AF831033

Asian Citrus Industry

The citrus industry plays a crucial role in the economies and agricultural landscapes of many Asian countries, contributing significantly to global citrus production and trade. This book provides a comprehensive analysis of the citrus sector across 14 major citrus-producing Asian nations, including China, India, Japan, Korea, Thailand, Bhutan, Nepal, Turkey, Pakistan, Bangladesh, Vietnam, Iran, Sri Lanka, and Indonesia. It presents an in-depth exploration of the historical evolution, production trends, key challenges, and opportunities shaping the citrus industry in Asia. Each chapter delves into country-specific aspects, including citrus production systems, major cultivars, post-harvest management, and production problems influencing the sector. It also highlights critical sustainability issues, including pest and disease management, technological innovations, and market-driven strategies that can enhance productivity and profitability for citrus growers.

By integrating a balanced blend of research insights, industry trends, and practical recommendations, *Asian Citrus Industry: Growth and Sustainability* serves as an essential reference for researchers, policymakers, agribusiness professionals, citrus growers, and students engaged in citriculture and horticulture. It provides valuable perspectives for those interested in understanding the diverse citrus ecosystems of Asia and exploring avenues for sustainable growth and international trade development.

Asian Citrus Industry

Growth and Sustainability

Edited by
Dilip Kumar Ghosh, Kwan Jeong Song,
Sangeeta Bhattacharyya, Subhra Saikat Roy
and Chandrika Ramadugu

CRC Press
Taylor & Francis Group
Boca Raton London New York

CRC Press is an imprint of the
Taylor & Francis Group, an informa business

Designed cover image: Shutterstock

First edition published 2026
by CRC Press
2385 NW Executive Center Drive, Suite 320, Boca Raton FL 33431

and by CRC Press
4 Park Square, Milton Park, Abingdon, Oxon, OX14 4RN

CRC Press is an imprint of Taylor & Francis Group, LLC

© 2026 Taylor & Francis Group, LLC

Reasonable efforts have been made to publish reliable data and information, but the author and publisher cannot assume responsibility for the validity of all materials or the consequences of their use. The authors and publishers have attempted to trace the copyright holders of all material reproduced in this publication and apologize to copyright holders if permission to publish in this form has not been obtained. If any copyright material has not been acknowledged please write and let us know so we may rectify in any future reprint.

Except as permitted under U.S. Copyright Law, no part of this book may be reprinted, reproduced, transmitted, or utilized in any form by any electronic, mechanical, or other means, now known or hereafter invented, including photocopying, microfilming, and recording, or in any information storage or retrieval system, without written permission from the publishers.

For permission to photocopy or use material electronically from this work, access www.copyright.com or contact the Copyright Clearance Center, Inc. (CCC), 222 Rosewood Drive, Danvers, MA 01923, 978-750-8400. For works that are not available on CCC please contact mpkbookspermissions@tandf.co.uk

Trademark notice: Product or corporate names may be trademarks or registered trademarks and are used only for identification and explanation without intent to infringe.

Library of Congress Cataloging-in-Publication Data
Names: Ghosh, Dilip Kumar editor | Song, Kwan Jeong editor | Bhattacharyya, Sangeeta editor |
Roy, Subhra Saikat editor | Ramadugu, Chandrika editor
Title: Asian citrus industry : growth and sustainability / edited by Dilip Kumar Ghosh, Kwan Jeong Song,
Sangeeta Bhattacharyya, Subhra Saikat Roy, and Chandrika Ramadugu.
Identifiers: LCCN 2025007696 (print) | LCCN 2025007697 (ebook) |
ISBN 9781032894669 hardback | ISBN 9781032953786 paperback |
ISBN 9781003584599 ebook
Subjects: LCSH: Citrus fruit industry–Asia
Classification: LCC HD9256.A2 A85 2026 (print) | LCC HD9256.A2 (ebook) |
DDC 381/.414304095–dc23/eng/20250813
LC record available at https://lccn.loc.gov/2025007696
LC ebook record available at https://lccn.loc.gov/2025007697

ISBN: 9781032894669 (hbk)
ISBN: 9781032953786 (pbk)
ISBN: 9781003584599 (ebk)

DOI: 10.1201/9781003584599

Typeset in Times
by Newgen Publishing UK

Contents

Foreword

The citrus industry has long been a cornerstone of global agriculture, contributing significantly to the economies of numerous countries. Most cultivated citrus varieties originated in Asia so a volume focusing on citrus cultivation in this region is particularly valuable to all citrus researchers and industry professionals. In Asia, citrus cultivation spans diverse agro-climatic conditions, reflecting a rich tapestry of genetic diversity, cultivation practices, and market dynamics. This book, Asian Citrus Industry: Growth and Sustainability, provides a much-needed analysis of the citrus sector in 14 major citrus-producing Asian nations, offering a detailed perspective on the scientific advancements, market trends, and sustainability challenges shaping the industry

As someone deeply engaged in citrus research, I recognize the importance of regional studies that contribute to global knowledge. This book not only examines country-specific insights but also fosters a broader understanding of citrus production systems and their evolving role in a rapidly changing global agricultural landscape.

I sincerely commend the authors for their insightful contributions and extend my congratulations to Dr. Dilip Ghosh and his team of editors for curating this outstanding volume. This book will undoubtedly serve as an indispensable resource for researchers, policymakers, and industry professionals alike.

I trust that this book will inspire further research and international collaboration, ensuring a resilient and prosperous future for the citrus industry.

Professor Mikeal L. Roose
University of California, Riverside

Citrus fruits hold immense significance in the agricultural economies of Asia, offering nutritional benefits, economic opportunities, and cultural heritage. However, citrus cultivation in the region faces multiple challenges, including climate change, pest and disease pressures, market volatility, and sustainability concerns. Addressing these challenges requires a concerted effort among researchers, industry stakeholders, and policymakers.

This book 'Asian Citrus Industry: Growth and Sustainability' is a collaborative endeavor by leading researchers and experts across Asia, offering an in-depth analysis of the citrus industry in 14 major citrus-producing nations of the region – Bangladesh, Bhutan, China, India, Indonesia, Iran, Japan, Nepal, Pakistan, South Korea, Sri Lanka, Thailand, Turkey, and Vietnam. It explores key research advancements, prevailing challenges, and emerging opportunities within each country, providing a comprehensive perspective on the region's citrus sector, fostering a better understanding of the opportunities and constraints faced by the citrus agribusiness. The book will serve as an essential reference for those involved in citrus research, production, and trade, ultimately contributing to the long-term sustainability of the Asian citrus industry.

Dr. Juliana Freitas-Astúa
Embrapa Cassava and Fruits
Brazilian Agricultural Research Corporation
Brazil

About the Editors

Dilip Kumar Ghosh, PhD, is the Director of ICAR-Central Citrus Research Institute (ICAR-CCRI), Nagpur, India. He earned his Master's and Doctoral degrees in Plant Pathology, specializing in Plant Virology, from ICAR-Indian Agricultural Research Institute, New Delhi, and has been associated with ICAR-CCRI since 1995. A distinguished Citrus Pathologist, Dr. Ghosh has developed PCR-based and LAMP/RPA-LFA-based rapid diagnostic kits for major citrus pathogens and developed novel management strategies for Huanglongbin. He identified *Candidatus* Phytoplasma cynodontis as the causal agent of Witches' Broom Disease of Acid Lime (WBDL) in India and was the first to report Hop Stunt Viroid infection in citrus in India. An FAO consultant, he has worked in Nepal and Bhutan and organized international training programmes. With 14 research projects, 76 research papers, 6 books, 480 NCBI GenBank Accessions and one US patent, he has earned numerous awards and fellowships. He has served as President of the Indian Phytopathological Society, as Executive Committee Member of the International Society of Citriculture and currently leads the Indian Society of Citriculture. In 2023, he convened the first Asian Citrus Congress in India, fostering global collaborations in citrus research.

Kwan Jeong Song, PhD, is a Professor in the Horticultural Science Major at Jeju National University and has served as Dean of Jeju Agriculture Meister College, Korea. He received a BA, MS, and PhD in the Department of Horticultural Science from Seoul National University, Korea. He served as Director of the Research Institute for Subtropical Agriculture & Biotechnology and Dean of the College of Applied Life Science at Jeju National University. He also served as President of the Korean Society of Citrus and Subtropical Climate Fruits and recently as President of the International Society of Citriculture and Convenor of the 15th International Citrus Congress. His research has focused on citrus genetics, molecular physiology, and polyploidization related to fruit set, fruit quality, and seedlessness. He has published more than a hundred articles and book chapters on these topics.

Sangeeta Bhattacharyya, PhD, is a Scientist in Agricultural Extension at the ICAR-Central Citrus Research Institute (ICAR-CCRI), Nagpur, India. She holds a B.Sc. (Hons.) in Agriculture from Visva-Bharati University, India, an M.Sc. in Agricultural Extension Education from ICAR-National Dairy Research Institute, India, and a Ph.D. from ICAR-Indian Agricultural Research Institute, New Delhi. Her research focuses on the adoption and impact assessment of scientific citrus farming technologies developed by the institute. She also examines gender dynamics in Indian citriculture, challenges faced by citrus growers in producing and exporting high-quality fruits, and the effectiveness of capacity-building programmes conducted by the institute. These programmes train citrus farmers, professionals, state horticulture officers, and citripreneurs in advanced citrus production technologies and related agribusiness ventures. Additionally, she is actively involved in demonstrating rejuvenation technologies for citrus orchards. Dr. Bhattacharyya has authored over 31 research papers, 11 book chapters, 33 popular articles, six training manuals, and one book.

Subhra Saikat Roy, PhD, is a Principal Scientist (Horticulture) at the ICAR-Central Citrus Research Institute (ICAR-CCRI), Nagpur, India. He earned his undergraduate degree from Uttar Banga Krishi Viswavidyalaya, followed by a Master's and Doctoral Degree in Horticulture from Bidhan Chandra Krishi Viswavidyalaya, India. A distinguished scholar, he was awarded the University Gold Medal (UG) and Certificate of Merit (PG) for academic excellence. With over 17 years of research experience, Dr. Roy has developed several innovative technologies. His research focuses on phytochemical profiling, bioprospecting, metabolomics, valorization, post-harvest technology, and citrus value addition. He has an extensive publication record, including over 50 research papers, 15 books, and

45 book chapters. Dr. Roy played a pivotal role as the Organizing Secretary of the Asian Citrus Congress – 2023. Throughout his career, he has been honoured with several prestigious awards in recognition of his outstanding contributions to horticultural science.

Chandrika Ramadugu, PhD, is a Project Scientist in the Department of Botany and Plant Sciences, University of California Riverside, Riverside, California, USA. She earned her Ph.D. from the Indian Institute of Science, Bangalore, India. Since 2005, Ramadugu worked exclusively on citrus in Riverside, California. Her interests range from citrus pathology, development of pathogen detection assays, horticulture, taxonomy, genetic transformations, study of diverse citrus germplasm, genomic studies, and development of short-term and long-term solutions to the citrus huanglongbing disease that has been a major concern for citrus industries in the USA, Brazil, Central and South American countries, and citrus growing regions in Asia. Currently, Ramadugu and her team are developing novel hybrids with genetic resistance/tolerance to the citrus huanglongbing disease through genome-assisted breeding of citrus cultivars with disease-resistant Australian limes. Ramadugu has published research articles in peer-reviewed scientific journals and collaborates widely with scientists from various research organizations. She is the leading principal investigator in several federally funded multi-institution research grants focused on finding solutions to economically significant citrus diseases.

Contributors

Umesh Kumar Acharya
National Citrus Research Program
Dhankuta, Nepal

Faruque Ahmed
Department of Agricultural Extension (DAE)
Farmgate, Dhaka, Bangladesh

R.M.N.T Amarsainghe
Agriculture Research Station
Department of Agriculture
Monaragala, Sri Lanka

Mehdi Azadvar
Kerman Agricultural and Natural Resources
Research and Education Center
Agricultural Research
Education and Extension Organization
Kerman, Iran

Muhammad Azam
Institute of Horticultural Science
Faculty of Agriculture, University of Agriculture
Faisalabad, Pakistan

H.M.U.A.G Jayalath Bandara
District Secretariat Office
Ministry of Agriculture
Nuwara Eliya, Sri Lanka

M.H.M. Borhannuddin Bhuyan
Citrus Research Station
Bangladesh Agricultural Research Institute, Sylhet, Bangladesh

Kurniawan Budiarto
Research Center for Horticulture
National Research and Innovation Agency of Indonesia
Cibinong Science Center
Jalan Raya Bogor, West Java, Indonesia

Mengji Cao
Citrus Research Institute of Southwest University
National Citrus Engineering Research Center
Beibei, Chongqing, China

Nirmala Friyanti Devy
Research Center for Horticulture
National Research and Innovation Agency of Indonesia
Cibinong Science Center
Jalan Raya Bogor, West Java, Indonesia

Himani Gammanpila
Extension and Training Centre
Department of Agriculture
Peradeniya, Sri Lanka

Sangay Gyeltshen
Agriculture Research Development Centre
Bajo, Bhutan

Zainuri Hanif
Research Center for Behavioral and Circular Economic
National Research and Innovation Agency of Indonesia
Cibinong Science Center
Jalan Raya Bogor, West Java, Indonesia

Ir Hardiyanto
Research Center for Horticulture
National Research and Innovation Agency of Indonesia
Cibinong Science Center
Jalan Raya Bogor, West Java, Indonesia

Seyed Mehdi Bani Hashemian
Citrus and Subtropical Fruits Research Center
Horticultural Science Research Institute
Agricultural Research
Education and Extension Organization Ramsar, Iran

Abuzar Hashempour
Citrus and Subtropical Fruits Research Center
Horticultural Science Research Institute
Agricultural Research
Education and Extension Organization
Ramsar, Iran

Mingyang He
Citrus Research Institute of Southwest University
National Citrus Engineering Research Center
Beibei, Chongqing, China

Phuangphet Hemrattrakun
Division of Horticulture
Department of Plant and Soil Sciences Faculty of Agriculture
Chiang Mai University
Chiang Mai, Thailand

Nguyen Van Hoa
Southern Horticultural Research Institute (SOFRI)
Tien Giang, Vietnam

Linhua Huang
Citrus Research Institute of Southwest University
National Citrus Engineering Research Center
Beibei, Chongqing, China

Mitsunori Iwasaki
Division of Citrus Research
NARO Institute of Fruit and Tea Science
Okitsu-nakacho, Shimizu, Shizuoka, Japan

Muhammad Jafar Jaskani
Institute of Horticultural Sciences, Faculty of Agriculture, University of Agriculture,
Faisalabad, Pakistan

Dong Jiang
Citrus Research Institute of Southwest University
National Citrus Engineering Research Center
Beibei, Chongqing, China

I. Kalubowila
Fruit Research and Development Institute
Department of Agriculture
Kananwila, Horana, Sri Lanka

Iqrar Ahmad Khan
Institute of Horticultural Sciences, Faculty of Agriculture, University of Agriculture,
Faisalabad, Pakistan

Nobuhiro Kotoda
Faculty of Agriculture
Saga University
Honjomachi, Saga, Japan

Chaireni Martasari
Research Center for Horticulture
National Research and Innovation Agency of Indonesia
Cibinong Science Center
Jalan Raya Bogor, West Java, Indonesia

Harun Meral
Institute of Science
Karamanoglu Mehmetbey University
Karaman, Türkiye
and
Department of Plant Production and Technologies
Food and Agriculture University
Konya, Türkiye

Daruni Naphrom
Division of Horticulture
Department of Plant and Soil Sciences Faculty of Agriculture
Chiang Mai University
Chiang Mai, Thailand

Ali Ozcan
Department of Chemistry
Kamil Ozdag Science Faculty
Karamanoglu Mehmetbey University
Karaman, Türkiye

Liangzhi Peng
Citrus Research Institute of Southwest University
National Citrus Engineering Research Center
Beibei, Chongqing, China

Ishara Perera
Fruit Research and Development Institute
Department of Agriculture
Kananwila, Horana, Sri Lanka

P.S.R Premarathne
Socio Economic and Planning Centre-subunit
Fruit Research and Development Institute
Department of Agriculture
Kananwila, Horana, Sri Lanka

Jamilur Rahman
Department of Genetics and Plant Breeding
Faculty of Agriculture
Sher-e-Bangla Agricultural University
Dhaka, Bangladesh

Md. Zahidur Rahman
Department of Genetics and Plant Breeding
Faculty of Agriculture
Sher-e-Bangla Agricultural University
Dhaka, Bangladesh

Faisal Shahzad
Institute of Horticultural Sciences, Faculty of Agriculture, University of Agriculture, Faisalabad, Pakistan

Chantalak Tiyayon
Division of Horticulture
Department of Plant and Soil Sciences Faculty of Agriculture
Chiang Mai University
Chiang Mai, Thailand

Mohan Bahadur Thapa
Nepal Horticulture Society
Lalitpur, Nepal

Nguyen Thi Kim Thoa
Southern Horticultural Research Institute (SOFRI)
Tien Giang, Vietnam

Vo Huu Thoai
Southern Horticultural Research Institute (SOFRI)
Tien Giang, Vietnam

Nangsel Tshomo
Agriculture Research Development Centre Bajo, Bhutan

Ali Tuncer
Institute of Science
Karamanoglu Mehmetbey University Karaman, Türkiye.

Development, Application and Research Center for Strategic Products (SARGEM) Konya
Food and Agriculture University
Konya, Türkiye

Dang Thi Kim Uyen
Southern Horticultural Research Institute (SOFRI)
Tien Giang, Vietnam

Xuefeng Wang
Citrus Research Institute of Southwest University
National Citrus Engineering Research Center
Beibei, Chongqing, China

Phuntsho Wangdi
Agriculture Research and Development Sub-Centre
Tsirang, Bhutan

Masashi Yamamoto
Faculty of Agriculture
Kagoshima University
Korimoto, Kagoshima, Japan

Kyung Uk Yi
Plant Resources & Environment Major
SARI
Jeju National University
Jeju, Korea

Yaohai Zhang
Citrus Research Institute of Southwest University
National Citrus Engineering Research Center
Beibei, Chongqing, China

Xiaochun Zhao
Citrus Research Institute of Southwest University
National Citrus Engineering Research Center
Beibei, Chongqing, China

Changyong Zhou
Citrus Research Institute of Southwest University
National Citrus Engineering Research Center
Beibei, Chongqing, China

Yan Zhou
Citrus Research Institute of Southwest University
National Citrus Engineering Research Center
Beibei, Chongqing, China

1 Citrus Industry of China

Changyong Zhou, Xiaochun Zhao, Liangzhi Peng, Dong Jiang, Yaohai Zhang, Linhua Huang, Mingyang He, Yan Zhou, Mengji Cao, and Xuefeng Wang

1.1 INTRODUCTION

China has a long citriculture history of over four millennia. In recent two decades, the citrus industry has been motivated by the Chinese government, including a plan on "Constructing advantageous citrus regions" and a policy on "The crucial task for poverty alleviation via industrial development", which have been facilitating its fast development. It took over two decades for China to reach the first 10 million tonnes (Mt) of production by 1997, then the time period the following each 10 Mt increase becomes less and less. China became the largest citrus production country in the world in 2007. In 2022, the total production and acreage of citrus were 60.04 Mt and 3.03 million hectares (Mhm^2) (National Bureau of Statistics of China, NBSC), or 48.77 Mt and 3.25 Mhm^2 (FAOSTAT), accounting for 29.33% and 30.79% of the world (ca 10% in 2000), respectively. China contributed over 60% of the incremental growth of citrus production to the world during the period of 2000–2022 (FAOSTAT). Since 2018, citrus has become the largest fruit industry, exceeding apple production in China. Take Guangxi Province as an example, whose citrus production and acreage reached nearly 19 Mt and 0.65 Mhm^2 in 2023, respectively. Therefore, citrus is of much economic importance to southern China's rural areas.

1.2 MAJOR VARIETIES/CULTIVARS

China is the primary centre of origin for many citrus species, such as *Citrus ichangensis*, *C. hongeensis*, *C. mangshanensis*, *C. daoxianensis*, *C. reticulata*, *C. sinensis*, *Fortunella*, and *Poncirus*. A large number of diversified citrus varieties are formed by the selection of various geographical and ecological conditions, and long cultivation history. A century ago, China was the main exporter of citrus varieties. With the development of the world citrus industry and the progress of science and technology, many excellent citrus varieties have emerged in various countries around the world, which has promoted the further development of the citrus industry. In the 1960s, due to the needs of the canned mandarin industry, the introduction and promotion of new varieties of Satsuma mandarins were strengthened in China. Several Satsuma mandarin cultivars such as Miyagawa Wase and Okitsu Wase etc. were introduced from Japan, and applied to a large-scale plantation, even more recently, some varieties such as Benimadona, Asumi, and Harumi, etc., have still received attention from growers in China. From the 1970s, the introduction of citrus varieties from overseas has gradually become institutionalized, standardized, and normalized.

Before the 1980s, the major citrus varieties were Satsuma mandarins; tangerines such as Ponkan, Nanfengmiju, and Bendizao; sweet oranges (SwO) such as Jincheng, Xinhuicheng, and Gailiangcheng; and Shatianyou pomelo, also kumquats. The Satsuma mandarins were the most popular cultivars, accounting for 60% of the total citrus production by then. After the 1980s, the plantation of navel SwO, pomelos such as Guanximiyou and Guangfengmajiayou, tangerines such

DOI: 10.1201/9781003584599-1

as Shatangju, Ponkan, Zaomi Ponkan, Nanfengmiju, Gonggan, Chazhigan, and mandarins such as Xingyumiju, Huangguogan, and recently, the late-/early-maturing varieties, etc., has been expanded fast. Some of them have been developed to a large scale, thus the production of mid-season varieties began to decline, so as to improve the variety structure with the rapid increase of late and early season citrus production to extend the period time of fresh fruits supplied to the market. The maturity of citrus has been further divided from early, mid, and late maturity to extreme early, early, mid, mid-late, late, and extreme late, and changed from a ratio of 1:8:1 at the end of 20th century to 0.2:1.5:5.0:2.0:1.0:0.3 by 2020. The promotion of new variety to the production was accelerated by the expansion of the citrus industry and demand of the market. The high total solid soluble content (TSS)/acid ratio, seedless, easy peeling, enriched flavour and aroma, and early or late-season varieties are favoured by consumers. The large number of high fruit quality, late and early ripening new varieties were rapidly applied to the production. Since the 1980s, more than 140 citrus varieties have been bred in China, among which 88% were developed from the selections of seedlings and bud mutants, 12% by cross-hybridization, biotechnology (cell fusion), and mutagenesis, or classified by traits, 79% are seedless, 21% with late or early maturity, including two rootstock varieties tolerant to iron deficiency in alkaline soil. The Act on Plant Variety Protection has been implemented since 1996, and ca 80 varieties/cultivars have been approved for protection in China now. Based on the data of 2022 from FAOSTAT, the proportion of mandarin (incl. tangerine and clementine) among the total production was 26.57% in the world, and 55.67% in China, whereas that of SwO was 45.95%, and 15.8%, respectively, showing a difference in consumption preference between the two major variety groups.

Based on the scale of cultivation acreage of over 667 hm^2 and 6,667 hm^2, ca 70 and 12 cultivars could be accounted in, respectively, which can be classified into several major groups:

1.2.1 Satsuma Mandarins

The planting area has declined from 0.8 Mhm^2 a decade ago to 0.47 Mhm^2 by now, it's suggested to be maintained at ca 0.2 Mhm^2 due to the characteristics of early maturity, good productivity, and easy planting.

1.2.2 Navel SwO

The planting area has risen steadily to at present 0.33 Mhm^2, and it is predicted that early-matured ones to have somewhat expanded space due to the good market.

1.2.3 Ponkan Tangerines

The plantation peaked at 0.27 Mhm^2 and went down to the current 0.1 Mhm^2.

1.2.4 Guanximiyou Pomelos

From the peak of 0.17 Mhm^2 a decade ago to the present 0.1 Mhm^2, which are mainly concentrated in the region of south 25°N latitude such as Pinghe County, Fujian Province.

1.2.5 Nanfengmiju Tangerines

The production area decreased from 0.1 Mhm^2 in the peak period to 53,000 hm^2, mainly concentrated in the traditional production region Nanfeng County and its surrounding areas of Jiangxi Province. It's predicted to be stable within 67,000 hm^2 because of its high storability, easy peeling, and rich aroma.

1.2.6 Shatangju Tangerines

The cultivated area is estimated to be more than 0.27 Mhm2, mainly distributed in Guangxi and Guangdong Province, especially in Guangxi, mainly due to HLB impact in Guangdong since 2006. In recent years, due to the excessive production led to the decline in economic benefit, the plantation is decreasing.

1.2.7 Bingtangcheng SwO

At present, the cultivation area is ca 67,000 hm^2, mainly distributed in Hunan and Yunnan Province, where the annual average temperature (AAT) is higher than 20°C, thus the typical characteristic of low acid with this variety type could perform well.

1.2.8 Late Ripening Hybrid Mandarins

The main varieties are Orah, Dayagan, Murcott, and Shiranui, etc., which have been developed rapidly within the latest 15 years, mainly distributed in Guangxi, Sichuan, Chongqing, Yunnan Province, and other frost-free areas, where the winter season temperature is requested to be not high and not low. The production area is about 0.15 Mhm^2, and still growing due to the good market.

In the past two decades, two rounds of variety upgrading each with a number of representatives have well adapted to the rapid development of citrus industry:

The first stage (2004–2013): Since 2004, Guanximiyou cultivars such as the red-fleshed, the yellow-fleshed, and Shanhongyou, etc., have been largely expanded from Pinghe to other producing areas in China. The promotion of medium and late-maturing citrus such as Shatangju, Dayagan, and Tarocco newlines, etc., led to the development of the late-maturing citrus industry in Chongqing, Sichuan, Guangxi, Yunnan, etc. Implementing several provincial projects in Chongqing, Sichuan, etc., has promoted the development of a number of high-quality SwO varieties. During this period, extending technologies such as wide-row cultivation, container-grown nursery trees, weed sod cultivation, fruit bagging, and canopy mulching, etc., have significantly improved the appearance quality of fruit.

The second stage (2014–present): Due to the strengthening of the awareness of variety registration and variety intellectual property protection, the number of variety registrations has surged, and the enthusiasm for independently breeding new varieties has been stimulated. The typical representative is the cooperation between the Citrus Research Institute (CRI) of Southwest University (SWU) (Former CRI, CAAS (Chinese Academy of Agricultural Sciences)) and a number of integrated enterprises, which has accelerated the breeding and extending of new varieties, such as Jinqiu Shatangju, Changye Xiangcheng, Huamei No.7, and Sunshine No.1, etc.

1.3 CULTIVATION ZONES

In the mainland China, around 1,000 counties within 19 provinces have been producing citrus, of which over 93.3% the production had been produced by nine provinces, including Guangxi, Hunan, Hubei, Guangdong, Sichuan, Jiangxi, Fujian, Chongqing, and Zhejiang in 2021 (NBSC). Citrus is mainly distributed in the areas between 18°5'N–33°43'N latitude, covering as far south as Qiongzhong County of Hainan Province, as far north as Hanzhong County of Shanxi Province, as far east as Changxing Island of Shanghai City, and as far west as Mêdog County of Tibet Autonomous Region, which is generally consistent with the natural geographical division of China. In 2003, the Ministry of Agriculture (MOA, nowadays reformed as MOA and Rural Affairs (MARA)), had put together the form of "Three superior belts with five bases" (the sweet orange belt of upper and middle reaches of Yangtze River, the mandarin belt of Zhejiang-Fujian-Guangdong, the navel SwO belt of south Jiangxi-south Hunan-north Guangxi, plus five bases (Lingnan late mandarin in Guangdong, Nanfengmiju in Jiangxi, extra-early mandarin in Yunnan, the Danjiangkou reservoir area mandarin in Hubei, and lemon in Sichuan and Yunnan)), followed by planning one more mandarin belt of west Hunan-west Hubei in 2008. The production in the region of Zhejiang-Fujian-Guangdong has gradually declined since the beginning of this century, which is mainly due to the increase of labour/land costs and the HLB problem, whereas that of the other three belts has increased dramatically, showing a tendency of inland/westward movement from east coast areas.

According to the vast diversified climate, soil, and citrus varieties, citrus production in China can be divided into five regions with characterized structures of citrus varieties.

1.3.1 South Subtropical Citrus Zone

This zone is composed of a large area of coastal plains and shallow hilly areas along the south of the Nanling Mountains and southeast coasts, which is roughly consistent with the 23°N latitude. The AAT is 19°C–23°C, the annual precipitation is 1,500–1,800 mm, the sunshine is 1,400-2,400 h, and the soil is mostly acidic. This area is very warm, with sufficient rain and light, and almost no frost. It is the most suitable area for the growth of all kinds of citrus varieties except late-matured ones. Citrus trees grow and develop fast, with early bearing, high yield, early maturing, and smooth and thin peel of fruit in light colouring, and low acid content. This zone is a very important traditional citrus-producing region in China. It is also one of the most abundant citrus variety resources areas, and rich in various types of native citrus varieties, such as mandarins: Ponkan, Nianju, Jiaogan, Gonggan, Fuju, Xuegan, Lugan, Shatangju, Mashuiju, Biangan, Shagan, Danju, and Chazhigan; SwO: Xinhuicheng, Anliucheng, Liucheng, Sihuicheng, Huazhoucheng, and Xiangshuicheng; pomelos: Shatianyou, Guanximiyou, Wendanyou, Pingshanyou, and Huazhoujuhong; others: kumquats, citrons, and limonias. The main cultivars are Shatangju, Orah, Gonggan, Satsuma, Mashuiju, Ponkan, Murcott, Jiaogan, Xuegan, Hongjiangcheng, Xinhuicheng, Anliucheng, Shatianyou, Guanximiyou, and kumquats of Jindan, Rong-an, Huapi, Cuimi.

1.3.2 Mid-Subtropical Citrus Zone

This zone is between the north 23°N and south 30°N. The terrain is complex and diverse and is mainly composed of hills and valleys. This area belongs to the mid-subtropical humid monsoon climate. The AAT is 16.5°C–19.5°C, the average extreme minimum temperature is from -1.0°C to -9°C, the annual precipitation is 800–1,400 mm, and the sunshine duration is 1,100–2,000 h. The soil is red soil, yellow soil, and purple soil. This area is the most extensive and important citrus distribution area in China. Due to the warm and humid climate, abundant and uniform distribution of precipitation, and distinct four seasons, it is suitable for the following main varieties: Satsuma, Ponkan, Shatianyou, SwO of Navel/Blood/Valencia, and kumquats. It is the origin and main distribution area of many species of trifoliate orange, kumquat, and citrus plants. In Hunan, Jiangxi, Fujian, Chongqing, Sichuan, Guangxi, and other places, the wild mandarin, Yichang papeda, junos, and Jindou (*Fortunella hindsii*) still exist in their natural habitat. The south of this zone is warm and humid, with light frost. It is the main production area of SwO, pomelos, and kumquat in China. Mandarins such as Satsuma, Ougan, Hongju, Tugan, Ponkan, SwO such as Jincheng/Taoyecheng/Navel/Blood, sour oranges for the usage of traditional medicine, pomelos such as Shatianyou/Liangpinyou/Sijipao, and lemons are widely distributed in this region. The north of this zone is less warm with frost. It is the most suitable area for the mandarin and kumquat. Although SwO and pomelos can be planted, they generally perform poorly. The main varieties are Satsuma, tangerines of Ponkan/Nanfeng/Hongju, hybrid mandarins and Huyou pomelo, kumquat, etc., and are also the main distribution areas of many ancient mandarin varieties, such as tangerines of Ruju/Zhuju/Tuju.

1.3.3 Northern Subtropical Citrus Zone

This area is located between 30°N and 33°N latitude. Due to the influence of topography, the citrus plantation areas are small and sporadic, mostly in valleys, basins, hills, and reservoir-side/lakeside/riverside plains. The AAT is 15°C–17°C, the average extreme minimum temperature is from -6.8°C to -3.5°C, the annual precipitation is 800–1,600 mm, and the sunshine is 1,400–2,200 h. The soil is mostly yellow soil, yellow-brown soil, and calcareous soil. Due to the regular frost in the winter, citrus can be injured by freeze. Citrus production used to be dominated by cold-tolerant mandarin landrace varieties, such as Zhoupigan, Huangpigan, Tuju, and Zhuju. At the current time, the main cultivars are Satsuma, Ponkan, Bendizao, and Zhuju. SwO and pomelos are only planted in areas with better microclimate conditions. Affected by the ecosystem, citrus plants in this zone generally grow slowly, with smaller fruits and darker colours, but higher acidity and later maturity.

1.3.4 Yunnan-Guizhou, Qinghai-Tibet Plateau Citrus Zone

This zone is located in the north latitude of 21°04'–29°23', a part of the Yunnan-Guizhou Plateau and the Qinghai-Tibet Plateau in southwestern China, which is affected by the Pacific and Indian Ocean monsoons at the same time, and belongs to the subtropical-tropical plateau humid monsoon climate. The dry and wet seasons are distinct, and the light and heat conditions are superior. It is suitable for the growth of various types of citrus. Citrus generally matures earlier and yields higher, with a bright colour, thick peel, rich flavour, and high TSS. Due to the high mountains and deep valleys, the climates of different altitudes at the same latitude are quite different. From the bottom of the valley to the top of the mountain, the climate varied from tropical to cold. Therefore, citrus is obviously distributed vertically with different types of varieties. Generally, citrus is planted below the altitude of 1,500–2,000 m, but mostly in small scale and scattered distribution. This area is also one of the citrus origin centres in the world. Many types of wild citrus resources have been discovered, which proves that China is the origin or one of the origins of papeda, mandarin, citrons, limonia, and other species. The main varieties cultivated are Bingtang, Jincheng, Navel, Ponkan, Satsuma, Huangguogan, Orah, Shatianyou, Dongshizaoyou, Menglunzaoyou, Mansailongyou, citrons, lemon, and limonia.

1.3.5 Tropical Citrus Region

This zone is in the south of 21°N latitude, including Leizhou Peninsula and Hainan. The AAT in this area is above 22°C, the annual precipitation is above 1,500 mm, and the annual sunshine is about 2,000 h. Summer is very long without winter and distinct four seasons. The climate is featured by high temperature and high humidity, adequate light, and frequent typhoons. This is a small citrus-producing area in China. Citrus grows rapidly, with early bearing, early ripening, high yield, and larger fruit, but light colour or green, low acidity. The main varieties are Ponkan, Jiaogan, Nianju, Hongjiangcheng, Shatianyou, Eureka lemon, and Meyer lemon.

1.4 MAIN TECHNOLOGIES AND PACKAGE OF PRACTICES

Improving technologies and practices is of much importance, changes within the last two decades in China in citrus cultivation from nursery, plantation, and protection, to post-harvest would be generally introduced hereafter.

1.4.1 Nursery

Trifoliate orange is the most common rootstock used in the region of acidic and neutral soil, occupying ca 80% of the total, followed by Hongju, which is mostly used in Sichuan, Chongqing, and other region of alkaline soil, and Goutoucheng in Zhejiang, Xiangcheng, Suanyou in some areas. Since 2000, over 90% of the newly planted orchards have used container nursery trees grafted onto Carrizo citrange in Chongqing, and with somewhat expansion to other areas. By 2010, over 100 modern citrus nurseries had been set up throughout the main citrus production provinces for expanding virus-free container nursery trees via the establishment of a 3-level virus-free propagation scheme partially supported by the state government, with the capability of producing over 100 million nursery trees, causing a continuous decrease in the proportion of open field budling production. The above practices followed in nurseries have a significant effect on controlling the virus and virus-like diseases such as tatter-leaf, and exocortis that had been prevalent in some regions before. For those graft-transmission diseases that also have insect vectors such as tristeza (*Citrus tristeza virus*, CTV), citrus yellow vein clearing (*Citrus yellow vein clearing virus*, CYVCV, invasive alien disease), and citrus Huanglongbing (HLB, *Candidatus* Liberibacter asiaticus, *C*Las), the virus-free propagation system cannot prevent them from reinfection in the field, such extra practices as cross-protection of mild strain and screenhouse protection in the early stage are needed. In the past decade, an isolated outbreak of severe stem pitting of CTV occurred in Ganzhou, Jiangxi Province.

1.4.2 Plantation

Citrus production areas in China are mainly distributed in the south and southwest of the Yangtze River. Citrus orchard on slope land accounts for more than 80%, and citrus orchards on flat and gentle slope lands are mainly distributed in southern and central Guangxi, the Sichuan basin, Guangdong, and some areas of Hunan and Jiangxi.

1.4.2.1 Soil Management

The vast majority in the south of the Yangtze River is red soil (mostly pH 4–6, over 1 m in depth), which is characteristic by depletion, acid, sticky, lacking organic matter, nitrogen, phosphorus, potassium, calcium, magnesium, and other nutrients, a small area in the mountain for citrus planting is yellow soil. The majority in Chongqing and Sichuan is purple soil (mostly pH 7.5–8.56, ca 0.4 m in depth), which is well-ventilated, permeable, and rich in mineral nutrients but difficulty to be absorbed by citrus, lacking organic matter. Since the 1990s, the annual tillage practice of citrus orchards (10–20 cm in depth) has decreased from 2–4 times down to 1–2 times or no tillage at all due to a shortage of labour.

1.4.2.2 Cultivation Mode

Contour terrace is a popular mode of land preparation with various widths from the usual 3–5 m to ca 10 m (slight slope), combined with a trench of 1–1.5 m in width, 0.6–1 m in depth, back-filled with soil improvement materials (turf, weeds, organic fertilizer, mineral phosphate fertilizer, and lime). Citrus are planted only after soil improvement materials are fully decomposed. If the land drainage is poor, then the ridge (2.5–3.5 m in width, 0.5–0.8 m in height) planting should be adopted. Sod cultivation with natural grass and artificially planted grass is mainly practiced in fully bearing orchards, maintained by mowing via portable or small self-propelled, occasionally large lawn mowers in the slight slope or flat orchards, whereas seasonal natural grass sod cultivation is to let the grass grow in the rainy season and kill the grass in the dry season. Artificial grass sod cultivation is to sow the high dense grass such as clover (*Trifolium repens*), vetch (*Vicia villosa*), rye (*Secale cereale*), and vulpia (*Vulpia myuros*) in the open space between the rows. The simple facility cultivation mode mainly includes a sunshade net in summer to prevent sunburn at high temperatures, and a canopy covering film or sunshade net to protect fruits from freezing damage in winter. In the northern region of Guangxi, a simple trellis is set up in winter for supporting film to prevent the kumquat fruit from cracking during the ripening period caused by rain and freezing damage. This can be used during the high cracking season for some varieties prone to cracking. The plastic greenhouse cultivation mode, accounting for only ca 0.1% of the total production area, is mainly distributed in the economically developed areas such as Zhejiang, Shanghai, and Jiangsu in the northern margin of citrus production, as well as in the suburbs of some big cities. The main purpose of greenhouse cultivation is to prevent frost damage and fruit cracking, which is also used to delay fruit harvesting so that varieties that matured in Nov. and Dec. can be delayed to harvest for the Spring Festival or sightseeing picking in Feb. or varieties that matured in Jan. and Feb. can be delayed to harvest in Apr. and Jun. There is a small size of experimental cultivation in Zhejiang, where Satsuma mandarin maturity could be adjusted to Jun.–Jul.

1.4.2.3 Weed Control

The annual rainfall in most citrus-producing areas is 1,300–1,800 mm in China, where weed barrier fabric is usually mulched on both sides of the trunk of a young tree (1–1.5 m in width), and removed in the 3rd or 4th year. Within the first three years of young orchards, shallow-rooted crops, including peanut, soybean, pepper, eggplant, sweet potato, carrot, watermelon, mustard, cabbage, radish, etc., are sometimes planted 30–40 cm outside the canopy drip line to suppress weed growth, increase orchard income, and improve the soil. Herbicides like glyphosate and glufosinate-ammonium are often sprayed once or twice a year to control weeds in young orchards.

1.4.2.4 Fertilization

Due to the difference in citrus varieties and soil types, the amount of fertilizer applied to citrus varies greatly in different regions and varieties. Usually, the 1–3-year-old orchard is mainly applied with nitrogen (N) fertilizer, supplemented by phosphate, potassium, and organic fertilizer. Taking medium-density cultivation as an example, 1–3-year-old trees are applied 50–100 g, 80–150 g, and 150–300 g of pure nitrogen per plant per year, respectively. The application amount of phosphorus is 1/3–1/5 of nitrogen, and potassium is 1/2–1/3 of nitrogen. The fertilization of the bearing tree mainly depends on the yield of fruit, in general, with the expected yield of 100 kg of fresh fruit, the application quantity is N 0.8–1.5 kg, P2O50.4–0.6 kg, K2O 0.6–0.8 kg, and MgO 0.1–0.3 kg. In the acidic red soil area, calcium content is low, and lime or calcium magnesium phosphorus is applied once every 1–2 years. The main nutrient source of citrus is chemical fertilizer, accounting for more than 2/3. The application of organic fertilizers has declined significantly in the past two decades, but there has been an increasing trend in recent years. A few orchards with good benefits mostly use organic fertilizers instead of chemical fertilizers. The third national soil general survey was carried out in 2022 in China, which is of much help for soil analysis of citrus orchards. Systematic balanced fertilization based on nutrient diagnosis of 12 leaf mineral elements began in the early 2000s in Chongqing, and large-scale leaf nutritional diagnostic fertilization has then gradually been promoted throughout the main citrus-producing provinces. However, this practice has been only conducted in a small proportion of citrus orchards, most orchards still rely on the experience for fertilization. With regards to micro-element deficiency in citrus orchards in China, over 2/3 and 1/3 of orchards are deficiency in zinc and magnesium, respectively. Furthermore, the deficiency of iron and manganese often occurs in alkaline purple soil and low calcium in red and yellow soil. The deficiency of zinc, iron, and manganese is mainly treated by soil application of organic fertilizer, zinc sulfate, zinc chloride, chelate iron and manganese, and foliar spraying of organic zinc and manganese compounds such as mancozeb, zinc and chelate zinc. The seasonal fertilization includes base fertilization in winter, sprouting fertilization in spring, fruit set fertilization in summer, and fruit growth fertilization in summer and autumn. Fertilization methods include soil trenching (or burrowing) fertilization, rotary tillage fertilization, broadcast fertilization, fertigation, soil injection fertilization, and foliar fertilization. Trenching fertilization for application of base fertilizer in most citrus orchards is by digging the trenches in winter, then burring the organic fertilizer, mineral phosphorus calcium fertilizer, and ground in the trenches. Rotary tillage fertilization is the application of organic or chemical fertilizer on the surface of land, and then rotary tillage with a rotary tiller for mixing the fertilizer with soil. Due to the increasingly serious problem of labour shortage, broadcast fertilization of nitrogen fertilizer or nitrogen-containing compound fertilizer during the growing season is also common. Fertigation is a fast-growing citrus orchard fertilization method since the new century, especially in the large-scale citrus groves managed by the company. Soil injection fertilization is another new method which fertilizer is dissolved in water and injected into a certain depth of soil distributed with dense roots through high pressure.

1.4.2.5 Flower and Fruit Management

The common method to inhibit the flower bud differentiation of 1–3-year-old trees is to increase the proportion and amount of nitrogen fertilizer, reduce the application of phosphorus and potassium fertilizer, increasing irrigation times to prevent soil drought in autumn and winter. To moderately cut the shoots before spring buds sprout is also an effective practice to reduce flowering and promote the vegetative growth. For the varieties that are ease to flower, 100–200 mg/kg gibberellin is commonly used at the young tree stage in autumn and winter, and complementing with foliar spray of urea for 2–3 times to inhibit the formation of flower buds, thereby reducing flowering. In order to promote flower bud differentiation, the often used practices include control of fertilizer application in summer and autumn, shoot twisting in autumn, partial root cutting in autumn, water and humidity control, foliar application of paclobutrazol, uniconazole, potassium dihydrogen phosphate from Sep. to Dec. A traditionally effective method of branch girdling in autumn is rarely used by now, mainly due to its difficulty in mastering the time and degree of girdling. The rainy season in most citrus-producing areas in China from Apr. to Jun., along with the temperature above 35°C in the occasionally sunny days, is not conducive to the fruit setting, in particularly for seedless varieties. In order to reduce the young fruit drop from April to June, 50–100 mg/kg gibberellin and cytokinin are usually sprayed once or twice during this season, and trunk girdling is also applied in some areas. In addition, the removal of excessive spring vegetative shoots, balanced fertilization, and foliar spraying fertilizer during flowering and young fruit periods are also often used to reduce young fruit drop. The commonly used method to reduce sunburn often encountered by Satsuma, Orah, Nanfengmiju, Murcott, etc. in southern China in summer, is to spray anti-sunburn materials onto the fruit surface, such as lime water. Orchard sunshade nets, fruit bagging, sod cultivation, etc., are also used to prevent sunburn.

1.4.2.6 Drought and Frost Prevention

Less than 20% of citrus orchards in China are equipped with drip or micro-spray irrigation systems. The occurrence of the main drought is normally in Jul.–Aug. in most areas, whereas in the dry and hot valley of Jinshajiang River is usually in Feb.–May due to the RH being lower than 20%, and in the south of Yangtze River is often in autumn and winter. Fully bearing orchards can basically survive the dry period without irrigation, but yield and fruit quality are often seriously reduced if severe drought is encountered. Such practices as manually watering trees via water pipes installed in orchards can somehow attenuate drought issues. In some winters, temperatures as low as -10°C to -12°C could be encountered in Zhejiang, Hubei, Shanghai, Shanxi, central, and northern Jiangxi, and northern Hunan, in particular within the low-lying hilly areas. Fruit frost damage could sometimes occur to late-matured citrus in Sichuan, Chongqing, and other areas when the minimum temperature drops to below -2°C, resulting in fruit drop, frostbite, and frost rot. To prevent citrus from freeze damage, planting the tree above the inversion layer of the slope, planting near large water bodies, building orchard protection forest, promoting the ageing of autumn shoots, soil irrigation before the freezing weather, tree trunk and large branches bandaging, canopy covering and other measures, are generally adopted to reduce freezing damage. In Sichuan, Chongqing, and other places, fruit bagging, fruit bagging plus tree canopy covering with film, and other measures, are generally adopted to prevent freezing damage of late-maturing citrus fruits. In addition, orchard smoke, spray anti-freezing solutions are also applied in some cases.

1.4.2.7 Pruning

Pruning usually is carried out in winter after fruit harvest, and in areas with freezing damage to the tree, the pruning is done in early spring. Late-maturing citrus is usually pruned after fruit harvest in spring. Some varieties of citrus with high fruit-setting rate will also be pruned in spring when the flower bud first appears to reduce the amount of flowers. Varieties such as Orah with very vigourous shoot growth also pruned in summer. Pruning is usually carried out manually, and mechanical pruning is rare in China.

1.4.2.8 Orchard Mechanization

The degree of mechanization of citrus orchards in China varies greatly from fully artificial orchards to highly mechanized orchards, most orchards are managed by combining a small portion of mechanization with manpower. More widely used machinery in orchards includes small trenchers, small rotary tiller, small lawnmowers, simple pipe irrigation facilities, constant pressure pipe spray systems, motor spray equipment, orchard ropeway transport systems, orchard rail transport systems, agricultural transport vehicles, and tricycles. However, even for small orchard machinery, there are not many orchards equipped with all of them, and many orchards only have some of the machinery.

1.4.3 Disease and Pest Control

There are ca 60 citrus diseases and pests commonly occurred in China. Due to chapter space limitations, thus taking HLB as an example for introduction, a few changes of control practices on the main or newly occurred ones within the last two decades would also be simply mentioned. HLB, spreading to over 340 counties via psyllids (*Diaphorina citri*, *Cacopsylla citrisuga*) by 2023, is the only left quarantine citrus disease in China since citrus canker was removed from the quarantine list in 2020. During the period of 2013–2018, about 50 million trees have been destroyed due to an outbreak of HLB in Ganzhou, Jiangxi. After fast developing in the recent decade, the superior navel SwO belt has been facing an upward trend of HLB epidemics. Meanwhile, the two HLB-free superior belts are facing much pressure induced by the northward movement of psyllids. In order to protect the Yangtze River citrus superior belt, MARA has set up a citrus quarantine-free zone in Chongqing and an HLB interception zone at Jinshajiang River valley of Pingshan County, Yibin Prefecture, Sichuan Province since 2007 and 2017, respectively. The main control strategy is based on three basic measures (application of HLB-free nursery trees, timely removal of HLB-infected trees, and large-scale joint prevention and control of psyllids). The classified management based on HLB incidence has been strengthened, complementing by "village rules and people's covenants", "technician-driven management", "highly dense planting", "windbreaks", and "killing summer shoots via chemicals" practices, which have been shown effective. In 2014–2018, MARA had annually convened a national on-site conference on HLB prevention and control in Guilin (2014), Ganzhou (2015–2017), and Guangzhou (2018), respectively. A national HLB alliance was also organized, formulating the HLB diagnostic criteria, prevention and control norms, releasing disaster relief funds, and setting up two state key research programmes, etc. "National Modern Agricultural (citrus) Industrial Technology System" and "National Collaborative Innovation Alliance for Integrated Prevention and Control of HLB for Citrus Superior Belts" have been established in 2007 and 2017, respectively, so as to strengthen the collaborative linkage throughout the country, effectively curbing the HLB harm. In order to prevent even large-scale outbreaks of HLB, the government of Guangxi released an Act for effective regulations on the prevention and control of HLB in 2019, which is the first provincial Act with this specific issue in China. A new invasive alien virus disease caused by CYVCV, mainly borne by citrus whiteflies (*Dialeurodes citri*), has been spreading fast throughout almost all citrus areas since its report in Yunnan Province in 2009, causing somehow damaging to lemon production in China, and its prevention and control measures mainly refer to those addressed for HLB. Much effort has been made to control citrus canker, e.g. three million trees were burned during 2002–2007 in Chongqing, so far only Chongqing still keeps it on the local quarantine list, whereas other citrus-growing provinces have treated it as a common disease since Nov. 2020 due to its widely spreading. Such fungi diseases with localized outbreaks as acute anthracnose (*Colletotrichum gloeosporidoides*) in Deqing of Guangdong, alternaria brown spot (*Alternaria alternata* tangerine pathotype, a new leaf-spotting disease) in Wanzhou of Chongqing, target spot (*Pseudofabraea citricarpa,* a new leaf-spotting disease) in Hanzhong of Shanxi, had occurred within the latest two decades. The control measures of fungi diseases usually combine agricultural practices with spraying fungicides, like those that are applied for common fungi diseases such as scab, black spot, and greasy spots, sometimes complementing with drone spraying due to the shortage of labour. The often used fungicides are mancozeb, azoxystrobin, prochloraz, and copper-based fungicides, etc. With regards to pest control, much emphasis has been made on pesticide reduction, so biological control practices have been widely applied, e.g. the predatory mite (*Neoseiulus barkeri*) has been extensively applied. In most areas, mites, scales, aphids, and leafminers commonly occur, whereas whitefly, thrip, and occasionally longicornia, leafminer beetle, etc., heavily attached citrus, in particular, fruit fly (*Bactrocera dorsalis*) spread fast to almost all citrus areas from the predominant coast areas due to its large range of host and climate warming change, *B. minax* has distributed in Sichuan, Hunan, Guizhou, Yunnan, Chongqing, Guangxi, Shanxi, and Hubei, so these two species of fruit fly have been removed from the quarantine list since 2009, the only left quarantine pest is *B. tsunamis*, which is limited in part counties with a certain low-temperature period in Sichuan, Guizhou, Yunnan, and Hunan. The practice package for pest control is based on IPM, and pesticides such as neonicotinoids, avermectins, decamethrin, acetamiprid, and petroleum spray oil emulsion, are still often used. In the recent decade, the practice of fruit bagging has become popular for protecting the production of pomelos, lemons, and late-matured varieties.

1.4.4 Post-Harvest

In China, the annual post-harvest loss of total production varies from 10% to 20% after a period of storage of 1–3 m. In order to reduce the loss from harvest to consumer table, technologies and practices for post-harvest have been paid much attention. In 2006, over 200 packing lines were built with a total treating capacity of 1,000 t/h, mainly distributed in Hubei, Jiangxi, Guangdong, Zhejiang, Fujian, and Hunan by then. Along with the fast development of the Chinese citrus industry, packing capacity has been significantly promoted in the past decade, covering 70% of total production for the fresh market.

1.4.4.1 Harvest

Although earlier fruit harvest can prolong the storage period, it is at the cost of somehow losing the original flavour of the variety, such practice presents a decreasing tendency but still existing, e.g. in order to avoid such practice, Gannan Navel SwO Association strongly recommended that growers could not harvest navel SwO fruits before Nov. 1 in 2023. On the other hand, since early this century, the successful practice of canopy mulching of kumquat during the winter season, so-called "Three avoid technology" (avoiding rain, frost, and snow), in Yangshuo County, Guilin Prefecture, Guangxi Province, has gained the high return of added value, which has stimulated strong demand for technologies of delaying harvest, on-tree storage technology has been being practiced in the past decade in particular in Sichuan, Chongqing via foliar spraying of 2,4-D and imazalil, and fungicides before fruit maturity, complementing with fruit bagging during the winter season, extending selling season of fresh fruit for 2–4 m.

1.4.4.2 Precooling

In general, precooling is applied mainly for the late ripening SwO in Sichuan and Chongqing from Feb. to May, and the early ripening mandarins in Yunnan from Jul. to Oct.

1.4.4.3 Sterilization

The most common post-harvest diseases are green mould (*Penicillium digitatum*) and blue mould (*P. italicum*), which are normally treated by fungicides such as imazalil, prochloraz, or iminoctadine tris (albesilate) in the packing house. The usage of microbial antagonists like yeasts is preferred, applied by dipping or spraying directly onto fruit. However, its commercialized application is still limited due to the lack of curative activity, high cost, and lower level of control efficacy when applied as stand-alone treatments. Therefore, the integrated practice package to enhance bio-control performance is still in demand.

1.4.4.4 Sorting

Accurate and non-destructive tools for online fruit quality assessment, monitoring, and early detection of pathological decay, are developed fast. Since 2014, Reemoon Technology Co., Ltd., one of the biggest suppliers of sorting equipment in China, offered an automated system to facilitate sorting fruits with different classification functionalities in one packing line based on Vis/NIR (Visible/near-infrared diffuse reflectance specta (structure-preserved error-correcting tree automata)) techniques, for sorting large amounts of citrus fruits based on automated and reliable inspection systems, with capable of examining fruit at a very high speed and measuring external properties, such as size, shape, surface texture, external defects, peel colour, and also the inner qualities with titratable acidity, TSS, pulp colour and segment drying. Their sorting systems (Figure 1.1) have been widely applied in citrus packing in China, and exported to more than 30 countries.

1.4.4.5 Packing

Wooden/carton boxes or plastic frames are still commonly used for packaging in China. Modified atmosphere packaging by using polyethylene or polyvinyl chloride films has been used for more than 40 years. In recent years, antimicrobial active food packaging has been encouraged in China.

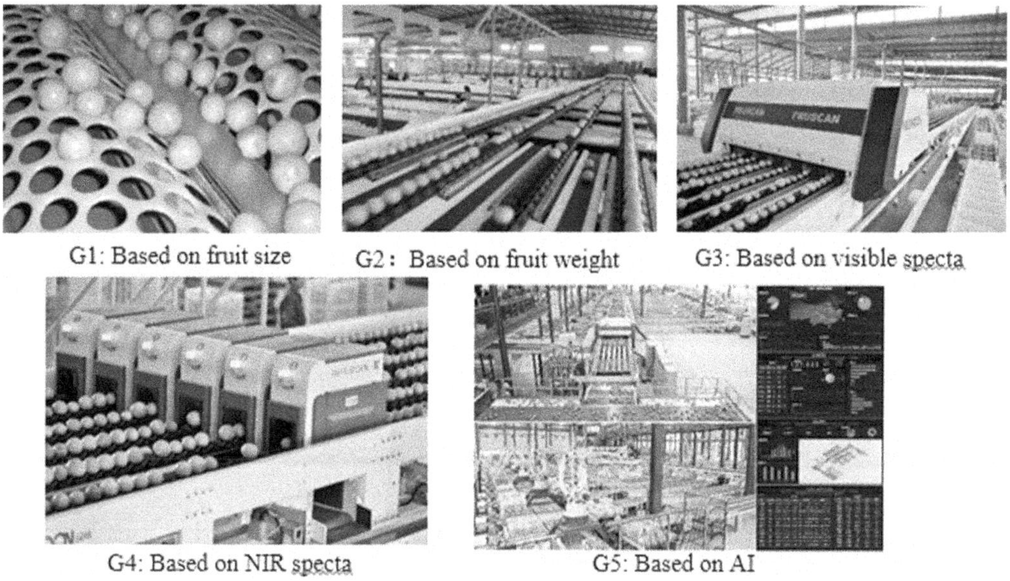

FIGURE 1.1 Five generations of fruit sorting systems made by Reemoon Technology Co., Ltd.

1.4.4.6 Storage and Transport

It's predicted that not more than 10% of the annual production in China will be stored for the long term. The cold storage temperature ranges from 4°C to 8°C for all citrus types or 12°C for lemons. The cold chain for storage and transport of citrus fruits, including precooling after package, cold storage prior to transport, cold storage in transit during shipment, and subsequent cold storage at wholesalers, retailers, and finally consumers, has been developed very fast in recent years, with a storage capacity of ca 220 Mm3, and refrigerated trucks of ca 380,000 units in 2022.

1.5 STATUS OF CITRUS INDUSTRY

1.5.1 Harvesting Window Of Major Cultivars

As China has a huge geographical span for citrus plantation, thus the climate diversity and abundant varieties have made a year-round supply of citrus fruits in China realized since a decade ago. The ripening period of major citrus varieties in China is shown in a sketch map (Figure 1.2). From the perspective of quantity, the proportion of mid and mid-late varieties occupies ca 70% of the total yield, including major ones such as Satsuma, Shatangju, Ponkan, Nanfengmiju, Newhall, Bingtangcheng, Guanximiyou. Satsuma mandarins cultivated via protected facilities and in the field in Huaning County, Yuxi Prefecture, Yunnan Province, could mature early in Jun. and Jul., respectively. Late-matured varieties such as Shiranui and Asumi could be marketed in Mar. and Apr. in Sichuan, and Orah in Jan. in Guangxi, and harvested in Jul. at the latest in Yunnan. The summer flowers of some early-matured varieties such as Miyagawa Wase, Beni madona could also be harvested in Feb. by taking advantage of climate in Yunnan.

1.5.2 Marketing Network of Citrus

In the planned economy era, the circulation of citrus mainly relied on the domestic supply and marketing cooperative system with densely covered network in China, though the total citrus production by then was very small, actually in short supply, no marketing pressure at all. In 1982, the reform of the household contract responsibility system started in China's rural areas, citrus plantation had then been based on small-scale peasant management, which was weak in marketing. Since 1983, the specialty crop tax had been addressed to citrus, which had gradually been cancelled

FIGURE 1.2 Harvesting window of major cultivars.

for all citrus-producing provinces by 2004, then no agricultural tax at all by 2006 throughout the country. Due to small-scale peasant management and information asymmetry, the transaction scale was usually small, mainly by wholesales of small vendors and roadside stalls for a long period. Unlike main grain crops, price regulation was not approached to cash crop citrus, therefore the development of citrus marketing network is much more lag behind than that for main grain crops, self-organizing for citrus marketing is of much importance. Initially, citrus specialized cooperative and citrus industry association have been practical in the main citrus-producing areas, followed by a model of "company + farmer households", of which the company acts as a bridge between farmers and markets, offering mutual benefits. Along with the fast development of the citrus industry, in particular within the latest 15 years, to build a modern marketing network has been becoming a real challenge towards the Chinese citrus industry. So far, ca 2,000 associations/economic organizations and 40,000 enterprises have been registered in somehow relating to citrus, however, the majority is still on a small scale, among which a few have become leading enterprises such as Beijing Huiyuan Juice Group Co., Ltd.; Zhongshan Yangshi Nanbei Fresh Fruit Co., Ltd.; Shenzhen Pagoda Industrial (Group) Co., Ltd.; Zhejiang Huangyan Canned Food (Group) Co., Ltd.; Zhejiang Wangbuliao Citrus Industry Cooperative; Chongqing Three Gorges Construction Group Co., Ltd.; Chongqing Mengtai Biotechnology Co., Ltd.; Hunan Chic Foods Co., Ltd.; Meizhou Lijinyou Agr. Co., Ltd.; and Hubei Qugu Foods Co., Ltd., etc. In order to improve the degree of farmers' organization and pricing power, since the beginning of this century, a few models of marketing management such as "company + base + farmer", "company + cooperative + base", and "cooperative + supermarket", have been demonstratively extending, thus reducing intermediate circulation even to citrus zero docking mode. Entrusting the acquirer to act as an agent is becoming a popular mode, and strengthening sales through alliance is also operating, e.g. Shenzhen Pagoda Industrial (Group) Co., Ltd. has established a market chain of ca 5,600 of franchise stores throughout the country since 2004. With the advent of the Internet, more and more people living in cities even towns in China choose to shop online via the most commonly used e-commerce platforms such as Taobao, Pinduoduo, JD, and Tmall, through which the proportion of citrus sold is still increasing. In Chongqing, e.g. there are 1472 citrus enterprises with e-commerce representatives such as "Navel SwO E-point Connection" in Fengjie (5,000 t sold in 2023), and "Tiannongbabu Agr. Tech. Co., Ltd." in Liangping, and "Chongqing Three Gorge Citrus Trading Center" in Zhongxian, etc.

1.5.3 Status of Export and Import of Citrus

China exports and imports several types of citrus products, mainly including fresh fruits, orange juice, and canned mandarin segments. The trade data of citrus fresh fruits within the last nine years in China is briefly reviewed (Figure 1.3).

1.5.3.1 Exportation of Citrus Fresh Fruits

The volume of exportation has kept relatively stable within the last decade. An average of 0.93 Mt (ca 2% of the total production) was exported to 87 countries with an average value of US$1,197.5 m. The top 10 destination countries ranked by Vietnam, Russia, Thailand, Malaysia, Philippines, Netherlands, Indonesia, Myanmar, Kazakhstan, and Kyrgyzstan, occupying 86% of the total annual export volume, whereas 69% of destination countries each averagely imported a small volume of less than 1,000 t per year. Among the main four types of citrus, the exportation proportion of volume and value, averagely ranked by mandarins for 70.81% and 73.94%, pomelos for 18.38% and 12.28%, oranges for 6.19% and 7.51%, and lemons for 4.54% and 6.20%, respectively. Vietnam was the major importer of 29.36% of mandarins and 67.52% of oranges, followed by the Netherlands for 38.58% of pomelos, and the Philippines for 28% of lemons. There was a significant difference in the volume of exportation between months, with over 70% of citrus fresh fruits exported from Oct. to next Jan.

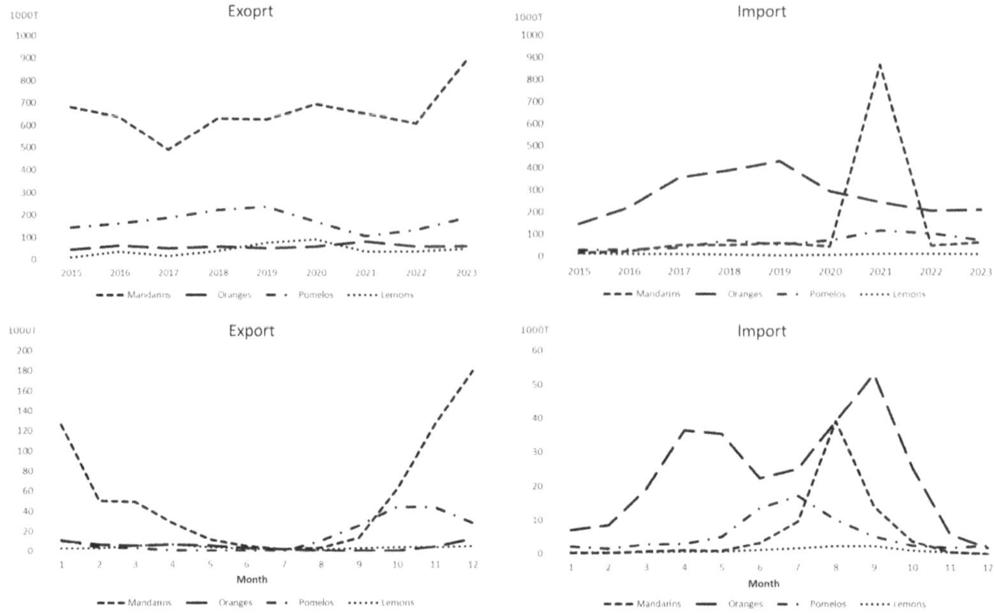

FIGURE 1.3 China's trade of citrus juice and canned mandarin segments in the past nine years.

1.5.3.2 Importation of Citrus Fresh Fruits

China imported ca 0.50 Mt of citrus fresh fruits annually with US$0.47 m in value from 27 countries. South Africa was the largest source for 52.68% of citrus fresh fruits imported to China, followed by Egypt and Australia for 17.7% and 11.01%, respectively. Among the main four types of imported citrus, oranges were the largest for 56.29% of total volume, mandarins, pomelos, and lemons accounted for 27.81%, 13.55%, and 2.35%, respectively. Although China is the largest citrus producer in the world, a very small proportion of fresh fruits is still in demand for importation to fill the insufficiency of the off-season market. Therefore, most of the citrus fruits imported to China were from countries located in the south hemisphere and during the period of Apr.–Sep., not primarily harvesting season in China. Citrus imported during this period occupied 70.4% of annually importing, in particular 50.74% concentrated from Jul. to Sep.

1.5.3.3 Trade of Orange Juice and Canned Mandarin Segments

In the past nine years, China has imported 0.77 Mt of orange juice with US$125.27 m in value, gradually increasing (84,500 t in 2020, 136,400 t in 2021, and 144,300 t in 2022), but only exported 15,902 t. Since the beginning of this century, China has become the main producer and exporter of citrus canned food, occupying ca 80% of the world trade. In the last decade, China has annually produced around 0.5 Mt of citrus canned food, of which ca 70% was used for exportation to over 60 countries, mainly to the USA (over 0.1 Mt), e.g. ca 0.28 Mt for exportation with US$371.2 m in value in 2022.

1.5.4 Status of Citrus Processing

Over 95% of China's total citrus production is consumed through fresh market, less than 5% used for processing.

1.5.4.1 Orange Juice

In the past two decades, the processing capacity of China's orange juice has gradually improved. China's annual output of citrus juice is less than 0.3 Mt (calculated as concentrated juice), while the demand is close to 1 Mt, and over 0.6 Mt demands importation. China has more than 10 large processing enterprises mainly distributed in Chongqing, Jiangxi, Sichuan, Fujian, and Yunnan Provinces, with more than 100 juicers, meaning an annual processing capacity of more than 1 Mt. However, the actual processing volume of citrus in China is far lower than its production capacity, mainly due to the insufficient supply of raw materials for processing left from the current fresh market.

1.5.4.2 Canned Mandarin Segments

China has continuously strengthened the advantages of the citrus canned food industry through upgrading of technology and equipment, e.g. applying enzymatic peel and de-capsulating technology instead of the traditional manual peel and acid-base capsule removal to produce citrus slices and fruit capsules, which meets the huge demand for citrus fruit raw materials for the explosive development of new tea in China.

1.5.4.3 Traditional Chinese Medicine and Snack Food

Since ancient times, China has used immature fruits of *Fructus aurantii*, *P. trifoliata*, etc. as traditional Chinese medicine for the treatment of lung diseases. Studies have confirmed that the content of synephrine, limonin, naringin, and neohesperidin in young or fallen citrus fruits is higher than that in matured fruits, and all of them have higher medicinal value, such as weight loss, immunity enhancement, prevention, and treatment of cardiovascular diseases, allergy, and inflammation. In addition, synephrine is the main functional component of *F. aurantii*, which in addition to its effects in regulating Chinese "qi" (vital energy) and promoting lung, also has significant weight loss function, replacing the diet drugs with serious side effects currently on the market. In Sichuan, Hunan, and Chongqing, there are citrus orchards dedicated to the cultivation and processing of traditional Chinese medicine to supply the domestic and foreign traditional Chinese medicine industry chain. Tangerine peel (namely Chenpi in Chinese) is made from the naturally dried citrus peel [the most used cultivar is called Chachigan (*C. reticulata* cv. chachiensis)]. In this process, it produces a special and pleasant aroma through microbial fermentation. Meanwhile, it has stronger medicinal value, especially for curing pneumonia and cough. Xinhui County, Jiangmen Prefecture, Guangdong Province, is one of the main producing areas of orange peel, through industrial restructure after the citrus industry was ruined by HLB since 2009, the development of ca 6,667 hm^2 orange peel industry had created a myth of RMB 6.6 billion in value in 2018.

1.5.4.4 Utilization of Citrus Peel Residue By-Products

China has gradually promoted the development of citrus intensive processing, thus extended the industrial chain with more added value. The whole fruit resource utilization can not only produce juice, canned food, beverages, jam, vinegar, wine, beer, etc., but also produce essential oil, pectin, dietary fibre, flavonoids, limonoids, molasses, etc. At present, over a hundred of citrus by-products via processing could be bought in the domestic market for dairy usage, though the proportion of citrus by-products among daily necessities is still very low. Overall, the development space for processing in particular intensive processing is still large in China. Taking a few cases as examples: Many Chinese processing enterprises of orange juice refined filtered pulp by-products for usage of cookie fillings or yogurt. A number of enterprises in Shandong and Chongqing extract pectin from the peel residue of SwO or lemon juice, and the annual output of pectin exceeds 1,000 t. Meizhou Lijinyou Agricultural Co., Ltd. annually extracts more than 1,500 t of naringin from immature pomelo fruits.

1.5.5 Status of Safety Inspection

The Citrus and Seedling Quality Inspection and Testing Centre (CSQITC) of MARA, has been established at CRI, CAAS since 1991, and well equipped with advanced detection instruments in the past two decades. As a third-party and non-profit technical service for social welfarc, CSQITC has conducted continuous years of risk investigation and safety assessment on the quality and safety status of citrus in nine major citrus production provinces, focusing on the three main risk factors, including fungal toxins such as citrinin, patulin, alternaria toxin, heavy metals such as nickel, lead, cadmium, arsenic, pesticides such as imidacloprid, feniconazole, probromophos, bifenthrin, etc. Risk assessment analysis showed that those factors were mainly concentrated on the fruit peel, but the content was extremely low and did not exceed the standard, e.g. pesticide residues accounted relatively low for 85%–96% of the entire fruit and extremely low in the flesh under the terms of restrictively sampling the whole fruit by the national limit standard request, more than 96% of fresh fruit samples met the national limit standard (GB 2763). Based on the residual behaviour of various pesticides during citrus processing studied by CSQITC, juicing was an effective way to reduce pesticide residues, and the dietary exposure risk of pesticide residues to the population was lower after juicing. Most fat-soluble pesticides were enriched in essential oils, while some pesticides might experience slight enrichment in peel residues, which could be reduced through microwave and ultrasonic treatments, adsorption, and distillation. Overall, pesticide residues are the main factor affecting the quality and safety of Chinese citrus, and the overall quality and safety of Chinese citrus remains at a high level, with guaranteed consumer safety. By now, 76 national and industry standards for citrus (excluding the detection method standards for pollutants and pesticide residues), have been released, including 11 pre-production standards for the production environment and seedlings, 16 in-production standards for production regulations, 40 post production standards for product grade and testing, and nine standards for processing technology and products. In addition, the main citrus-producing areas have formulated over a hundred quality standards for nursery trees and commercial fruits, as well as supporting local standards for orchard construction, budling propagation, cultivation, protection, harvest, storage and preservation, based on the development of local characteristic varieties.

1.5.6 Status of Citrus Research and Extension

In China, scientific research on citrus was initiated by Profs. WU Gengmin, ZENG Mian, and ZHANG Wencai, et al. in the 1930s. Currently, many universities and research institutions are conducting the research on citrus. As indicated by the number of publications on citrus research, China published 3,358 research articles in 2011–2020, ranked the first in the world.

1.5.6.1 Major Institutes/Universities/Platforms

CRI, CAAS (now CRI, SWU, in Beibei, Chongqing) found in 1960 was the only state level citrus specialized institute by then. In 1992, CRI, HZAU (in Wuhan, Hubei) was established by MOA, aiming at the development of citrus biotechnology. The main citrus-producing provinces have set up their own CRIs / Fruit Tree Res. Inst. (FTRI) / Horticulture Res. Inst. (HRI), etc., mostly attached to the provincial Academy of Agricultural Sciences (AAS), including CRI, Zhcjiang AAS (in Huangyan); FTRI, Chongqing AAS (in Jiangjin); FTRI, Guangdong AAS (in Guangzhou); FTRI, Fujian AAS (in Fuzhou); HRI, Hunan (in Changsha); HRI, Sichuan AAS (in Chengdu); Fruit and Tea Res. Inst., Hubei AAS (in Wuhan); HRI, Guangxi AAS (in Nanning); CRI, Guangxi Academy of Specialty Crop Sciences (in Guilin); Navel Orange Res. Inst. of Jiangxi (in Ganzhou). Other universities somehow engaged in citrus research are those located at the main citrus-producing areas, such as Zhejiang Univ., SWU, South China Agr. Univ., Hunan Agr. Univ. (HNAU), Jiangxi Agr. Univ., Guangxi Univ., Fujian Agr. and Forestry Univ. (FAFU), and Gannan Normal Univ., etc. More recently, more and more units, such as Plant Protection Inst., CAAS, Inst. of Microbiology, Chinese Academy of Sciences (CAS), etc., joined the HLB research issue. In those major citrus-producing prefectures, some even counties, local governments also set up their own citrus specialized research units. However, non-governmental organizations (NGOs) for citrus research are seldom registered in China. In the past two decades, a large number of platforms such as various-level key labs, and research centres, have been set up for accelerating the innovation activities for citrus research, mostly attached to the above-mentioned research organizations, just taking three state level platforms set up by the Ministry of Science and Technology (MOST) as examples: National Key Laboratory for Germplasm Innovation & Utilization of Horticultural Crops (hosted by HZAU), National Citrus Engineering Research Centre (hosted by CRI, SWU), and National Navel Orange Engineering Research Centre (hosted by Gannan Normal Univ.).

1.5.6.2 Significant Progress in Breeding

Through the efforts of generations, great progress in many aspects of citrus research has been achieved in the past 60 years, particularly from this century, China has been becoming one of the leading countries in citrus research in the world. In 1978, the triploid plantlets had been successfully produced by culturing the endosperms of *C. grandis* and *C. sinensis* at CRI, CAAS. More than 140 new citrus varieties have been released from breeding programmes since the 1980s. In the past 15 years, significant progress has been achieved in citrus genome research. The genomic sequence of *C. sinensis* was assembled in 2012, and omics technology has been fully applied for mining the functional genes in citrus, together with other advanced breeding technologies such as molecular marker-assisted breeding, shortening juvenile phase in breeding process, and gene editing, laid a foundation for genomics-assisted breeding in citrus. The breeding of new varieties is being accelerated with the progress of technology, average release of new varieties increased from one variety per year in the 20th century to 3 and 7 varieties per year in each decade of this century, respectively.

1. **Germplasm**: Nature habitats of wild species of papeda and mandarin were discovered in Sichuan, Hunan, Hubei, Guangxi, and Yunnan, and several new species of citrus such as *C. honghensis*, *C. mangshanensis*, *C. daoxianensis*, and *Poncirus polyandra*, were identified. Many elite cultivars, such as Nanfenmiju, Jincheng, Shatianyou, Bendizaoju, and Dahongtiancheng, were identified from a large number of landraces. A new type of rootstock, Ziyang Xiangcheng (*C. junos*) was identified to be tolerant to alkaline stress and iron deficiency, which is one of the most important rootstocks in China. Further exploration discovered the mini-citrus, a mono-embryonic early flowering germplasm of *F. hindsii*. The homozygous inbred lines of mini-citrus were developed by HZAU for use as an ideal model plant for functional genomic research in citrus. In the early 1980s, National Citrus Germplasm Repository (NCGR) was established at CRI, CAAS in Chongqing. By 2020, 1,990 accessions of citrus germplasms were preserved *ex situ* in NCGR, including more than 400 citrus germplasm resources from 17 countries mostly introduced in 1972–1990. The collection covers 24 representative species and 14 varieties of germplasms from *Citrus* and its eight relative genera of the Rutaceae family, among which, 73.5% of accessions are indigenous. Over 800 accessions have been investigated for phenotypic traits, and genotyped at the whole genome-wide level for SNP/Indel identification, 213 were selected as the representative germplasm for gene function study, and 128 were applied as core parental germplasm for breeding. Citrus germplasm is also preserved *in vitro* in China. In the early 1990s, HZAU began studying *in vitro* preservation of citrus. A citrus callus bank of 101 accessions was set up before 2007. A system of cryopreservation of citrus *in vitro* was also established with over 90% of regeneration rate after cryopreservation of shoot tips. Over 500 accessions of virus-free citrus germplasm are preserved in virus-free germplasm repository in CRI, CAAS. Some citrus germplasm resources are also preserved in several provincial- even prefecture-level citrus germplasm repositories.

2. **Varieties**: Systematic selection of citrus varieties in China began in the 1930s, mainly in the region of current Sichuan and Chongqing. More than 10 improved SwO varieties were selected and some of them were the important cultivars in Chongqing, Sichuan, Hubei, and other provinces. Since the 1950s, the citrus breeding has gradually become systematic, standardized and institutionalized. Many varieties were selected from the variants of landraces in the 1950s–1980s. Some of those varieties such as Shatianyou and Yuhuanyou pomelo, Nanfengmiju and Ponkan tangerines, Guoqing No. 1 Satsuma mandarin, Jincheng and Taoyecheng sweet orange, Yangshuo kumquat, etc., are still in large production in China. Since the 1990s, more and more varieties were bred by selection of bud mutation and hybridization. Many new varieties were selected from the bud mutation of cultivars, such as red flesh pomelo Shanhongyou and Hongmianyou from Guanximiyou, early and late-maturing blood orange from Tarocco, SwO Jinhong, Jinyu, and Jinmi from Bingtangcheng, Satsuma mandarin Ekan No.2 and Huayi No.1, etc. Much more attention has been paid for hybrid breeding since the beginning of this century. Elite germplasms with the characteristics such as male sterile, seedless and monoembryony have been selected as the major breeding parents. Many mandarin varieties such as Dayagan, Jinqiu Shatangju, Q Ju, Yangguang No. 1, Huamei No.7, Shangpin No.2, and Hongyun Xianggan, etc., have been bred through cross breeding. Complementing with shortening juvenile phase technology, which could promote the early flowering of hybrid offspring from 8 years to 2 years, the period of time required for hybridization breeding processing was significantly reduced. In China, the regional trial is required for register of new variety before releasing for promotion in production. The suitable planting region of new varieties is based on their adaptability to ensure the best performance in production.
3. **Genes**: In the latest 15 years, the remarkable progress has been achieved in citrus molecular genetics in China. Large number of genes controlling important traits have been identified. HZAU have sequenced and assembled high-quality genomes of some citrus species such as Valencia sweet orange, pomelo, *F. hindsii* and trifoliate orange, and constructed transcriptome, variome and epigenome databases of these species. Meanwhile, CRI, SWU has established a comprehensive database of citrus genomic variations. The key genes related to citric acid accumulation in citrus fruits was identified in the domestication regions of cultivated mandarins, providing the genetic evidence for the domestication history of mandarin. A MITE transposon insertion was found to be present in the promoter region of *CitRWP* gene, and was completely co-separated with the polyembryonic trait. The colour of citrus pulp and peel was found to be closely related to the content of flavonoids and carotenoids, and regulated by the gens of *CsMADS6*, *LCYb*, and *CCD4*. The R2R3 type *MYB* gene, a regulatory gene for anthocyanin synthesis in citrus fruits, was activated by the insertion of a copia retroposon into the promoter region of this gene, resulting in a significant colour change in blood orange varieties. The key genes regulating fruit ripening, such as ethylene receptor genes *ETR3*, *MADS-RIN*, *PHYF*, have been discovered and identified. The fruit quality of citrus is mainly affected by the ratio of sugar and acid, a series gene related to the contents of sugar and citric acid in citrus fruit were identified. Some genes related to the forming and development of flower organs were identified, such as *APETALA1*, *LEAFY*, *CiFT*, *MADS-box*, *miR156*, etc., which could be used for promoting early flowering of transgenic plants. Some genes such as *LYP2*, *FLS2*, *NBS-LRR*, *CsWAKL08z*, *CsPrx25*, *CsWRKY22* and *CsWRKY61*, were found to be associated with the resistance to canker. Knock-out the canker susceptibility gene *CsLOB1* significantly enhanced the resistance of sweet orange to canker. The work also discovered many genes related to the abiotic stress tolerance in citrus. Genes such as *CsCBF1*, *PtrICE1*, *CsPIF8*, *ERF6*, *PtrERF109*, *PtrERF108*, *COR*, *PtrBAM1*, *PtrA/NINV7*, *PtrGST U17*, *PtrRafS*, etc., were found to be related to cold

tolerance, *Dehydrins*, *LEA*, *CiFRI*, *CiBZR1*, *PtrABF3*, *BMA3*, *PtrADC*,*PtrMYB*, *WRKY70*, *CsNF-YA5*, *HATs*, and *HDAC* were linked to drought tolerance, and*FcABF2*, *FcWRKY40*, and *CrNCED1* were responsible for salt tolerance.

1.5.6.3 Significant Progress in Protection

In the past 40 years, especially the latest two decades, significant research progress in citrus protection has been made, in particular the construction of nationwide virus-free propagation scheme, quarantine-free zone, and HLB interception zone, etc., has laid a profound base. In the past decade, a few ongoing projects have also been aiming to protect the fruit out-appearances, thus to increase high-quality fruit rate. Taking a few recent researches as examples:

1. **Virus**: A series of rapid diagnosis techniques have been set up, six new citrus-associated viruses have been identified in cultivated citrus in China through high-throughput sequencing. 44 virus isolates of the *Closteroviridae* family in wild citrus habitats were identified via metatranscriptomic analysis to reveal that frequent horizontal gene transfer, gene duplication, and changes in expression strategies contribute to the complexity and diversity of functional genomes. Since these newly discovered viruses from wild citrus may become pathogenic factors, monitoring their pathogenic and epidemiological characteristics is crucial for global citrus production.
2. **HLB**: A few key functions of *C*Las effectors have been discovered, e.g. SDE3 and SDE4405 could suppress autophagy-mediated immunity to promote infection; AGH17470 could induce plant immunity, whereas AGH17488 could enhance host susceptibility, etc. The vitellogenin-like protein of *D. citri* could significantly increase *C*Las titer, and the miR-275 (microRNAs) etc. of *D. citri* hijacked by *C*Las could improve the fecundity. The ongoing State Key R&D Plan Project (Study on catastrophe mechanisms and sustainable prevention and control technology of HLB) is putting the following issues on the priority, such as screening/generating of small molecular compounds/nanodrugs with potential curing effectiveness, creating materials against *C*Las via transgenic technology and/or gene-editing technology, real time monitoring of psyllids via AI technology, etc.
3. **Fungi**: 18 species of *Botryosphaeriaceae* have been identified as causing branch diseases across major citrus production areas in China, which were pathogenic to *C. reticulata* shoots *in vitro*, with *Lasiodiplodia* being the most aggressive genus both *in vitro* and *in vivo*. Microbes such as *Pantoea* asv90 and *Methylobacterium*asv41 were discovered to inhibit the pathogen via phyllosphere microbiome analysis, calling "cry for help" strategy, and *Sphingomonas* spp. also play a role in iron competition, suggesting that iron competition plays a key role in their disease suppression mechanisms. These findings provide new insights into plant-microbe-pathogen interactions and offer promising strategies for controlling plant diseases by manipulating microbial communities.
4. **Pest**: A few key microRNAs have been identified to function as regulators, e.g. miR-1-3p had an intermediary function required for male sex determination in early embryogenesis of *Bactrocera dorsalis*, miR-275/305 cluster was essential for maintaining energy metabolic homeostasis in *B. dorsalis*, a miR-9b-mediated signal cascade could control both wing dimorphism and development of *Aphis citricidus*. These researches highlight the extensive function of microRNAs, and also provide potential targets for RNAi-based pest control. The gut microbes of citrus pests also attracted attention, a dual oxidase gene (BdDuox) played an important role in regulating the homeostasis of the gut bacterial community in *B. dorsalis*. In addition, the gut microbiota of *B. dorsalis* could promote the host resistance to low-temperature stress, providing new targets for pest control. The ongoing projects are mainly focusing on the interaction amongst pathogen-plant-vector / pathogen-plant-microbe, the function of

pathogen proteins and effectors, clarifying the key signal pathways and regulatory networks of citrus response to pathogen infection, etc.

1.5.6.4 Significant Progress in Other Aspects

Significant progress in cultivation has also been made to especially adapt to the hilly and mountainous region, the layout and construction of orchards have basically reached the ecological and environmental protection standards, the waste of water and fertilizer has been greatly reduced, the input of cultivation and management labour has been reduced by more than two-thirds, the demand in drones and mechanized planting have been largely stimulated. Along with the enlargement of orchard average scale and improvement of technologies addressed, the per unit yield and high-quality fruit rate have reached the world average level. The goal of a year-round supply of fresh fruits has reached in China. Research over the past two decades has also made significant progress in screening antagonistic microorganisms, elucidating antimicrobial mechanisms, and improving biocontrol efficacy, e.g. cold atmospheric plasma (CAP), ozone, GRAS inorganic or organic salts, and natural antifungal compounds have been used to control green mould and blue mould. As well as in the construction of safety and health limit standard system, it is gradually aligning with international standards. In the 1970s, China initially established the relevant pesticide residue limit standards (GB n53-1977). Currently, GB 2763-2021 [National Food Safety Standard – Maximum Residue Limits (MRLs) for Pesticides in Food], has addressed 217 pesticides related to citrus and 537 MRLs for citrus fruits. So far, 145 active ingredients of pesticides are registered for citrus, including 75 insecticides and acaricides, 57 fungicides, 15 herbicides, and 18 plant growth regulators in China. At present, among the MRLs developed by CAC and major citrus trading partner countries in China for citrus, there are 122 pesticide MRLs in China that need to be developed or improved, such as boscalid, carbaryl, clothianidin, fludioxonil, methoxyfenozide, pyraclostrobin, and spinetoram. There is a gap between the residue limit standards of some high-risk pesticides and those of foreign countries, and their rationality is being studied, such as profenofos, methidathion, triazophos, carbendazim, etc (Liang et al. 2022).

1.5.6.5 Extension

The extension of citrus technology in China is led by MARA via a network composed of four levels (province, prefecture, county, and township). The agricultural technology extension centres or stations are established at all levels to undertake extension and implementation of agricultural technologies. Since 2007, "National modern agricultural industrial technology system" has been established by MARA, implementing with "Technical envoy system" organized by MOST, to promote technology extension. These two systems are composed of selective citrus experts throughout the country to participate in technical guidance and training supported by state specialized project funds. Since 2015, in order to complete the task of poverty alleviation and prevent return to poverty in the poverty-stricken areas, especially the western minority areas, "Partner assistance system" organized by Organization Department of the CPC Central Committee has also been helping technology extension. The other teams include the experimental demonstration bases and courtyards of experts voluntarily supported by universities and research institutes, and the grassroots technical extension team composed of salesmen from numerous agricultural supplies and sales stores throughout the countryside. In addition, some well-managed citrus orchards are often visited and learned by surrounding growers to promote and extend their technology. There are many means for extension, the most common form is delivered by both the scientific and technical experts and local staffs in charge of extension, or organized by cooperatives or associations, through on-site visiting and training, distributing information materials such as easily understanding sheets, brochures, audios, and videos, etc. In the latest decade, new media such as QQ, WeChat, TikTok, small APPs, internet meetings, etc., have been becoming popular forms for extension.

1.6 CHALLENGES AND STRATEGIES

Although the volume of citrus production in China has become the largest throughout the world since 2007, the per unit yield and high-quality fruit rate have just reached the world average level by now. In order to accelerate the process of transforming into a strong citrus production country, many challenges are still faced by the citrus industry in China. **Firstly**, along with the northward movement of citrus psyllids from southern China, HLB outbreak risk in more areas has been increasing, and the pressure to prevent and control other pests and diseases has also been increasing, whereas more stringent standards have been applying for ecological improvement and environmental protection, the conflict between them is needed to be solved by more advanced green technology, etc. **Secondly**, along with the transferring of rural labour to cities driven by urbanization policy, the citrus industry has been facing a shortage of strong growers, increasing labour/land costs, furthermore, the gap in science and technology between extension staff and the rural labour force has been enlarging, thus labour-saving cultivation technology and grassroots specialized service teams are needed to be strengthened. **Thirdly**, the majority of growers with small-scale operations are still weak in marketing with less pricing power and information asymmetry, whereas the proportion of citrus production used for processing and exportation in China is still very low, even that for post-harvest handling of citrus fruits is relatively low, and extreme early-/extreme late-matured varieties are still very much in demand, so there is a large space left for growers and operators to get benefits from growing new varieties, upgrading high-quality fruit rate, and enlarging transaction scale via modern market system, etc. Taking the current supply situation of orange juice as an example to explicate a good window period for China's orange juice processing industry: For recent years, global orange juice has been in serious demand, mainly due to a great reduction in the USA and Brazil, resulting in the trade price of concentrated orange juice doubled in 2023 in contrast to that in 2022, even two times increase of price for China's domestic concentrated orange juice. This trend will be hard to ease within a decade. In other words, the current situation should be the best period for China to vigorously develop the citrus processing industry, and the current average per capita consumption of orange juice worldwide is 23 times that in China, thus the domestic potential market for orange juice is pretty large too.

For the above-faced challenges, China government has been implementing a series of strategies such as "Construction advantageous agriculture regions", "Industry supports agriculture", "Agricultural supply-side structural reform", "Rural revitalization", "Promote the integrated development of the primary, secondary, and tertiary industries", and "The belt and road initiative", etc., which have been facilitating the development of the citrus industry. Developing appropriately scaled citrus operations, improving commercial services and technology extension, stepping up vocational skills training for rural citrus growers, intensively processing citrus products, and converging citrus with tourism, wellness, and traditional culture, etc., are strongly encouraged and supported by all levels government, which will promote the steady development of citrus industry in China. With regards to the science and technology issue, breeding of consumer-favoured new varieties with characteristics such as unique and rich in flavour and anthocyanins, easy peeling, seedless, etc., upgrading high-quality fruit rate, standardizing efficient and green production models with appropriately scaled operations, etc., are the striving directions for further development of citrus industry in China. Taking post-harvest handling as an example to explicate the potential space for further development: In order to optimize treatments to control fruit quality and reduce post-harvest losses, using innovative plastic films to contain water loss during storage and transit will find greater application, together with innovative washing and packaging techniques, including the use of protective atmospheres and passive modulation systems for the atmospheres inside packages. Green techniques such as physical treatments and bio-control techniques, will be preferred in the post-harvest processing. Also, the development of combined approaches, such as multi-sensors, advanced imaging technology analyses, and AI tools, to monitor quality markers in real time will significantly improve the competitiveness of the entire supply chain in China. Alongside, the cold chain is still not very widespread for citrus fruits in China, thus there exists much-developing space for citrus industry in this aspect as well. As a matter of fact, most of the above-mentioned challenges must have been met by many countries, in particular those with similar citriculture environments such as Japan and Korea, who had dealt well with those issues some decades ago. We believe that the faced science and technology issues will be solved step by step by scientists, in particular with the cooperation of international citrus community.

1.7 CONCLUSION

This chapter briefly reviews the development process of the Chinese citrus industry in the past 60 years, particularly in the last two decades when the fast-forward button was pressed by the Chinese government. Other than a big pace made for the volume of citrus production, much progress has also been achieved in various aspects of China. Over 140 new citrus cultivars have been released for citrus industrial structure adjustment, a year-round supply of fresh fruits has been reached, and the industrial chain continues to extend. Over 95% of citrus fruits produced in China are consumed in the domestic fresh market, loose-skin mandarins and tangerines are still the main part, over 80% in 2000 and ca 60% by now in China. The amount of fresh fruits and canned food for exportation in the latest decade has kept stable at ca 1 Mt and 0.3 Mt, respectively, mainly mandarins/tangerines. The currently faced challenges and strategies are also briefly introduced in the chapter, there still exists developing space for orange juice processing and exportation, etc. in China. We trust that breeding will be further accelerating with the help of omics technology, particularly for breeding processing-used cultivars, etc. In general, many common problems would be encountered by most citrus-producing countries, among which some could be solved through international cooperation. Since China's opening up to the world in 1978, the pace of internationalization has stridden forward for the citrus industry as well, nowadays many citrus scientists in China have overseas study experience, mainly through cooperative projects and scientist/student exchange programmes, in particular with developed countries. We believe that win-win collaborations amongst Asian countries will bring a bright shared future for the Asian citrus community, various means of learning with each other are available for the exchange of citrus resources and technologies, even for the formulation of new trade policies. We believe that "Belt and road initiative" strategy raised by the Chinese government will benefit the citrus industry in Asian developing countries. Currently, many Asian countries are facing the challenge of HLB outbreak, for those areas where HLB occurrence is getting close by or at a low epidemic level, it is suggested to put prevention first, treasure the window period, don't be misled by "easily curing of HLB" saying, at least not for now.

REFERENCES

Alam, M.A., Subhan, N., Rahman, M.M. et al (2014). Effect of citrusflavonoids, naringin and naringenin on metabolic syndrome and their mechanisms of action. *Advances in Nutrition,* 5(4):404–417.

Bai, J., Tan, X., Tang, S. et al (2024). Citrus p-synephrine improves energy homeostasis by regulating amino acid metabolism in HFD-induced mice. *Nutrients* 16(2):248.

Cámara, M.A., Cermeño, S., Martínez, G. et al (2020). Removal residues of pesticides in apricot, peach and orange processed and dietary exposure assessment. *Food Chemistry* 325:126936.

Chen, H.M., Li, Z.A., Wang, X.F. et al (2014). First report of Citrus yellow vein clearing virus on lemon in Yunnan, China. *Plant Disease* 98:1747.

Chen, T., Ji, D., Zhang, Z. et al (2021). Advances and strategies for controlling the quality and safety of postharvest fruit. *Engineering* 7(8):1177–1184.

Chen, Y., Pan, H., Hao, S. et al (2021). Evaluation of phenolic composition and antioxidant properties of different varieties of Chinese citrus. *Food Chemistry* 364:130413.

China, Seed, Industry. (n.d.). *Big Data Platform.* http://202.127.42.47:6009/Home/BigDataIndex (in Chinese).

China Southern Fruit Tree Magazine (2004). *Chinese citrus practical and technical literature compilation (1972–2002).* Chongqing Press, Chongqing (in Chinese).

Dahro, B., Li, C.L., Liu, J.H. (2023). Overlapping responses to multiple abiotic stresses in citrus: From mechanism understanding to genetic improvement. *Horticulture Advances* 1:4.

Deng, X.X. (2022). A review and perspective for citrus breeding in China during the last six decades. *Acta Horticulture Sinica* 49(10):2063–2074 (in Chinese).

Deng, X.X. (2023). *Citrus variety in China* (2nd edn.). China Agriculture Press, Beijing (in Chinese).

Deng, X.X., and Li, L. (2010). *Production technology of citrus standard orchard.* China Agriculture Press, Beijing (in Chinese).

Deng, X.X., and Peng, S. (2013). *The citrusology*. China Agriculture Press, Beijing (in Chinese).

Dong, C., Zhao, Q.Y., Li, J. et al (2021). Residual dynamics of pyraclostrobin in citrus juice processing. *Chinese Journal of Pesticide Science* 23(6):1219–1226 (in Chinese).

Du, J., Wang, Q.Y., Shi, H.W. et al (2023). A prophage-encoded effector from "*Candidatus* Liberibacter asiaticus" targets ASCORBATE PEROXIDASE6 in citrus to facilitate bacterial infection. *Molecular Plant Pathology* 24:302–316.

Du, J., Wang, Q.Y., Zeng, C.H. et al (2022). A prophage-encoded nonclassical secretory protein of "*CandidatusLiberibacter asiaticus*" induces a strong immune response in Nicotiana benthamiana and citrus. *Molecular Plant Pathology* 23:1022–1034.

FAOSTAT. (n.d.). *Database of Food and Agriculture Organization of the United Nations.* www.fao.org/faostat/en/#home

Feng, Y.R. (2019). *Study on processing factors of pesticide residues of five common preservatives in orange and orange juice*. Dissertation, Tianjin Agricultural College (in Chinese).

GACC. (n.d.). *Database of General Administration of Customs of the People's Republic of China.* http://stats.customs.gov.cn/ (in Chinese).

Han, K., Wang, Y.M., Zhou, H.Y. et al (2022). Residual behavior of cyenopyrafen in the process of orange juice. *Food and Fermentation Industries* 48(12):24–29 (in Chinese.).

Hao, Y., Kang, J., Guo, X. et al (2023). p.H.-responsive chitosan-based film containing oregano essential oil and black rice bran anthocyanin for preserving pork and monitoring freshness. *Food Chemistry* 403:134393.

Huang, D., Yuan, Y., Tang, Z.Z. et al (2019). Retrotransposon promoter of ruby1 controls both light- and cold-induced accumulation of anthocyanins in blood orange. *Plant, Cell & Environment* 42(11):3092–3104.

Huang, Y., Xu, Y.T., Jiang, X.L. et al (2021). Genome of a citrus rootstock and global DNA demethylation caused by heterografting. *Horticulture Research* 88:69.

Li, P.D., Zhu, Z.R, Zhang, Y.Z. et al (2022). The phyllosphere microbiome shifts toward combating melanose pathogen. *Microbiome* 10:56.

Li, Q., Qi, J., Qin, X. et al (2020a). CitGVD: A comprehensive database of citrus genomic variations. *Horticulture Research* 7:12.

Li, Q., Qin, X. Qi, J. et al (2020b). CsPrx25, a class III peroxidase in *Citrus sinensis*, confers resistance to citrus bacterial canker through the maintenance of ROS homeostasis and cell wall lignification. *Horticulture Research* 7:192.

Li, Y.J., Xu, J.B., Zhao, X.P. et al (2021). The dissipation behavior, household processing factor and risk assessment for cyenopyrafen residues in strawberry and mandarin fruits. *Food Chemistry* 359:129925.

Liang, D.W., Zhang, Y.H., Wang, C.Q. et al (2022) Comparative analysis of the national and international standard of maximum residue limits for pesticides in citrus. *Food and Fermentation Industries* 48(3):273–283.

Liu, Q.Y., Zhang, S., Mei, S.Q. et al (2021). Viromics unveils extraordinary genetic diversity of the family Closteroviridae in wild citrus. *PLoS Pathogens* 17:e1009751.

Liu, Y.L., Ke, L.L., Wu, G.Z. et al (2017). miR3954 is a trigger of phasiRNAs that affects flowering time in citrus. *Plant Journal* 92: 263–275.

Long, Q., Du, M., Long, J. et al (2021). Transcription factor WRKY22 regulates canker susceptibility in sweet orange (*Citrus sinensis* Osbeck) by enhancing cell enlargement and *CsLOB1*expression. *Horticulture Research* 8:50.

Lu, S., Zhang, Y., Zheng, X. et al (2016). Molecular characterization, critical amino acid identification, and promoter analysis of a lycopene β-cyclase gene from citrus. *Tree Genetics & Genomes* 12:106.

Lu, S., Zhang, Y., Zhu, K. et al (2018). The citrus transcription factor CsMADS6 modulates carotenoid metabolism by directly regulating carotenogenic genes. *Plant Physiology* 176(4):2657–2676.

Lu, S.W., Zhang, Y., Zhu, K.J. et al (2018). The citrus transcription factor CsMADS6 modulates carotenoid metabolism by directly regulating carotenogenic genes. *Plant Physiology* 176(4): 2657–2676.

Mahmoud, A.M., Hernández, B.R.J. et al (2019). Beneficial effects of citrus flavonoids on cardiovascular and metabolic health. *Oxidative Medicine and Cellular Longevity* e5484138.

National Agricultural Technology Extension Service Center of China (2010). *Simplified cultivation techniques of fruit trees.* China Agricultural Press, Beijing (in Chinese).

NBSC. (n.d.). Database of National Bureau of Statistics of the People's Republic of China. https://data.stats.gov.cn/easyquery.htm?cn=C01 (in Chinese).

Nian, X.G., Luo, Y.R., He, X.Y. et al (2024). Infection with '*Candidatus Liberibacter asiaticus*' improves the fecundity of Diaphorina citri aiding its proliferation: A win-win strategy. *Molecular Ecology* 33:e17214.

Omeroglu, P.Y., Celik, B.A., and Alibasoglu, E.K. (2022). The effect of household food processing on pesticide residues in oranges (*Citrus sinensis*). *Foods* 11:3918.

Palou, L., Valencia-Chamorro, S.A., and Pérez-Gago, M.B. (2015). Antifungal edible coatings for fresh citrus fruit: A Review. *Coatings* 5:962–986.

Peng, A., Chen, S. Lei, T. et al (2017). Engineering canker-resistant plants through CRISPR/Cas9-targeted editing of the susceptibility gene *CsLOB1*promoter in citrus. *Plant Biotechnology Journal* 15(12):1509–1519.

Peng, J.Y. (2023). *Study on cost control of Chongqing GLA Fruit Limited Company.* Dissertation, Southwest University (in Chinese).

Peng, L.Z. (2001). *Cultivation, storage and transportation technology of high quality navel orange.* China Agriculture Press, Beijing (in Chinese).

Peng, L.Z. (2007). *Citrus orchard construction and management.* Chongqing Press, Chongqing (in Chinese).

Peng, L.Z. (2009). *Technical manual of citrus disaster prevention and mitigation.* China Agriculture Press, Beijing (in Chinese).

Peng, L.Z. (2013). *Technical guide for safe production of sweet orange.* China Agriculture Press, Beijing (in Chinese).

Peng, T., Yuan, Y.Z., Huang, A.J. et al (2023). Interaction between the flagellum of *Candidatus* Liberibacter asiaticus and the vitellogenin-like protein of *Diaphorina citri*significantly influences *C*Las titer. *Frontiers in Microbiology* 14:1–12.

Peng, W., Yu, S.N., Handler, A.M. et al (2020). miRNA-1-3p is an early embryonic male sex-determining factor in the Oriental fruit fly *Bactrocera dorsalis. Nature Communications* 11:932.

Qian, Y.Z., and Li, P.W. (2018). *Agricultural product quality and safety: Introduction.* China Agriculture Press, Beijing (in Chinese).

Raza, M.F., Wang, Y.C., Cai, Z.H. et al (2020). Gut microbiota promotes host resistance to low-temperature stress by stimulating its arginine and proline metabolism pathway in adult *Bactrocera dorsalis. PLoS Pathogens* 16:e1008441.

Saini, R.K. Ranjit, A., Sharma, K. et al (2022). Bioactive compounds of citrus fruits: A review of composition and health benefits of carotenoids, flavonoids, limonoids, and terpenes. *Antioxidants* 11(2):2.

Shang, F., Niu, J.Z., Ding, B.Y. et al (2020). The miR-9b microRNA mediates dimorphism and development of wing in aphids. *Proceedings of the National Academy of Sciences of the United States of America* 117:8404–8409.

Shi, H.W., Yang, Z.H., Huang, J. et al (2023). An effector of "*Candidatus* Liberibacter asiaticus" manipulates autophagy to promote bacterial infection. *Journal of Experimental Botany* 74:4670–4684.

Shi, J.X., Gong, Y.A., Shi, H.W. et al (2023). "*Candidatus* Liberibacter asiaticus" secretory protein SDE3 inhibits host autophagy to promote Huanglongbing disease in citrus. *Autophagy* 19:2558–2574.

Shi, W.J. (2014). *Study on the determination methods of alternaria toxins in citrus and fruit Juices.* Dissertation, Southwest University (in Chinese).

Strano, M.C., Altieri, G., Allegra, M. et al (2022). Postharvest technologies of fresh citrus fruit: Advances and recent developments for the loss reduction during handling and storage. *Horticulturae* 8:612.

Sun, J.C., Wang, D.L., Liu, C.R. et al (2021). Research progress of citrus on-tree storage. *Chinese Fruits* 7: 1–6, 10 (in Chinese).

Sun, L.M., Zhang, J.Z., Mei, L., and Hu, C.G. (2014). Molecular cloning, promoter analysis and functional characterization of APETALA 1-like gene from precocious trifoliate orange (*Poncirus trifoliata* L. Raf.). *Scientia Horticulturae* 178(23):95–105.

Talon, M., Caruso, M., and Gmitter, F.G. (2020). *The genus* Citrus. Woodhead Publishing.

Wang, N.Q., Peng, L.Z., Chun, C.P. et al (2012). Study on virgin soil nutrient in southern Jiangxi province. *South China Fruits* 41(5):1–4 (in Chinese).
Wu, G.A., Terol, J., Ibanez, V. et al (2018). Genomics of the origin and evolution of Citrus. *Nature* 554:311–316.
Wu, J., Zhang, S. Atta, S. et al (2020). Discovery and survey of a new mandarivirus associated with leaf yellow mottle disease of citrus in Pakistan. *Plant Disease* 104:1593–1600.
Wu, Z., Ma, Y., Xiong, H. et al (2023). Simultaneous determination of spiropidion and its five major metabolites in sweet orange fruit and various processing by-products using ultra-high performance liquid chromatography-tandem mass spectrometry. *Food Research International* 174:113498.
Xiao, C., Wang, M., Peng, T. et al (2021). Analysis of hotspots and frontiers of Chinese citrus research based on WOS and CiteSpace in the past decade. *China Fruits* 214(8):36–42 (in Chinese).
Xiao, X.E., Wang, W., Crous, P.W. et al (2021). Species of *Botryosphaeriaceae* associated with citrus branch diseases in China. *Persoonia-Molecular Phylogeny and Evolution of Fungi* 47:106–135.
Xie, J.F., Chen, H., Zheng, W.P. et al (2022). miR-275/305 cluster is essential for maintaining energy metabolic homeostasis by the insulin signaling pathway in *Bactrocera dorsalis*. *PLoS Genetics* 18:e1010418.
Xu, Q., Chen, L.L., Ruan, X., et al (2013). The draft genome of sweet orange (*Citrus sinensis*). *Nature Genetics* 45:59–66.
Xu, Y.Y., Zeng, R.F., Zhou, H. et al (2022). Citrus FRIGIDA cooperates with its interaction partner dehydrin to regulate drought tolerance. *Plant Journal* 111(1):164–182.
Yao, L.X., Zhou, X.C., Peng, Z.P. et al (2006). Soil nutrient fertility in citrus orchards of Guangdong Province. *Chinese Journal of Soil Science* 1:41–44 (in Chinese).
Yao, Z.C., Wang, A.L., Li, Y.S. et al (2016). The dual oxidase gene *BdDuox* regulates the intestinal bacterial community homeostasis of *Bactrocera dorsalis*. *The International Society for Microbial Ecology Journal* 10:1037–1050.
Ye, L.X., Zhang, J.X., Hou, X.J. et al (2021). A MADS-box gene *CiMADS43* is involved in citrus flowering and leaf development through interaction with CiAGL9. *International Journal of Molecular Sciences* 14, 22(10):5205.
Yi, H., and Liu, H. (2022). Distribution characteristics of citrus varieties and selection of varieties suitable for planting in China. *China Fruits* 1:1–7 (in Chinese).
Yi, X.T., Zhao, C.B., Li, Y.F. et al (2019). Characteristics analysis of Marco-elements and mid-elements in citrus leaves in Guangxi Province. *Journal of Fruit Science* 36(02):153–162 (in Chinese).
Yue, Y., Wang, B., Xi, W. et al (2023). Modification methods, biological activities and applications of pectin: A review. *International Journal of Biological Macromolecules* 253, 127523.
Zhang, S., Huang, A., and Zhou, X. (2021). Natural defect of a plant rhabdovirus glycoprotein gene: A case study of virus–plant co-evolution. *Phytopathology* 111:227–236.
Zhang, S., Yang, C. Qiu, Y. et al (2024). Conserved untranslated regions of multipartite viruses: Natural markers of novel viral genomic components and tags of viral evolution. *Virus Evolution* 10:veae004.
Zhang, S.S., Wang, X.F., He, J. et al (2023). A Sec-dependent effector, CLIBASIA_04425, contributes to virulence in '*Candidatus* Liberibater asiaticus'. *Frontiers in Plant Science* 14:1–10.
Zhang, X.K., Zhao, J. F., Wang, Y. et al (2022). Molecular characterization of four Citrus tristeza virus isolates that cause severe symptoms in navel orange in Ganzhou, China. *Tropical Plant Pathology* 47:509–520.
Zhang, Y.H., Liu, C.H., Wang Q. et al (2019). Identification of *Dialeurodes citri* as a vector of *Citrus yellow vein clearing virus* in China. *Plant Disease* 103:65–68.
Zheng, X.J., Zhu, K.J., Sun, Q. et al (2019). Natural variation in CCD4 promoter underpins species-specific evolution of red coloration in citrus peel. *Molecular Plant* 12: 1294–1307.
Zhou, C.Y. (2020). The status of citrus Huanglongbing in China. *Tropical plant pathology* 45(3): 279–284.
Zhou, C.Y. et al (2020). *China's fruit science and application—Citrus*. Shaanxi Science and Technology Press, Xi'an (in Chinese).
Zhou, K., and Ye, Y. (2005). *Fruit trees of China—Citrus*. China Forestry Press, Beijing (in Chinese).
Zhou, W., Hu, M., Zang, X. et al (2020). Luteolin attenuates imiquimod–induced psoriasis-like skin lesions in BALB/c mice via suppression of inflammation response. *Biomedicine & Pharmacotherapy* 131, 110696.
Zhou, X.W., Cheng, C.F., LV, B. et al (1991). Study on the content index of mineral nutrient elements in Jincheng leaves. *Journal of Southwest University* (natural sci. ed.) 1:19–24 (in Chinese).
Zhou, Y., Chen, H.M., Cao, M.J. et al (2017). Occurrence, distribution and molecular characterization of citrus yellow vein clearing virus in China. *Plant Disease* 101:137–143.

Zhou, Y., Zhou, C.Y., Li, Z.A. et al (2008). Mild strains cross-protection against stem-pitting Tristeza of sweet orange. *Scientia Agricultura Sinica* 41(12):4085–4091 (in Chinese).

Zhu, C.Q., Zheng, X.J., Huang, Y. et al (2019). Genome sequencing and CRISPR/Cas9 gene editing of an early flowering mini citrus (*Fortunella hindsii*). *Plant Biotechnology Journal* 17(11):2199–2210.

Zhu, J., Liu, B.Y., Li, S.J. et al (2020). Effects of 2,4-D on keeping citrus fruit fresh on the tree and the underlying mechanisms. *Acta Horticulturae Sinica* 47(11):2086–2094 (in Chinese).

Zhu, K., Chen, H., Mei, X. et al (2023). Transcription factor CsMADS3 coordinately regulates chlorophyll and carotenoid pools in *Citrus hesperidium. Plant Physiology* 193(1):519–536.

Zhu, P.P., Li, Y.F., Qiu, J.Y. et al (2019). Correlation between soil and leaf nutrient element contents in citrus orchards of Yuxi City, Yunnan Province. *Journal of Fruit Science* 36(12):1658–1666 (in Chinese).

Zhuang, Y.M. (1994). *Citrus nutrition and fertilization*. China Agriculture Press, Beijing (in Chinese).

2 Citrus Industry of India

Sangeeta Bhattacharyya, Subhra Saikat Roy, and Dilip Ghosh

2.1 INTRODUCTION

On a global scale, citrus (*Citrus* sp.) is widely recognized as one of the most important perennial fruit crops. India is the third largest citrus producer globally, following China and Brazil. Citrus cultivation in India spans across 11.15 lakh (1.11 million) hectares, making it the third most cultivated fruit after mango and banana. The annual production of citrus stands at 14.57 million tonnes (Ministry of Agriculture and Farmers Welfare, Government of India, 2023). The genus *Citrus* belongs to the family Rutaceae. Citrus fruits have distinct aroma, taste and are storehouses of several important nutrients like vitamin C, phytochemicals, and dietary fibres which are of vital importance in human health improvement due to their antioxidant properties and protection from various chronic diseases (Liu et al., 2012).

2.2 STATUS OF CITRUS CULTIVATION IN INDIA

A diverse array of species can be found within this citrus genus across the globe, but in India, the commercially significant ones that are produced include mandarins (*Santra*), sweet oranges (*Mosambi*), acid limes (*Nimbu*), lemons, grapefruits, and pummelos. Mandarins (*Citrus reticulata* Blanco) are the most abundantly produced cultivars in India (Table 2.1) and these fruits secure a special place in the dietary platter of Indians mostly as fresh fruit. Different varieties of mandarins are found to exist throughout the length and breadth of the country. Nagpur mandarin is widely cultivated in the Vidarbha region (Nagpur and Amravati Divisions) of Maharashtra and also some parts of Madhya Pradesh. Kinnow mandarin is grown in Abohar and Ganganagar regions of Punjab and parts of Haryana. In Karnataka Coorg mandarin is popular while Khasi mandarin dominates in North-East India.

2.3 GENETIC DIVERSITY OF CITRUS IN INDIA

Citrus mainly comprises of five cultivated genotypic groups viz. mandarin, sweet orange, pummelo, grapefruit, limes, and lemons. However, there are different types and cultivars in all these genotypic groups. Apart from identified species, some hybrids have also been developed.

2.3.1 Botanical Classification of Citrus

The genus citrus belongs to the order Geraniales, suborder Geraniineae, and the family Rutaceae. The Rutaceae family is divided into six subfamilies, and citrus comes under subfamily

DOI: 10.1201/9781003584599-2

TABLE 2.1
Area and Production of Citrus in India (Area in '000 Ha Production in '000 MT)

	2021–22		2022–23		2023–24	
Citrus	**Area**	**Production**	**Area**	**Production**	**Area**	**Production**
Lime/lemon	312.66	3776.35	312.37	3786.72	318.13	3886.46
Mandarin	454.18	6302.55	459.91	6178.07	466.87	6149.69
Sweet orange	228.35	3895.43	231.45	3927.07	237.01	3814.15
Other citrus fruits	103.11	782.25	96.70	713.37	92.64	725.43
Citrus Total (1–4)	1098.30	14756.58	1100.43	14605.22	1114.65	14575.72

Source: Ministry of Agriculture and Farmers Welfare, Govt. of India (2023).

TABLE 2.2
Botanical Classification of Citrus

Kingdom:	Kingdom : Plantae
Division:	Division : Embryophyta Siphonogama
Sub-division:	Sub-division : Angiospermae
Class:	Class : Dicotyledonae
Order:	Order : Geraniales (21 families)
Suborder:	Suborder : Geraniineae (12 families)
Family:	Family : Rutaceae (7 subfamilies)
Subfamily:	Subfamily : Aurantioideae (2 tribes)
Tribe:	Tribe : Citreae (3 subtribes)
Sub-tribe:	Sub-tribe : Citrinae (3 groups, 13 genera)
Group C:	Group C : Citrus fruits (6 genera)
Genus:	1. *Poncirus*
	2. *Fortunella*
	3. *Citrus*

Aurantioideae. Further Aurantioideae is subdivided into two tribes *viz.*, Clauseneae and Citreae. The tribe Citreae is subdivided into three sub-tribes, namely Triphasiinae, Balsamocitrinae, and Citrinae. The sub-tribe Citrinae comprises the genus Eremocitrus, Microcitrus, Clymenia, Fortunella, Poncirus, and Citrus. Different botanists have classified citrus in various ways in their times. However, there are many ambiguities on them and none of them would justify the classification yet. Albeit, the classification of citrus by Swingle and Tanaka has been largely followed. Tanaka was liberal and assigned 145 species; whereas, Swingle was conservative and assigned only 16 species and eight botanical varieties in genus *Citrus*. The botanical classification of citrus is listed in Table 2.2.

2.3.2 Distribution of Citrus Germplasm in India

India is a huge genetic reservoir of citrus germplasm. Different species of citrus and their availability in India are depicted in Table 2.3.

TABLE 2.3
Indigenous Citrus Germplasm and Their Distribution in India

S.N.	Common name	Species name	Status	Vernacular name	Distribution
1	Mandarin	*C. reticulata* Blanco	Cultivated	Santra, Soh-niamtra, Soh-umkhudai, Sohsyiem, Naga santra, Narengi	All citrus growing regions
2	Sweet orange	*C. sinensis* (L.) Osbeck	Semi-wild and edible	Soh-niangriang, Soh-bitara, Tasi, Athugudi	All citrus growing regions
3	Acid lime	*C. aurantifolia* L.	Cultivated	Kagzilebu, *Elumbichai,* Abhayapuri lime, Kaimganj lime	All citrus growing regions
4	Lemon	*C. limon* (linn.) Burmann	Cultivated	Pati lebu, Assam lemon, Godha pati, Kata jamuri, Jora tenga, Elachi lebu, Naga changney, Soh- systong, Pani-jamir	Assam, Arunachal Pradesh, Nagaland, Meghalaya, Mizoram
5	Pummelo	*C. grandis* Osbeck syn. *C. maxima* (Burm.) Merr.	Cultivated, home gardens	Dawadi, Nowgong, Byrni, Zemabawk, Aizawil, Nai chakotha, Devanahalli, Gagar, Aijal	All over the country
6	Grapefruit	*C. paradisi* Macfad	Cultivated	Saharanpur special, Serbial, Champa umtambi	Limited scale in Karnataka, Punjab
7	Rough lemon	*C. jambhiri* Lush	Wild, occasionally cultivated	Soh-myndong, Kata jamir, Soh-jhalia, Sindhuri, Nemutenga, Kachai lemon, Mitha tulia, Kotangiri, Renuka lemon, Jatti Khatti, Jullandhari Khatti, Jambheri Kodur, Khatta Patiala, MooguNimbe	Almost all citrus growing regions of India
8	Citron	*C. medica* Linn.	Occasionally cultivated, wild	Pinchipunia, Bira jora, Mitha jora, Sipa-egra, Singking, Tashing, Soh manong, Haijange, Naya-changney, Kadarangai, Yemekaipuli	All citrus growing regions of the country
9	Karna Khatta	*C. karna* Rafin.	Wild, inedible	Soh-sarkar	Mizoram, Meghalaya (Southern slope of Khasi hills)
10	Sweet lime	C. *limetta* (Risso) Lush.	Occasionally cultivated and edible	Mitha kagzi	Hills of Assam, Mizoram, Arunachal Pradesh
11	Sweet lime	*C. limettioides* Tanaka	Occasionally cultivated and edible	Mitha nimbu, Kagzi-wak-ma, Sakka nimbu, Sarbatichikna	Hills of Assam, Mizoram, Arunachal Pradesh
12	Sour orange	*C. aurantium* L.	Wild, cultivated and inedible	Karan jamir, Gandha- Hantara, Narthangai, Kaipuli, Herale	Tamil Nadu, Andhra Pradesh, Assam, Meghalaya, Nagaland, Arunachal Pradesh
13	Sour pummelo	*C. megaloxycarpa* Lush.	Wild and inedible	Bortenga, Huknatenga, Holongtenga, Jimir tenga, Ambilbed	North-Eastern (Mainly Assam and Tripura) and several other parts of the country
14	Ginger citrus	*C. assamensis* Dutta and Bhattacharya	Wild, occasionally cultivated	Ada jamir	Barak valley of Assam and lower Jaintia hills of Meghalaya

(*continued*)

TABLE 2.3 (Continued)
Indigenous Citrus Germplasm and Their Distribution in India

S.N.	Common name	Species name	Status	Vernacular name	Distribution
15	Melanesian papeda	*C. macroptera* Montr	Wild, home gardens and inedible	Satkara, Tithkara, Soh- Kwit, Hampur-arong, Chamarbaphang, Heiribob	Assam (Cachar, Karimgunj and Karbi-Anglong districts), Meghalaya (Khasi and Garo hills), Tripura and Manipur
16	Ichang Papeda	*C. ichangensis* Swingle	Wild and inedible	Ketsa-shupfu	Nagaland and adjacent hilly areas of Assam
17	Khasi Papeda	*C. latipes* (Swingle) Tanaka	Wild and inedible	Soh-Kymphor-shrieh, Soh-shyrkhoit	Khasi hills of Meghalaya and Manipur
18	Indian Wild Orange	*C. indica* Tanaka	Wild and inedible	Mamang Narang	Some pockets of Assam, Nagaland, Manipur and Meghalaya (Garo hills) and Mizoram
19	King mandarin	*C. nobilis* Lour	Wild, edible	Jeneru-tenga	North-Eastern India
20	Kitchili	*C. maderaspatana* Hort. Ex. Tan.	Wild, occasionally cultivated, edible	Kitchli, Vadalapudi, Belladikithuli	Malnad region of Karnataka and Guntur district of Andhra Pradesh
21	Gajinima	*C. pennivesiculata* Tan.	Wild, occasionally cultivated	Gajanimma, Baduvapulli	South India
22	Calamondin	*C. madurensis*Lour.	Wild		Northeastern India
23	Cleopatra mandarin	*C. reshni Hort. Ex.Tan*	Wild	Laddu, Sylhet, Billi-kichili, Karpuratenga Naga Santra, Kodakithuli, Kokni, Soh-seim	All over the country
24	Kaula mandarin	*C. reticulata* var. *crenatifolia*	-	-	North-Western India
25	Indian grapefruit	*C. rugulosa Hort. Ex. Tan.*	Wild	Attani	North-Western India
26	Nakoor lime	*C. nakoor Singh*	-	Nakoor lime	Malnad of Karnataka
27	Kafir lime	*C. hystrix* DC	Wild		Home gardens of Andhra Pradesh and Tamil Nadu
28	Hill lemon	*C. pseudolimon Tanaka*	Semi-wild, edible	Galgal, Soh-long, Kil-Kil	North-Eastern region and foothills of Western India
29	Sadaphal	*C. semiperflorens Tan*	-	Sadaphal	Several parts of India
30	Citron hybrids	*C. limon* var. *limonum*	Occasionally cultivated, edible	Gandhraj lemon, Godhapti-lebu, Kata jamuri, Elachi-lebu	All over India
31	Rangpur lime	*C. limonia* Osbeck	Wild, occasionally cultivated	Marmalade orange, Kusai lime, Sylhet lime	North-Eastern and Central India

(Updated from Singh and Chadha, 1993; Sharma et al., 2004).

2.3.3 Commercial Cultivars

There are certain genotypes in citrus, which are directly exploited as commercial cultivars. However, some improvement programmes at different places across the world evolved very fine commercial cultivars through selection, hybridization, mutation breeding etc. citrus improvement programmes. The different commercial genotypes and cultivars of India are as follows:

2.3.3.1 Sweet Orange (*C. sinensis* (L.) Osbeck)

Commercially, it is the largest produced citrus type. Sweet orange is believed to be a natural hybrid of pummelo and mandarin. Over the centuries of their cultivation many varieties have evolved to suit particular climatic conditions and the individual preference of the local population.

2.3.3.2 Mandarin (*C. reticulata* Blanco)

Like that of other citrus species the precise origin of the mandarin is far from certain but is believed to be either north-east India or south-west China.

2.3.3.3 Grapefruit (*C. paradisi* Macfad.)

It has come to existence in the West Indies. The grapefruit's name alludes to clusters of the fruit on the tree, which often appear similar to that of grapes. It is believed to be a cross of Pummelo and Orange. The quality of grapefruit, like that of most citrus varieties, is influenced by several factors, particularly climate and rootstock. The pigmented grapefruit varieties derive their colour from the pigment lycopene.

2.3.3.4 Pummelo (*C. grandis* Osbeck syn. *C. maxima* (Burm.) Merr.)

The pummelo is one of the original citrus species from which the rest of cultivated citrus hybridized. After a Captain Shaddock of an East India Company ship introduced it to Barbados, the fruit was called "shaddock" in English.

2.3.3.5 Acid Lime (*C. aurantifolia* L.)

Limes comprise a varied group of types which are so different from one another in tree and fruit characteristics that they have been given separate species status. It is convenient to divide them into two broad groups: the acid or sour limes of which there are two important types, and the sweet or acidless limes.

2.3.3.6 Lemon (*C. limon* (L.) Burmann)

Lemon trees grow vigorously and become large if not pruned. They are more cold–sensitive than oranges, grapefruit, or mandarins. Lemons will produce a series of flowering flushes throughout the season. See Table 2.4.

2.4 AGRO-CLIMATIC ZONES AND SOIL TYPES IN INDIA WHERE CITRUS IS CULTIVATED

The rich biodiversity of citrus fruits is spread across India. Major states of the country where the commercial cultivars are mostly grown have been depicted in Table 2.5. The citrus cultivars have adapted themselves to a wide range of climatic conditions, i.e., from cool temperate conditions to tropical humid conditions. They mostly grow vigorously in deep, well-drained soils. In general, saline, calcareous, light soils are not good for growth of citrus plants.

TABLE 2.4
Commercial Citrus Cultivars of India

Commercial Cultivars (local name)	Scientific name	Popular types/varieties	Illustration
Mandarin (*Santra*)	*Citrus reticulata* Blanco	Nagpur mandarin, Kinnow mandarin, Darjeeling mandarin, Khasi mandarin, Coorg mandarin, Arunachal Orange, Tamenglong Orange, PAU Kinnow-1, Sikkim mandarin, Clementine mandarin, Daisy, W. Murcott, NRCC mandarin Seedless – 4 (N-4 seedless), Mudkhed Seedless, etc.	
Sweet Orange (*Mosambi*)	*Citrus sinensis* Osbeck	Mosambi, Cutter Valencia, Katol Gold, Jalna Sweet Orange, Sathgudi, Jaffa, Pineapple, Hamlin, Blood Red Malta, Malta, Washington Navel, Natal, Pera, Westin, Phule Mosambi, Pusa Round, Pusa Sharad, etc.	
Acid Lime (*Nimbu*)	*Citrus aurantifolia* Swingle	NRCC Acid Lime-7, NRCC Acid Lime-8, PKM-1, Balaji, Pramalini, Sai Sharbati, PDKV Bahar Lime, Indi Limbe, Jai Devi (PKM - 1), Vikram, PDKV Chakradhar, PDKV Trupti, Pusa Abhinav, Pusa Udit, Phule Sharbati, Rasraj, etc.	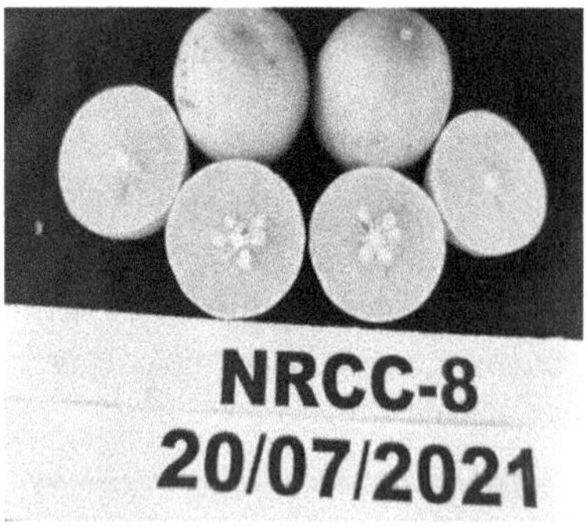
Lemon	*Citrus limon* (L.) Osbeck	Assam Lemon, Pant Lemon-1, Baramasi Lemon, Konkan Lemon, Punjab Baramasi lemon, PAU Baramasi lemon -1, Kagzi Kalan, Cardamom Lemon, Jaintia Lemon, Gandharaj lemon, Meyer lemon, Eureka Lemon, etc.	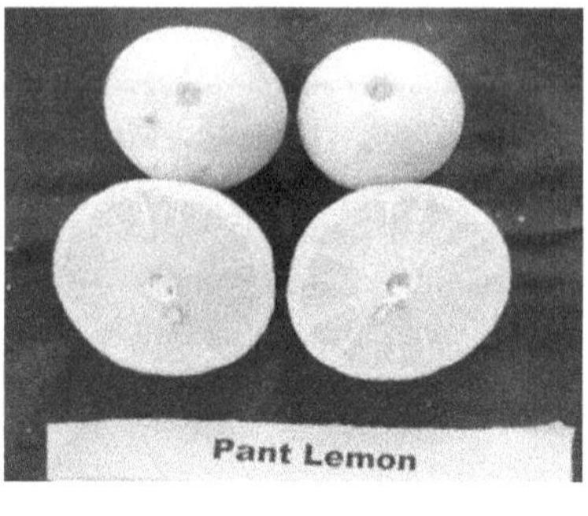

(*Continued*)

TABLE 2.4 (Continued)
Commercial Citrus Cultivars of India

Commercial Cultivars (local name)	Scientific name	Popular types/varieties	Illustration
Pummelo	*C. grandis* Osbeck syn. *C. maxima* (Burm.) Merr.	NRCC-Pummelo-5, US-Pummelo-145, Devanahalli Pomello, Arka Anantha, Arka Chandra, Pusa Arun, etc.	
Grapefruit	*Citrus paradise* Macfadyne	Flame Grapefruit, NRCC Grapefruit-6, Duncan, Marsh Seedless, Red Blush, Foster, Flame, Rio Red, Star Ruby, etc.	

TABLE 2.5
Climatic Zones and Soil Types in India Where Citrus Is Cultivated

Name of the citrus fruit	Major states/climatic zones	Soil type required
Sweet orange	Maharashtra, Andhra Pradesh, Punjab, Rajasthan, Uttar Pradesh	Deep well drained loamy soils are the best for the cultivation of sweet orange. Heavy soils, if well drained, yield good crops but cultivation becomes difficult. The pH of soil should be 6.5 to 7.5 and EC of water should be less than 1.0.
Nagpur mandarin Khasi mandarin Kinnow, Darjeeling mandarin, Coorg Mandarin	Maharashtra, Madhya Pradesh, Rajasthan, North-Eastern Himalayan region (NEH), Punjab, Uttar Pradesh, West Bengal, Sikkim, Andhra Pradesh, Karnataka and Tamil Nadu	Generally grows well in black cotton soils with varying clay content (30–60%). Medium deep soils of 3–4 ft depth with 30–40% clay can also sustain good productivity (Ladaniya et al., 2019)

(*continued*)

TABLE 2.5 (Continued)
Climatic Zones and Soil Types in India Where Citrus Is Cultivated

Name of the citrus fruit	Major states/climatic zones	Soil type required
Acid lime	Andhra Pradesh, Karnataka, Maharashtra, Gujarat, Madhya Pradesh, Tamil Nadu, Odisha	Well drained, medium deep (1–1.5 m) soil with a pH of 6.5 to 7.5 is ideal (Ladaniya et al., 2014).
Grapefruit	Andhra Pradesh	Grapefruit trees grow best in well-draining soil that keeps extra moisture from destroying their roots. These trees also like to grow in loamy soil, a mixture of mostly sand and silt with a small amount of clay.
Lemon	Gujarat, Andhra Pradesh, Uttar Pradesh, Assam, Karnataka	Sandy loam soil is known for being well-draining while also retaining sufficient moisture. Lemon trees are also best grown in soil with a pH between 5.5 and 6.5, which can be adjusted using lime or sulphur.
Pummelo	Andhra Pradesh, Assam, NEH	Pummelo thrives equally in clay, loam or sand with a highly acidic and highly alkaline pH. Regardless of the soil type, pummelo should be provided with good drainage and water at least once a week.

2.5 SIGNIFICANT TECHNOLOGIES AND PACKAGE OF PRACTICES FOLLOWED IN CITRUSFARMING IN INDIA

From nursery management, orchard establishment, nutrient and water management to plant protection and post-harvest management; citrus cultivation requires intense care. Since it is a perennial crop, hence rootstock selection and good quality planting material for orchard establishment plays a highly crucial role in determining the health and longevity of the orchard in later years. Hence, scientific farming is the only way to get enhanced production, productivity and ultimately profit in this venture. Research institutes, agricultural universities, State Agriculture and Horticulture Departments, Government bodies and even private agri-players are in their pursuit of technology development, assessment, refinement and transfer to farmers' field. ICAR-Central Citrus Research Institute, Nagpur, Maharashtra is a premiere institute dedicated to citrus research in India. The institute, established since 1985, has developed several notable technologies which have revolutionized the citrus farming scenario of India. Along with CCRI technologies related to citrus farming developed, standardized by other institutes and practiced throughout India have been discussed below:

2.5.1 High Yielding Varieties

New varieties of grapefruit, pummelo, and sweet orange were developed and released/recommended by ICAR-CCRI, Nagpur:

- **NRCC Mandarin Seedless-4:** N-4 seedless is a natural mutant identified by Central Citrus Research Institute, Nagpur and released for commercial cultivation during 2015. It is commercially seedless variety (0–2 seeds per fruit). The plants are upright and medium large in nature. The average fruit weight is 145 g with an average of ten segments per fruit. It has the TSS of 10.44 ^{0}Brix with acidity of 0.72%. The fruits are attractive and colour turns to deep orange colour during ripening.

- **Flame Grapefruit:** The grapefruit is seedless, red fleshed and has a good blend of acid and sweet and bitter taste. The productivity is about 25 tonne per hectare (6×6 m distance) which increases to double in high density (4×4 m) cultivation.
- **NRCC Pummelo–5:** Fruit has excellent sweet taste and has a little tinge of guava flavour. It is red fleshed, juicy and fruit weighs about 800 g. The yield is high and a grown-up plant bears over 400 fruits. The yield could be 32 tonnes per hectare from about 400 plants at 5 × 5 m spacing. Pummelo has naringin and other flavonoids apart from lycopene and has therapeutic value.
- **NRCC Grapefruit– 6:** This variety yields around 400 fruits/plant, which is about 152 kgs/plant or 42 tonnes/hectare.
- **Pummel US-145:** The fruit is white fleshed and weighs about 700–800 g with sweet taste. The fruit has 15 segments which are very juicy. The per hectare productivity is about 40–50 tonnes. Fruit mature in August and can be harvested up to October. This is an early fruit of the citrus group since Nagpur mandarin comes to market in October.
- **Cutter Valencia:** This is a tight jacket orange and considered as the best fruit world over for processing. Fruit matures in the last week of November and can be harvested up to February. It can be used for table purpose also. The yield is 21 tonnes at 6×6 m spacing and can be increased up to 30 tonne at 5×5 spacing. The juicy fruit has sweet taste and pleasant aroma.
- **NRCC Acid Lime-**7: NRCC acid lime -7 is a clonal selection identified by ICAR-Central Citrus Research Institute, Nagpur and released by the Institute during 2014. The tree is vigorous, spreading, and high yielding (54 t/ha in more than 10-year-old plant at 6 x 3 m spacing). The fruits are round, light yellow in colour and yield good quality fruits. It bears the fruit inside the canopy. The average fruit weight is 48 g. It has high juice content. The number of seeds per fruits are 8.45, acidity 7.05% and total soluble solids is 7.93%. It is suitable for juice and processing.
- **NRCC Acid Lime-**8: NRCC acid lime-8 is a clonal selection identified by ICAR-Central Citrus Research Institute, Nagpur and released by the Institute during 2014. The tree is vigorous, spreading and high yielding (59 t/ha in more than 10-year-old plant at 6 x3 m spacing). It is a cluster bearing type (2.9 fruits/cluster). The average fruit weight is 50 g. It has high juice content (51.5%). The fruits are round in shape and the rind is thin. The number of seeds per fruit is 8.31, acidity 7.1% and total soluble solids 7.96%. It is suitable for juice and processing.

2.5.2 Rootstocks

The presently used rootstocks Rough lemon and Rangpur lime have wide adaptability across the country. Rangpur lime produces medium to high-quality fruit, tolerant to drought and acidic soils. It has a deep root system and is also tolerant to Tristeza. It has been proven to be the best available rootstock for sweet oranges (Mosambi and Sathgudi). It has produced good results for Coorg mandarin also. Rough lemon is vigorous, produces high yield, tolerant to Tristeza and Exocortis and tolerant to high pH. It is widely used for Nagpur and Kinnow mandarins. Alemow (*C. macrophylla*) produces medium canopy and is tolerant to *Phytophthora* diseases (Ladaniya et al., 2019). Both these rootstocks (Rangpur lime and Rough lemon) are susceptible to *Phytophthora* root rot diseases. Alemow has given very good results as rootstock for Nagpur mandarin and acid lime in a long-term experiment at CCRI, Nagpur. It is being commercially distributed to nurserymen and growers.

2.5.3 Disease-Free Containerized Planting Material

This is a very significant and important step ahead to establish the citrus industry on scientific lines. The growing medium in the nursery is disinfected (growing medium should be free from soil-borne pathogen such as *Phytophthora, Fusarium, Pythium*) and mother trees are indexed regularly to remove any diseased mother tree (Indexing for pathogen like Tristeza, Exocortis, Ring spot, citrus mosaic, and greening). Such work has been started by ICAR-CCRI, Nagpur during 2000-2001 for the first time in Vidarbha region of Maharashtra, India. Now several such nurseries have started on commercial scale under Govt. sector. The private sector is slow to take up this technology/protocol but initiative has been taken by some companies.

2.5.4 Cultural Practices

2.5.4.1 Planting System

Planting is generally done during July to September in most parts of India. Pit making with FYM and top soil (1:1 ratio) is done in summer. Generally, spacing of 6 m x 6 m or 5 m x 5 m is recommended for planting in a square system (conventional) which provides plant density of 277 and 400 plants per area respectively. In the high-density system of planting (HDP) of Nagpur mandarin, spacing of 6 m x 3 m (row to row 6 m and plant to plant 3 m) spacing is recommended accommodating 555 plants per ha that doubles the plant density and hence the yield. The higher yield (Table 2.6) is due to restriction in root growth and vegetative growth, which increases flowering and fruiting. Though additional cost incurred is for regular pruning and plant protection but the close spacing provides higher returns in initial years sometimes up to 15 years (Ladaniya et al., 2019). While planting care is taken that rootstock union remains at least 6 inches above ground. In Ultra High Density Planting (UHDP) the spacing is 4 x 2 m and the density is 1,250 plants/ha (Ladaniya et al., 2021).

Fruit quality is not adversely affected due to HDP and UHDP (Table 2.7). Peel thickness (2.74–2.87 mm), juice content (43.88–45.71%), total soluble solids content (TSS) (9.31–9.87%) and juice acidity (0.72 –0.78%) were reported to be normal (Ladaniya et al., 2021).

2.5.4.2 Nutrient Management

Citrus is considered a highly nutrient responsive crop. Balanced fertilization is considered one of the effective ways of not only cutting the cost of production, but increasing the farmers' income as

TABLE 2.6
Comparison of Conventional Planting System, HDP and UHDP

Spacing (No. of plants/ha)	First crop (4th year after planting)	Second crop (5th year after planting)	Third crop (6th year after planting)	Fourth crop (7th year after planting)	Fifth crop (8th year after planting)	Average yield of first five years
			Conventional density			
6X6 m (277)	1.19	5.10	14.65	20.85	25.92	13.54
			High density			
6X3 m (555)	3.61	13.20	36.18	57.16	68.20	35.67
			Ultra-high density			
4X2 m (1250)	6.40	25.34	64.98	71.04	71.50	47.85

TABLE 2.7
Fruit Qualities in Different Planting Systems

Plant spacing and density	Peel thickness (mm)	Fruit juice content(%)	TSS (%)	Titrable acidity (%)
Conventional planting 6X6 m (277 plants/ha)	2.87	45.71	9.55	0.78
HDP 6X3 m (555 plants/ha)	2.79	45.33	9.87	0.75
UHDP 4X2 m (1250 plants/ha)	2.74	43.88	9.31	0.72

TABLE 2.8
Fertilizer Schedule for Nagpur Mandarin

Nutrient source	Total fertilizer (g/tree/ year)	Soil application						Foliar application					
		Ambia (g/plnat)			Mrig (g/plant)			Ambia			Mrig		
		Apr	Aug	Nov	Sept	Nov	Jan	Apr	Aug	Nov	Sept	Nov	Jan
Urea	1300	433	433	434	433	433	434						
Single Super phosphate	1260	630	630		630	630							
Muriate of Potash	180		180			180							
$FeSo_4$	100	50	50		50	50		0.50%	0.50%	0.50%	0.50%		0.50%
$MnSO_4$	100	50	50		50	50		0.50%	0.50%	0.50%	0.50%		0.50%
$ZnSO_4$	100	50	50		50	50		0.50%	0.50%	0.50%	0.50%		0.50%

well. The complete scheduling of macro- and micronutrients for Nagpur mandarin through soil and foliar application has been summarized in the Table 2.8. These fertilizers doses can be supplemented further with additional application of N, P, and K (may be 100–200 g/plant) depending on yield obtained and leaf nutrient status. This depends on soil type and yield required.

Apart from the above-mentioned chemical fertilizers, farm yard manure (FYM) is applied by mixing in soil of the plant basin before the monsoon (after a few light showers). Ten Kg of well decomposed FYM is applied to one year old plant per year while it is 20,30,40 and 50 kg for the 2,3,4- and 5-year-old plants in traditional method (Ladaniya et al., 2019).

2.5.4.3 Water Management

Citrus is sensitive to water deficit and water logging. During rains the drainage provision is required to remove the excess water from root zone. The dry season extending from October to May coincides with comparatively more critical growth stages of citrus, which includes leaf expansion, bloom, fruit set and development. Hence, during this dry season, water management assumes critical importance. Frequency of irrigation determines tree health and production. In the traditional method of irrigation through the ring or basin method, during winter, the irrigation interval may be from 10–15 days and in summer it is much shorter (after every 4–5 days). During summer, the top soil dries up quickly and light irrigations are advisable. Bearing trees are watered at shorter intervals than trees which are non-bearing. Irrigation during harvest/maturity periods is avoided because it reduces TSS and acid. Under Central Indian conditions, in Nagpur mandarin, 28–30 irrigations are sufficient (during crop cycle of 8–9 months) for proper growth and productivity in surface method (basin or ring method) of irrigation. In surface method or traditional method of irrigation losses of water are very high during flow of water from source to the plant due to higher evaporation in open field. But in the drip system these losses are avoided. Drip irrigation and fertigation are being used in many citrus plantations keeping in view the water scarcity. It allows citrus cultivation in hilly areas, on poor soils and thus can allow expansion of citrus to new areas. General output of this technology is increased yield per unit water use and better fruit quality.

2.5.4.4 Canopy Management and Pruning

Citrus trees including mandarins are evergreen and if neglected they tend to overgrow and produce dense unmanageable canopy. Interior branches become unproductive due to lack of sunlight and diseases of pest infestation. Hence, regular pruning is required to remove dead wood in citrus trees. Once bearing starts, pruning almost becomes necessary as the deadwood quantity increases. In plantations with 6 m x 6 m and 6 m x 3 m spacing, pruning operation can be easily mechanized with tractors and pruners available as on today. In general, care is taken never to remove more than 10–15% of the total shoots at a time in small trees (Ladaniya et al., 2019). The structure of a tree should be carefully seen before they are pruned. Pruning is done from outside and top twigs and branches. The top scaffold branch is pruned first, followed by the second and finally the first. For each scaffold, pruning starts from the secondary branch and then the side shoots. Diseased as well as heavily insectinfested branches are removed. Branches and shoots which are in the wrong position, or growing in the wrong direction (undesirable branches and shoots which are in a cross or overlapping manner or at close angle) are removed from the young age. Pruning secateurs are used on young trees for heading back and thinning out cuts. Heading back promotes the growth of lower buds and increases branching.

2.5.4.5 Flowering Regulation

Nagpur Mandarin is an evergreen tree fruit crop with two distinct commercial seasons of flowering and fruiting. Any one crop either *Ambia* or *Mrig* is recommended at a time on the plant. Both crop if taken together can exhaust the plant and may result in early decline. The types of cropping of Nagpur Mandarin practiced in Central India are given in Table 2.9.

Although leaf water potential (as measured by pressure chamber apparatus) in excess of -3.5 MPa has been standardized for imposing water stress in Nagpur mandarin for inducing flowering, for all practical purposes for inducing uniform flowering in Nagpur mandarin it is necessary to impose water deficit stress (< 30% clay content) 30–40 days in medium soils (<40% clay content) and 40–50 days in heavy soils (>50% clay content). The symptoms of the adequate water stress are discernible on the plants in the form of temporary wilting of leaves, leaf curling and senescence of old leaves (Ladaniya et al., 2019).

In the soils having high clay (more than 50%) in the subsurface, it is advisable to use soil application of paclobutrazol @ 9–12 g/plant in the month of November for *Ambia bahar* and in the month of April for *Mrig bahar* followed by two irrigations within 8–10 days before imposing water stress. Alternatively, chlormequat chloride @ 4 ml/l can be used as foliar spray twice at 15 days interval at the beginning of stress period.

For the *Ambia* crop of Nagpur mandarin, the date of alleviation of stress is crucial for ensuring economic final fruit retention. The earlier the stress is alleviated (in the first week of January), the better is the fruit retention and final yield. After alleviation of water deficit stress either by irrigation water (in January) for *Ambia* flowering or by arrival of rains (in June) for *Mrig* flowering a foliar application of GA_3 10 ppm along with urea 1% (1 g GA_3 + 1 kg urea in 100 litres of water) may be given for expression of full potential of flowering. In heavy bearing Nagpur mandarin orchards, for *Mrig bahar* crop regulation, NAA application @ 300–400 PPM concentration can be done

TABLE 2.9
Ideal Weather Conditions for Flower Emergence in Nagpur Mandarin

	Cropping Season	
Parameters	**Ambia/Ambe(spring)**	**Mrig (Mansoon)**
Period of water Stress	Nov– Dec– Jan	April–May
Quantum of water Stress (Mpa)	-3.5 Mpa	-3.5 Mpa
Period of Flowering	Jan–Feb–March	June–July
Temperature at the time of Flower initiation (C)	Minimum Temperature >12 °C maximum Temperature 27–32 °C	Minimum Temperature >20 °C maximum temperature 27–32 °C
Humidity (%)	30–80	>70
Periods of Fruit harvesting	Oct–Dec	Feb–Mar–April

immediately after fruit set for fruit thinning. All the foliar application of micronutrients may be done in the months of March–April for *Ambia* crop and August–September for *Mrig* crop when the age of the fully expanded new flush leaves is 2–3 months and they are photo-synthetically most efficient (Ladaniya et al., 2019). For increasing fruit size, treatment combinations with monopotassium phosphate 2%, DAP 2%, 2,4- D 10 ppm and $GA_3$10 ppm can be alternatively used for increasing the fruit size of Nagpur mandarin and sweet oranges.

In case of acid lime, mostly *Mrig bahar* (June–July bloom) is taken as the natural stress. However, *Hasta bahar* (October–November bloom) fruits fetch a high price in the market in summer. Use of growth regulators like chlormequat chloride (CCC) at 2000 ppm concentration as foliar spray or paclobutrazol 6 g soil application applied twice at 15–20 days interval during last week of August and in the first fortnight of September induces profuse flowering and fruiting in acid lime in the month of October–November thus ensuring the availability of fruits in the months of March through May (Ladaniya et al., 2014).

2.5.4.6 Integrated Management of Diseases and Insect-Pests

Earlier, the major problem faced was Tristeza and die-back. The rootstocks used (such as sweet orange, sweet lime) were susceptible to Tristeza. These were replaced with Tristeza tolerant rootstocks *viz.* Rough lemon and Rangpur lime. But these rootstocks are susceptible to *Phytophthora* diseases (Gummosis, collar rot, and root rot). The integrated approach that can be used effectively to manage *Phytophthora* diseases includes 1) new tolerant rootstocks, 2) improved planting method on raised beds, 3) *Trichoderma* inoculated planting material for new plantations, 4) Use of *Trichoderma* in grown-up plants/orchards, 5) Drip system of irrigation, 6) regular use of Bordeaux paste and Bordeaux mixture, 7) use of fungicides viz. metalaxyl, mancozeb, and Fosetyle Al, 8) careful cultural operations and intercropping to minimize injury to roots.

The HLB (Huanglongbing) or Greening has emerged as a new serious threat that is wide spread. For HLB management the three-pronged strategy has been quite successful.

1. Healthy planting material from insect-free nursery
2. Elimination of symptomatic tree and
3. Insecticide treatment to control psyllid proved very effective to contain spread.

Fruit sucking moth, fruit flies, bark-eating caterpillar and trunk borer are the other important pests. Bio-rational control methods need to be focused to reduce chemical use and amount of residues in fruits. Mass trapping, biological control, selective pesticide applications, area-wide management in case of fruit flies need to be used in combination as a part of the IPM programme.

Among insect-pests trunk borer particularly in NEH region, bark-eating caterpillar in Central India, citrus psylla in North-West India and fruit flies and fruit sucking/piercing moths in most parts of the country pose a threat to citrus industry. Effective management practices include integrating cultural, chemical and biological control measures which are cost-effective would lead to sustainability. Area-wide (AW) –IPM has been reported to be more effective and environmentally friendly for pests like fruit flies. Increasing cost of chemical control measures increase cost of production. Keeping in view the final price scenario during bumper crop season, it is necessary to develop low-cost insect-pest and disease management practices. This would lead to less pollution and also less chemical residue in fruit. The organic cultivation practices are gaining acceptance as they are low cost and such fruit fetch high price in market. Once established with proper certification, the organic cultivation could be the boon for the citrus industry of the country.

2.5.4.7 Mechanization

Use of machines and automation is increasing in the citrus sector and has been useful in addressing cultivation/workmen related problems particularly irrigation/fertigation, pruning of trees, spraying and harvesting, particularly in case of fruit destined for processing. In the case of closely planted citrus in row formation, it is easy to adapt to mechanization for spray application, pruning and harvesting.

2.5.4.8 Organic Citrus Cultivation

Organic citrus cultivation in India is gaining momentum as farmers and consumers increasingly recognize its environmental and health benefits. Organic citrus farming involves the use of bio-fertilizers, compost, green manure, and natural pest control methods instead of synthetic agrochemicals. This approach enhances soil fertility, reduces pesticide residues, and promotes biodiversity. North-east Indian states such as Arunachal Pradesh, Assam, Manipur, Meghalaya, Mizoram, Nagaland, and Sikkim are gradually adopting organic practices, supported by initiatives like the National Programme for Organic Production (NPOP), Participatory Guarantee System (PGS) and Mission Organic value Chain Development for North-eastern Region (MOVCDNER). The demand for organic citrus fruits is rising in both domestic and international markets due to their superior quality and chemical-free nature. However, challenges such as pest and disease management and yield stability need to be addressed through research, extension services, and policy support to ensure sustainable growth in India's organic citrus sector.

2.5.4.9 Post-Harvest Management and Processing

Systematic and pioneering research on fresh citrus handling in order to develop complete protocol of handling was started at ICAR-CCRI, Nagpur during 1989–1990. Now technologies and expertise for maturity standard of fruit, harvesting stage, mechanized handling, wax coating and grading, packaging, pre-cooling, storage at refrigerated and ambient condition of *Mosambi*, Nagpur mandarin and acid lime are available. ICAR-CCRI has given consultancy services and technology transfer on commercial scale for handling of these citrus fruits. Several commercial units for mandarin fruit handling are operational in Vidarbha under the private sector. Many cold storages are operational and storing Nagpur mandarin fruit during March, April, and May to fetch off-season high prices. The institute is continuously engaged since 1990–1991 in educating, training and capacity building of all concerned in the field of fresh citrus handling. In fact, the establishment of Kinnow packing lines (packing houses) in Punjab and Rajasthan started in 1995–1996 after convincing studies of mechanized handling for Nagpur mandarin at ICAR-CCRI, Nagpur during 1991–1994. ICAR-CCRI demonstrated, propagated and popularized concept of mechanized handling of citrus fruit first time for citrus varieties in the country. Now more than 95% Kinnow is brushed and/or wax coated before grading and packaging in corrugated boxes/cartons. These handling practices have brought down post-harvest losses considerably.

In post-harvest management, processing and value addition, ICAR-CCRI, Nagpur has developed and standardized the following technologies:

- Developed maturity indices for Nagpur mandarin fruits.
- Standardized the mechanized processes of sorting, washing, wax coating and size grading for Nagpur mandarin, sweet orange, and acid lime.
- Optimized degreening technique for colour improvement in citrus.
- Identified suitable coating materials and their doses for extending the storage life of Nagpur mandarin and sweet orange fruits under ambient and refrigerated conditions.
- Standardized the pre-cooling technique for Nagpur mandarin to minimize the post-harvest losses during storage.
- Developed low-cost evaporative cool chamber (7.5 ft length x 6 ft width x 7 ft height) with 1.5-tonne fruit storage capacity to store citrus fruit up to 3 weeks.
- Designed corrugated fibre board-based container and standardized the packaging method in vented polyethylene for long-distance transport of Nagpur mandarin.

TABLE 2.10
Potential Value-Added Products of Citrus Fruits Developed by ICAR-CCRI

S.N.	Citrus cultivar	Potential value-added product	
1.	Mandarin	Ready-to-serve beverage	Juice concentrate
	(*Citrus reticulata*)	Carbonated drink	Spray dried powder
		Sweetened juice	Ice-cream
		Squash	Barfi
		Marmalade	Candy
		Jam	Jelly bites
		Jelly	Pastel gummies
2.	Sweet Orange	Ready-to-serve beverage	Sweetened juice
	(*C. sinensis*)	Squash	Juice concentrate
		Candy	Spray dried powder
3.	Lime, Lemon and Kachai Lemon	Ready-to-serve beverage	Jelly
	(*C. aurantifolia, C. limon* and *C. jambhiri*)	Carbonated drink	Pickle
		Energy drink	Juice concentrate
		Squash	Jelly bites
		Cordial	Pastel gummies
4.	Pummelo	Ready-to-serve beverage	Blended juice
	(*C. maxima* or *C. grandis*)	Squash	Candy

- Standardized juice clarification technique for Nagpur mandarin and sweet orange.
- Optimized the drying technique for sweet orange peel for maximum retention of bioactive compounds.
- Optimized pectin extraction from mandarin peel.
- Developed edible film using acid lime peel.
- Standardized extraction methods for peel essential oil from Nagpur mandarin, sweet orange, acid lime, lemon, pummelo, and grapefruit.
- Developed poultry feed prepared from de-oiled Nagpur mandarin processing waste. The feed, contains nearly 8% protein, 20% sugar and 8% fibre, can be mixed in broilers diet @ 10% by replacing maize.
- Accomplished the phytochemical characterization of indigenous citrus germplasm of North-east India.

Fruit processing is an emerging sunrise sector in India with immense growth potential. Citrus fruits, known for their diverse flavours and aromatic profiles, offer vast opportunities for value addition and product diversification. However, despite this potential, less than 5% of citrus fruits in India undergo commercial processing. The ICAR-Central Citrus Research Institute has pioneered the development of a wide range of value-added citrus products and has successfully commercialized some of these technologies with citripreneurs and industry stakeholders (Table 2.10). These innovations are designed to be environmentally sustainable and economically viable, making them well-suited for adoption by small to medium scale processing industries.

2.5.5 Technologies Developed By ICAR-Central Citrus Research Institute, Nagpur

Some of the technologies developed and standardized by ICAR-CCRI, Nagpur, have been enlisted below which have revolutionized the citrus farming scenario of India

- Shoot tip grafting technique for cleaning of planting material from viruses and virus-like pathogens
- Micro-budding technique for retrofitting of nursery phase
- Production of disease-free planting material and bud wood certification
- High density planting for Nagpur mandarin and acid lime
- Raised bed planting system for Nagpur mandarin and sweet orange
- Leaf sampling technique for reliable foliar analysis and leaf nutrient standards for optimum production
- Fertilizer requirement for pre-bearing and bearing Nagpur mandarin and acid lime
- Weed control in citrus orchards up to 300 days in a year using herbicides
- Sensor-based surface and subsurface micro-irrigation system for Nagpur mandarin
- Control of fruit drop in citrus
- Flowering regulation in Nagpur mandarin
- Integrated management of insect-pests in citrus
- Mass multiplication technique of predator *Mallada desjardinsi* (Navas) for bio-control citrus insect pests
- Molecular diagnostics of major virus diseases and citrus Huanglongbing disease
- Integrated management of *Phytophthora* causing foot and root rot and post-harvest diseases
- Mechanical sorting, washing, waxing and size grading of citrus fruits
- Prolonging of harvest of Nagpur mandarin using gibberellic acid
- Maturity standards for citrus fruits
- Degreening technique for Nagpur mandarin and sweet orange
- Evaporative cool chamber for field level storage of citrus fruits
- Packaging of citrus fruits using vented corrugated boxes for distant transport
- Citrus-based value-added processed products and functional foods

2.6 STATUS OF CITRUS INDUSTRY OF INDIA

The Indian citrus industry with its variety of cultivars has a wide harvesting window which keeps the industry in business in most months of the year. This characteristic makes the Indian citrus industry unique in the world. A glance into the harvesting months of the commercial cultivars of citrus in the country would help us in understanding better. The following sub-sections will highlight the present status of the citrus industry through EXIM and production data, citrus processing sector and marketing network existing in the country.

2.6.1 Harvesting Window of Major Cultivars of Citrus

2.6.1.1 Harvesting Month of Mandarin

Nagpur mandarin is recognized among the premium class of mandarins grown worldwide and occupies the first position in India in terms of its market demand. The mandarin here grows twice a year. The fruit available from October to December is *Ambia* (spring blossom) which has a slightly sour taste. It is followed by the sweeter *Mrig* crop in February. Normally, farmers go for either of the two varieties. Nagpur mandarin is one of the best mandarins in the world. Production of this fruit crop in the central and western parts of India is increasing every year. *Mrig* crop (monsoon blossom) which matures in February–March has great potential for export since arrivals of mandarin fruit in the international market are very less during this period.

First introduced in Punjab, it's a high-class hybrid of Willow leaf and King. Due to greater performance in terms of fruit quality and farming feasibility, it has gained steady commercial importance among graded mandarin varieties. Fruits are found in deep orange colour with medium in size, 12–24 seeds and with rich juice sacks. Harvesting time is January–February.

In the north-eastern Himalayan region of India, the harvesting of mandarins (Khasi mandarin, Tamenglong orange, Arunachal orange, Sikkim mandarin and Darjeeling mandarin, etc.) are done in November–December which is sometimes extended up to mid of January.

2.6.1.2 Harvesting Month of Sweet Orange

The main season of blossom is August–September, which comes to fruiting in March which is having a very good market price. The second season practiced for blossom is November–December where fruits come to harvest in the months of July–August.

2.6.1.3 Harvesting Month of Acid Lime

In general, maturity of acid lime depends largely on certain factors including nutrition, farming techniques, climatic condition, moisture availability etc. Plants start bearing fruits from 2½ – 3 years after planting. Among all citrus groups of fruits, acid lime takes a shorter maturity period of 5–6 months. In India, harvesting period for acid lime includes two seasons i.e. from July–September and November–January. A third blossom can be practiced which is ready for harvest during the summers and fetches high market price.

2.6.2 Domestic Market

The agencies involved in the marketing are grower, pre-harvest contractors, commission agents, wholesale buyers, processing industries, exporters, retail traders, and consumers. The normal practice for the growers is to auction the orchard to contractors at the flowering stage. Sometimes, the original contractor passes on the orchard to sub-contractor and makes some profit in the process. This means, the producer has to depend on commission agents, brokers, and middlemen, who, very often get more profits with lesser efforts in shorter time as compared to the producers. In such cases commission agents are assured of their commission, whereas the producer's income depends upon the supply, quality, and demand. Hence, this practice is not very useful to the growers, and also to the consumers. If the growers and producers can manage to sell their produce directly to processing industries, exporters, retailers, etc. avoiding the involvement of the middlemen, citrus cultivation would then truly prove profitable for the growers.

Sometimes, due to heavy production and lack of cold storage facilities, a glut is created in the market, which leads to lesser remunerative prices of the fruit for the growers. To avoid this, the many State Departments of Horticulture organize primary co-operative societies in the states with an objective to eliminate the services of the middlemen and to get more remunerative prices of the fruit for the farmers. At the co-operative societies, post-harvest infrastructure facilities will be provided such as pre-cooling, cold storage, refrigerated vans for transport in the areas of major centres of production and establishment of the processing units for producing juice, lime squash etc. pre-cooling and cold storage facilities will help the growers to store the produce during the season and supply in off-season to the consumers and processing units so as to obtain reasonable rates besides enhancing the shelf life of the fruits from 4–16 weeks depending upon the kind of fruits stored. By using refrigerated vans the losses during transport can be reduced.

2.6.3 Export Market

2.6.3.1 Status of Export of Citrus from India

India is the third highest producer of citrus in the world and successfully exports the fruit to different countries of the world (Table 2.11). Bangladesh is the biggest market for Nagpur mandarins. Nagpur and Amravati districts of Maharashtra together produce 7 lakh MT of Nagpur mandarins on 1.26 lakh hectares.

TABLE 2.11
Export Status of Citrus (Fresh/Dried) from India to Other Countries (2023-2024)

HSN code: Commodity (fresh & dried)	Quantity	Value	Major importing countries
8051000: Orange	69861.71	20.34	Bangladesh, Nepal, Bhutan, UAE
8052100: Mandarin	30.98	0.02	Kuwait, Nepal
8052900: Other Oranges and Similar Citrus Hybrid	684.33	0.32	UAE, Kuwait, Oman
8054000: Grapefruit and Pomelo	3.54	NA	Rusia, UAE
8055000: Limes and Lemons	10024.37	4.48	Nepal, UAE, Bhutan
8059000: Other Citrus Fruits	5432.48	1.70	Bangladesh, Nepal
Total	**80604.92**	**26.86**	

Source: DGCIS Annual Data (2024).

2.6.3.2 Essentials for Export of Citrus

Certain points should be kept in mind before proceeding for export. The minimum requirements for export have to be met by growers before they can prepare their crop for export as fresh fruit. Grades A and B of mandarins are eligible for export with sizes 7–8 cm and 6–7 cm respectively. The minimum maturity requirements for export of Nagpur mandarin are when its juice content is 40%, minimum total soluble solids content (TSS) is 10^0 Brix, minimum sugar content is 80% of TSS, TSS/acidity ratio is 14 and colour is greenish yellow in Ambia season and orange in Mrig season. These are possible to achieve only when Good Agricultural Practices (GAP) are followed by citrus growers right from production to harvest. While exporting, fruits must be intact and free of bruising, cuts and rotting. The development and condition of citrus fruit must be such as to enable them to withstand transport and handling and to arrive in satisfactory condition at the place of destination.

2.6.3.3 Uniqueness and Potential of Export of Nagpur Mandarin

As Nagpur mandarin enjoys two fruiting seasons in Vidharbha and also coupled by the world-renowned GI tag, it has a greater edge in export potential than other citrus fruits. But Indian citrus growers are yet to tap this potential to their advantage. At the same time there are various constraints and weaknesses that plague citrus export from India. Amongst them, lack of availability of exportable quality fruit is a major weakness. Only 10–20% of fruits, particularly Nagpur Mandarin, produced in the country fulfil the quality requirement for export. India has a limited varietal base of Nagpur mandarin and more varieties need to be developed as per market demand. There is a lack of sufficient infrastructure like packing houses, pre-cooling units and transport vehicles for timely delivery of fresh fruits and prevention of post-harvest losses. Diseases and insects infestation remain a major constraint to citrus crop. Diseases like *Phytopthora* root rot, greening, twig blight, CTV, citrus yellow mosaic, citrus canker and infestation of leaf miner, psylla, mites, thrips, scales, fruit sucking moth, fruit flies, and aphids pose a major hurdle to citrus cultivation. Lack of export promotion and market development campaigns is also a weakness of our system. Marketing and advertising campaigns for citrus fruits are essential to sensitize citrus growers about the strength of Nagpur mandarin and opportunity of its export. With more free trade opening up and insurance covers for exporters, there exists a wide range of opportunities for export (Kumar et al., 2023).

2.6.4 Contribution of Private Sector

In the Citrus Industry of India, private sector companies are also contributing in promotion/expansion for example Jain Irrigation systems Ltd., Jalgaon. They have introduced five sweet orange varieties from Brazil. Their nursery practices are different than what commonly seen in India particularly with respect to growing media used and container used to grow the plant. The rootstock seedlings and budlings (micro-budded plants) are grown under greenhouse with controlled temperature and RH condition. These are advancements and very healthy plants are being provided. This needs to be added with more details. Similarly, Sahyadri Farmers Producers Company Limited, Nashik, has made remarkable strides in citrus processing while also importing citrus varieties from Spain, enhancing market competitiveness. In addition, commercial citrus nurseries such as Seven Star Fruits Pvt. Ltd., Mumbai, and Pratima Biotech Pvt. Ltd., Raipur, have played a significant role in increasing the availability of disease-free, high-quality citrus planting material. Their contributions have strengthened the citrus cultivation in India, supporting both small-scale farmers and commercial citrus growers.

2.7 STATUS OF CITRUS RESEARCH AND EXTENSION ACROSS INDIA

2.7.1 Research in Citrus

An autonomous organization called the Indian Council of Agricultural Research (ICAR) exists under the Ministry of Agriculture and Farmers Welfare, Government of India which has established research institutes for every crop throughout the country. Indian Council of Agricultural Research-Central Citrus Research Institute (ICAR-CCRI), Nagpur is the premiere institute conducting research on citrus fruit crop. The institute develops technologies for overall improvement of productivity and profitability of citrus farmers across India and provides sustainable solutions to the problems of citrus growers through extension and farm advisory services (ICAR-Central Citrus Research Institute, 2024). The institute was established in 1985 and since then it has been engaged in development of improved production technologies of citrus ranging from disease-free planting material production technology and nursery management techniques to post-harvest handling of fruits. ICAR-CCRI technologies have transformed the citrus production scenario by benefitting farmers of all scale (Bhattacharyya et al., 2023)

ICAR-All India Coordinated Research Project (AICRP) was initiated during 4thfive year plan on July 31, 1971, with AICRP on Tropical fruits and AICRP on Sub-tropical fruits. In 2013, these two programmes were merged and renamed as "ICAR-AICRP on Fruits" and it is currently operated from the ICAR-Indian Institute of Horticultural Research, Bengaluru, Karnataka. Under AICRP on Fruits, in Citrus, at present, 30 different experiments are being conducted under six different themes *viz.* 1. Management of genetic resources and varietal improvement, 2. Planting density, propagation and rootstocks, 3. Nutrient and Water management, 4. Growth, development and value addition, 5. Pest management (Insect pests and nematodes), and 6. Disease management which is functional at 13 different centres viz. 1. Dr. PDKV, Akola, Maharashtra; 2. Central Horticultural Experimental Station, Chettali, IIHR, Karnataka; 3. Punjab Agricultural University, Ludhiana, Punjab; 4. ICAR-Central Citrus Research Institute, Nagpur, Maharashtra; 5. College of Horticulture and Forestry, CAU, Pasighat, Arunachal Pradesh; 6. Horticultural College and Research Institute, TNAU, Periyakulam, Tamil Nadu; 7. Mahatma Phule Krishi Vidyapeeth (MPKV), Rahuri, Maharashtra; 8. ICAR Research Complex for Eastern Region, Regional Centre, Ranchi; 9. Citrus Research Station, Dr. YSRHU, Tirupati; 10. Citrus Research Station, AAU, Tinsukia, Assam; 11. Agricultural Research Station, Sriganganagar, SK. RAU, Rajasthan; 12. ICAR-Indian Agricultural Research Institute, New Delhi; and 13. Horticulture Research and Extension Centre, Vijayapura, UHS, Karnataka. The various activities taken under AICRP (F) – Citrus are listed below.

- A total of 386 citrus germplasm have been collected and conserved in the field gene banks of Pasighat, Sriganganagar, Ludhiana, Rahuri, Tirupati, and Tinsukia. A rootstock hybridization programme is also under taken at PAU, Ludhiana to develop Phytophthora tolerant rootstocks.

- In varietal evaluation trails, Nagpur mandarin clones/varieties, Kinnow mandarin clones, Coorg mandarin clones, and Khasi mandarin clones under mandarin group; Malta clones, Mosambi clones/varieties and sathgudi clones under sweet oranges; acid lime clones and varieties, grapefruit varieties, and Pummelo varieties are being evaluated for yield, and quality.
- Citrus rootstocks viz., NRCC-2, NRCC-4, NRCC-5, NRCC-6, CRH-12, and Alemow are evaluated for Phytophthora tolerance and its effect on yield and quality of Nagpur mandarin, Kinnow mandarin, Khasi mandarin, Mosambi and Sathgudi varieties. Jatti Khati, Rough lemon and Rangpur lime rootstocks serve as controls.
- Long-term experiments on nutrient management under high density planting; Residual and cumulative effect of nutrients in citrus; Standardization of stage-wise requirements of nutrients for mandarin, sweet orange, and acid lime is taken to determine the optimum doses of fertilizers.
- Organic production of citrus for mandarin and sweet orange is undertaken to develop the organic package of practices.
- Impact of climate on crop phenology, productivity and incidence of pest and diseases under varying climatic conditions is undertaken for mandarin, sweet orange, and acid lime.
- Status of emerging pests and their natural enemies in citrus is undertaken to identify new emerging insect pests and their natural enemies.
- Screening of promising root stocks against root rot is undertaken to screen the different rootstocks against root rot.
- Integrated management of citrus greening disease is undertaken to develop a total package of the management of citrus greening in Mandarin, sweet orange, and acid lime.

Apart from different AICRP centres, research work on citrus is also being undertaken at ICAR-Indian Agricultural Research Institute, New Delhi (Plant Pathology, Horticulture and Post-harvest technology divisions). Some other SAUs like Punjab Agricultural University, Ludhiana;Tamil Nadu Agricultural University, Coimbatore; Assam Agricultural University, Jorhat and Central Agricultural University, Imphal are also presently involved in carrying research work and technology development in areas of citrus improvement, production, protection and post-harvest management aspects. ICAR Research Complex for NEH Region, Umiam, Meghalaya with its six regional centres has been providing research back up for citrus crop in the NEH region. The institute has developed an effective management package for the rejuvenation of declined Citrus orchards. Central Horticultural Experimental Station, Chettalli under the ICAR-Indian Institute of Horticultural Research, Bengaluru is working on horticultural crops especially on "Coorg mandarin" in Coorg region of Karnataka.

2.7.2 Citrus Extension Service Providers

In ICAR, the institute which works in the domain of citrus fruits is the ICAR-Central Citrus Research Institute (ICAR-CCRI) located in Nagpur, Maharashtra. The institute develops technologies for overall improvement of productivity and profitability of citrus farmers across India and provides sustainable solutions to the problems of citrus growers through extension and farm advisory services (ICAR-Central Citrus Research Institute, 2015). Apart from ICAR-CCRI, state agriculture universities, Farm Science Centres (Krishi Vigyan Kendras or KVKs), state agriculture department machinery, private consultancies or nurseries, or input dealers etc. are also involved in providing extension or advisory services to citrus growers of their respective regions.

2.8 CHALLENGES OF CITRUS INDUSTRY OF INDIA AND STRATEGIES AS WAY FORWARD

The citrus industry of India is facing several challenges but nonetheless with the dedicated research, extension, technology development, assessment and transfer; the industry is progressive steadily coupled by the efforts of citrus growers, farmers, exporters, entrepreneurs, and other stakeholders. A glance at major challenges but the strategies for countering them and making our way forward have been discussed below:

- **Varieties:** In citrus, the cultivars need to be seedless, because consumers prefer seedless fruit. The efforts are on to develop seedless varieties. There is a need to develop early and late maturing mandarin cultivars for marketing in August–September and also during April–May.
- **STG:** Still there is a long way to go on this count because it is capital intensive (infrastructure-wise) and very skilled manpower (to do indexing of mother trees for pathogen) is required. The main threat today to the citrus industry of the country is Greening or Huanglongbing. This pathogen is spreading due to citrus psyllids and through infected planting material in nurseries. Large-scale training and awareness programmes throughout citrus growing belts of the country are essential. Among diseases of acid lime, citrus canker is a main threat to plantations all over the country and chemical control is the only solution for this disease at present. Development of canker resistant variety would reduce cost of production to a great extent. Among insect pests, citrus leaf miner is important pest in nurseries and orchards particularly in young emerging and expanding leaves. Citrus psylla is seen in all mandarin growing areas and is responsible for spreading greening disease.
- **Rootstocks:** Keeping in view a wide range of biotic and abiotic stresses, cultivar and region-specific rootstocks need to be developed. Rootstocks like Carrizo, Troyer, Alemow, C-35 citrange and Volkamer lemon need to be tried for tolerance to drought, flood, saline and acidic conditions and high fruit quality and higher yields. Multiple tolerance to nematodes, *Phytophthora* and Tristeza and Exocortis are important characteristics for any rootstock. Hybrid rootstocks (Rough lemon x Trifoliate) are being evaluated. Tetraploids need to be evaluated for their tolerance to abiotic and biotic stresses, fruit yield, quality, and canopy spread. Development of a breeding technology to facilitate development of tree-size controlling rootstocks is required in Advanced Citrus Production System (ACPS). The somatic hybrid Changsha mandarin (*C. reticulate* x Trifoliateorange) has been found to be a cold hardy, dwarf that produces high yield.

  Tolerance to salts could be one area of research where we need to evaluate a large number of available rootstocks to tackle this abiotic stress. Cleopatra mandarin can be tried as an interstock and Rough lemon as rootstock for Kinnow. Cleopatra mandarin has been found to be salt tolerant and can be used in salinity affected areas. For the north-eastern region, Rubido x trifioliate and its hybrids can be tried. Performance of Khasi mandarin on Carrizo citrange has been found to be good at Nagaland. Carrizo has shown tolerance to flooding and moderate tolerance to Fe chlorosis caused in calcareous soils also.
- **Pruning:** For very close spacing, special machines for pruning, spraying and harvesting will have to be developed. Tree canopies will have to be restricted for these operations to be smoothly and easily conducted. It has been observed that high-density plantation reach a plateau of economic sustainability after some years and it becomes no longer beneficial in terms of benefit: cost ratio as the yield may decrease with increase in woody parts of the trees. It again depends on the bearing habit of the scion variety.
- **Post-harvest Management:** There is a long way to go on this front in the case of Nagpur mandarin. The Nagpur mandarin fruit of advanced stage of maturity gets puffy and it becomes

difficult to handle such fruit on the packing line. Also, such fruit cannot be stored for long. There is a need to increase awareness among growers and all concerned that healthy uninfected fruit from the orchard has long shelf life. Pre-harvest infection takes place in the orchard itself and post-harvest fungicide application/ treatment is not much effective in such fruit. Therefore, pre-harvest sanitary practices, good orchard health and pre-harvest fungicidal sprays play a great role in bringing down post-harvest losses due to rotting during marketing. Careful post-harvest handling is equally important to avoid injuries and infection during transportation and marketing. Great developments have taken place in the country during the last 20 years due to subsidy and promotion of post-harvest infrastructure by government agencies. But again lot needs to be done to utilize this infrastructure and create awareness about post-harvest fragile nature of fruit among traders and growers.

- **Technologies for processed products**: In citrus-based processed product market, price factor and quality assurance go hand in hand and in a bid to make profit, often quality gets jeopardized. Citrus fruit juice is the first ranking industry worldwide. However, it is not well organized in most countries. It is a partnership of producer, and all in the supply chain to work in tandem towards the goal of consumer satisfaction. This requires for all to behave more responsibly for all possible efficiencies to reduce cost and provide better quality.
- **Promoting citrus for human health:** Citrus is synonymous with health. With increasing life style stress, public in general prefer safe residue free healthy fruit. Citrus production with GAP would go a long way in health-conscious societies / countries. Requisite publicity of health-promoting properties of citrus needs to be done on wide scale and more often.
- **Soil health:** This is becoming a serious issue with intensive production practices and the need to increase productivity. Use of bio-fertilizers, microbial consortiums, organic manures and supplements amendments are going to play an important role in maintaining soil fertility. Recycling of organic waste/residue generated in orchard (like dry leaves, grasses, weeds, pruned wood etc.) through composting and vermicomposting is the key and integrated farming practices like animal husbandry, green manuring, use of VAM and P, K, and Zn solubilizing microbes and cultivation of N fixing leguminous intercrops would define sustainability through higher fertility of citrus orchard soils.
- **Managing water resources:** Irrigation resources are limited in small farms. Higher water holding capacity with higher organic matter in soil and drip, subsurface or deficit irrigation techniques (partial root zone drying) will be required as water is going to be the most precious resource in citriculture and agriculture in general. Soil water conservation measures and rain water harvesting would help in ground water recharging and increased water availability.
- **Adoption and mitigation of climate change:** Frequency of extreme climatic events like hailstorm, excess rains, drought, and excessively long dry spells are leading to abiotic and biotic stresses. With increased temperature and relative humidity insect-pest incidence pattern is changing. With continuous rains for 8–10 days in some years, incidence of *Phytophthora*root rot has been found to be increased. Cultivation of citrus under hail nets to protect the trees and crop may become a common thing in the future.

2.8.1 Way Forward

In the efforts of developing seedless fruits, NRCC Nagpur mandarin seedless–4 has been developed and released by ICAR-CCRI, Nagpur. The seedless Kinnow has also been developed at Punjab Agricultural University, Ludhiana. ICAR-CCRI, Nagpur has also developed processing technologies for sustainability, food safety and social compliance. A greater collaboration, scientific discussion and research in association with changes at policy level both national and at international platforms is necessary to transform the Indian citrus industry.

As India mostly exports its citrus produce to Asian countries, strengthening the Asian citrus industry through collaborations, scientist/student exchange programmes, and new trade policies is a commendable goal that can lead to economic growth and improved agricultural practices across the continent. Here are some steps and strategies that Asian citrus-producing countries can consider to strengthen Asian citrus scenario at global level:

- **Establish a Regional Citrus Council:** Creating a regional organization or council specifically dedicated to the citrus industry, with participation from all Asian citrus-producing countries. This council can serve as a platform for information sharing, policy development, and collaboration on research and development.
- **Exchange Programmes for Scientists and Students:** Implementing exchange programmes for citrus scientists and students across Asian countries. This will facilitate the sharing of knowledge, research findings, and best practices in citrus cultivation, disease management, and post-harvest handling. Scholarships and research grants can support these initiatives.
- **Joint Research Projects:** Collaborating on research projects to address common challenges faced by the citrus industry, such as pest and disease management, climate change adaptation, and varietal improvement. Jointly funded research initiatives can yield innovative solutions that benefit all participating countries.
- **Technology Transfer:** Establishing mechanisms for the transfer of agricultural technologies and best practices between countries. This can include the exchange of agricultural machinery, equipment, and infrastructure, as well as training programmes for farmers and agricultural workers.
- **Standardization and Certification:** Developing and adopting common standards for citrus production, quality control, and certification across the region. Harmonized standards will facilitate trade and ensure product quality, making Asian citrus products more competitive in international markets.
- **Trade Agreements and Policies:** Formulating new trade policies and agreements among Asian countries to reduce trade barriers, tariffs, and non-tariff barriers related to citrus products. This can stimulate intra-regional trade and enhance market access for Asian citrus in other parts of the world.
- **Infrastructure Development:** Investing in transportation and logistics infrastructure to improve the supply chain for citrus products. Efficient transportation networks, cold storage facilities, and distribution systems will reduce post-harvest losses and increase market reach.
- **Market Promotion:** Collaborating on marketing campaigns to promote Asian citrus products both domestically and internationally. Highlight the unique qualities and flavours of Asian citrus varieties to capture consumer interest.
- **Capacity Building:** Supporting capacity-building programmes for citrus growers, processors, and exporters. Training in sustainable farming practices, value addition, and export readiness can enhance the competitiveness of Asian citrus products.
- **Environmental Sustainability:** Collaborate on environmentally sustainable citrus production practices to minimize the industry's ecological footprint. This includes promoting organic farming methods, water management, and biodiversity conservation.
- **Risk Management:** Developing joint risk management strategies to deal with common challenges like climate-related risks, pest outbreaks, and market fluctuations. Establishing contingency plans and risk-sharing mechanisms can help countries mitigate potential losses.
- **Institute-Industry-Government Collaboration:** Encouraging collaboration between government agencies and the private sector within each country to ensure a coordinated effort in implementing these strategies.

- **Regular Summits and Conferences:** Organizing regular citrus industry summits and conferences to bring together stakeholders, share progress, and set future goals for the development of the Asian citrus industry.

By implementing these strategies and fostering a spirit of cooperation and mutual benefit, Asian countries can strengthen their citrus industry, enhance food security, and contribute to the economic development of the region. This is in turn would benefit trade exchanges for India as well.

2.9 CONCLUSION

India with her vast biodiversity of citrus cultivars and germplasm, holds immense opportunities for research and scientific collaborations in citrus sector. Again, with the increasing sensitivity towards, health, nutraceuticals, natural immunity boosters etc. especially after the global pandemic Covid-19, citrus fruit cultivation and citrus-based processed products can be well considered as a rising sun industry. Hence it is high time for cultivators, industrialists, entrepreneurs, exporters to work hand in hand with scientists and policy makers to make ground breaking changes in citrus industry of India. Sustainable citriculture can open new domains of sustainable food systems which are essential in today's scenario of ever-increasing population, shortage of agricultural land coupled by climate change. However, a joint action from all stakeholders can address these challenges and transform Indian and overall global citrus industry.

REFERENCES

Bhattacharyya, S., Sonkar, R. K., Tagde, C., and Marghade, J. 2023. From Nursery to Orchards: ICAR-CCRI Technologies Transforming the Livelihood of Citrus Growers. *Indian Citriculture*, 1(2):108–116.

DGCIS (Directorate General of Commercial Intelligence and Statistics), India. 2024. Annual Export. Accessed on 18th March, 2025.

ICAR-Central Citrus Research Institute. 2015. Vision Document 2050.https://ccri.icar.gov.in/site/assets/files/1318/ccrinagpur_vision2050.pdf

ICAR-Central Citrus Research Institute. 2024. Director's Message. Available at: https://ccri.icar.gov.in/ccringp/ (accessed on 11 Jan 2024).

Kumar, D., Bhattacharyya, S., and Ghosh, D. 2023. Assessing the Export Potential of Nagpur Mandarin: the Promising Citrus Fruit of Central India. *Current Science* 124(7): 782–784.

Ladaniya, M. S., Huchche, A. D., Das, A. K., Marathe, R.A., Behere, G.T., Rao, C. N. and George, A. 2021. *High Density Cultivation of Nagpur Mandarin*. Extension Bulletin No. 67, ICAR-Central Citrus Research Institute, Nagpur.

Ladaniya, M. S., Srivastava, A. K., Huchche, A. D., Singh, I. P., Das, A. K., Anavrat, V., Rao, C. N., Shirgure, P. S., and Venkatramana, K. T. 2014. *Improved Cultivation Practices for Acid Lime*. Extension Bulletin No. 54, ICAR-Central Citrus Research Institute, Nagpur.

Ladaniya, M. S., Srivastava, A. K., Vijayakumari, N., Huchche, A. D., Singh, I. P., Ghosh, D. K., Sonkar, R. K., Das, A. K., Kumar, D., Rao, C. N., Shirgure, P. S., George, A., and Kommu, K. K. 2019. *Good Agricultural Practices for Nagpur Mandarin*. Extension Bulletin No. 63, ICAR-Central Citrus Research Institute, Nagpur.

Liu, Y. Q., Heying, E., and Tanumihardjo, S. E. 2012. History, global distribution and nutritional importance of citrus fruits. *Comprehensive Reviews in Food Science and Food Safety* 11: 530–545.

Ministry of Agriculture and Farmers Welfare, Government of India. 2023. *Area and Production of Horticultural Crops for 2022–23* (Final). https://agriwelfare.gov.in/en/StatHortEst.

Sharma, B. D., Hore, D. K., and Gupta. S. G. 2004. Genetic resources of Citrus of north-eastern India and their potential use. *Genetic Resources and Crop Evolution* 51: 411–418.

Singh, H. P., and Chadha, K. L. 1993. Genetic Resources of Citrus. In: *Advances in Horticulture. Vol. 1: Fruit Crops Part 1*, ed. K. L. Chadha and O. P. Pareek, 95–121. Malhotra Publishing House, New Delhi, India.

3 The Citrus Industry of Japan

Nobuhiro Kotoda, Masashi Yamamoto, and Mitsunori Iwasaki

3.1 INTRODUCTION

3.1.1 The History Of Citriculture In Japan

Citrus fruits are cultivated all over the world, from temperate to tropical regions. In Japan, they are cultivated from Okinawa to the northern Kanto (Tokyo and the surrounding area) region at the northern limit. Citrus fruit production currently stands at approximately 1.022 million tonnes, more than that of apples (0.612 million tonnes), making it the number one fruit species cultivated domestically [Ministry of Agriculture, Forestry and Fisheries (MAFF), 2021]. In Japan, Shiikuwasha (*Citrus depressa* Hayata) has grown wild in the Ryukyu Islands since ancient times, and Tachibana [*C. Tachibana* (Makino) Tanaka] has grown wild in Kyushu, Shikoku, and southern Honshu (Tanaka 1926). These are the only two citrus species that are unique to Japan, so the citrus fruits currently on the market have either been introduced from overseas or have developed through natural or artificial hybridization among the introduced species/cultivars and the two wild species. See Figure 3.1.

The Kojiki (Records of Ancient Matters) (712 AD), compiled during the Nara period (710-794 AD), contains the story of "Tokijiku no Kagu no Ko no Mi" (meaning "always fragrant fruit"), which was brought back from the land of Tokoyo (the place is unknown) by Tajima Mori in 61 AD at the command of Emperor Suinin. This "Tokijiku no Kagu no Ko no Mi" was written as "Ima Tachibana" ("Ima" means "Present" or "Modern"). This story is little more than a legend, but according to Dr. Tyozaburo Tanaka, there is a theory that this fruit is different from the Tachibana that grows wild in Japan. On the other hand, it has been pointed out that Komikan [Kinokuni (*C. kinokuni* hort ex. Tanaka)] has been cultivated in the Kyushu region since ancient times. There is no doubt that Daidai (*C. aurantium* L.) and Komikan were introduced to Japan at a fairly ancient time.

Japan's oldest collection of poetry, the Manyoshu (late 7th to late 8th century), contains approximately 4,500 poems, including 72 poems about Tachibana, indicating that Tachibana was already widely familiar to Japanese people at that time. Even in present-day Japan, we often see Tachibana trees planted on the grounds of temples and shrines. The trifoliate orange [*Poncirus trifoliata* (L.) Raf.] and Yuzu (*C. junos* Seibold ex Tanaka), which were mentioned in Manyoshu and Shoku Nihongi (797 AD), respectively, were also widespread throughout Japan during the Nara and Heian periods (794–1185 AD). In addition, the name "Kuen" (citron) is recorded in the Honzo Wamyo (918 AD), Japan's oldest pharmacological encyclopedia, suggesting that citron was introduced to Japan before the 10th century. Furthermore, regarding kumquats (*Fortunella* spp.), which enjoy a certain level of popularity even in modern Japan, the name "kumquat" appears in "Teikin Orai" [late 14th to 15th century; the late Nanboku-cho period (1337–1392 AD) to the early Muromachi period (1336–1573 AD)], one of the textbooks used in temple schools in Edo period (1603 to 1868).

DOI: 10.1201/9781003584599-3

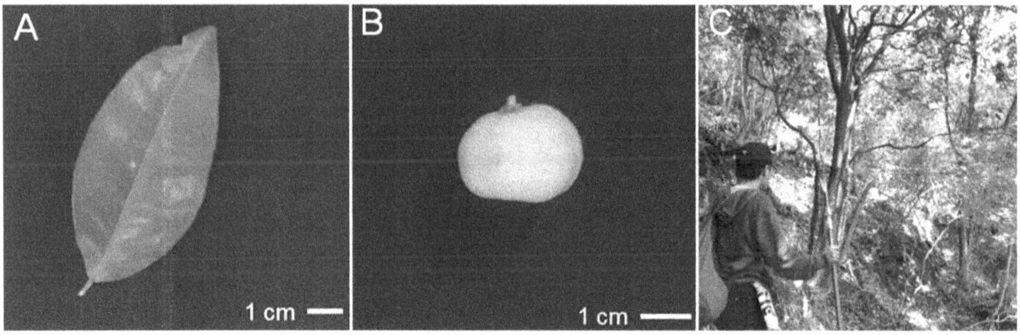

FIGURE 3.1 Representative leaf (A) and fruit (B) of the Tachibana tree (C) growing naturally on the slope in Izu Peninsula. The bars represent 1 cm.

In the Middle Ages (after the end of the Muromachi period), citruses such as Kunembo (*C. nobilis* Lour.), Oobenimikan (*C. tangerine* hort. ex Tanaka), Buntan (pummelo; *C. maxima*), and sweet orange (*C. sinensis*) were introduced to Japan. The Komikan tree over 800 years old exists and is said to have been planted there in 1157 (the second year of the Hogen era). This tree was designated a national natural monument in 1937. Recent genome analysis has suggested that Kunembo is one parent of Satsuma mandarin (*C. unshiu* Marcow.) (Shimizu et al. 2016). The Satsuma mandarin is native to Japan, and an old tree was discovered in Takanosu, Nagashima-cho, Izumi-gun, Kagoshima Prefecture, and was estimated to be about 300 years old. As it was a grafted tree, the original tree is thought to be even older. The 300-year-old tree has already died, and the oldest Satuma mandarin currently is the third-generation grafted tree.

At the end of the 16th century, cultivation of Sakurajima Komikan (small mandarins produced in Sakurajima) began in Kagoshima Prefecture, and Kinokuni mandarins in Hiroshima, Wakayama, and Shizuoka Prefectures. Natsudaidai (*C. natsudaidai* Hayata), which also plays an important role in the process of forming Japanese citrus cultivars, is known to have originated in Yamaguchi Prefecture in the mid-Edo period (around 1700 AD). In the Edo period, it is said that a certain merchant, Bunzaemon Kinokuniya (late 17th century to early 18th century), brought Kinokuni mandarins (called 'Kishu mikan') from Arida in Wakayama, the place of which is used to be called 'Kinokuni' or 'Kishu' that are both old names of Wakayama, to Edo and amassed great wealth. At that time, Satsuma mandarins were also cultivated, but because they were seedless, they were disliked by the Japanese, who valued the prosperity of descendants and were hardly cultivated.

In the Meiji era, when the national isolation system of the Edo period was abolished, various cultural artefacts were imported from the West, and citrus fruits such as sweet oranges, navel oranges, and lemons were introduced from Europe and the United States, and Ponkan (*C. reticulata* Blanco) from Taiwan, from the second half of the 19th century. Also, from the Meiji era onwards, the cultivation of Satsuma mandarins, which had little been cultivated during the Edo period, began in earnest. In Japan, Kinokuni had been the main citrus fruit since the Edo period, but in the 20th century, large-scale mandarin cultivation progressed, mainly in Wakayama and Shizuoka Prefectures, and Satsuma mandarins spread nationwide. Citrus cultivation gradually increased, but production dropped significantly during World War II. However, from the 1950s (1955 onwards), just before the start of the postwar period of high economic growth, a large increase in Satsuma mandarin production began.

3.1.2 Recent Citrus Production

The production of Satuma mandarins was 1 million tonnes in 1964, 2 million tonnes in 1968, 3.56 million tonnes in 1972, and peaked at 3.66 million tonnes in 1975 (Figure 3.2). The number of cultivars of Satsuma mandarin increased significantly as early maturing cultivars such as 'Miyagawa

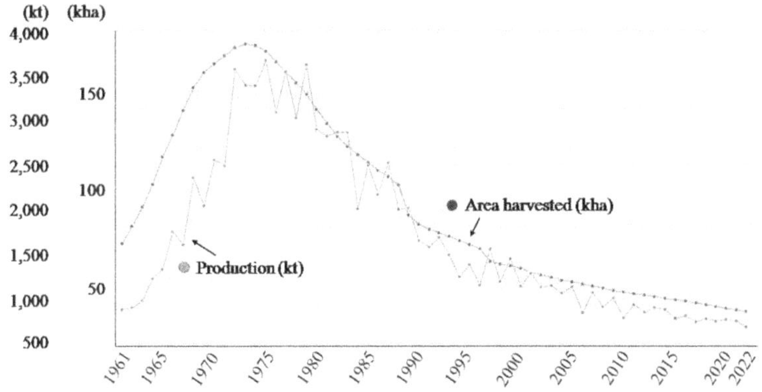

FIGURE 3.2 Area and production quantities of Satsuma mandarins in Japan. (FAOSTAT, 1960~2022).

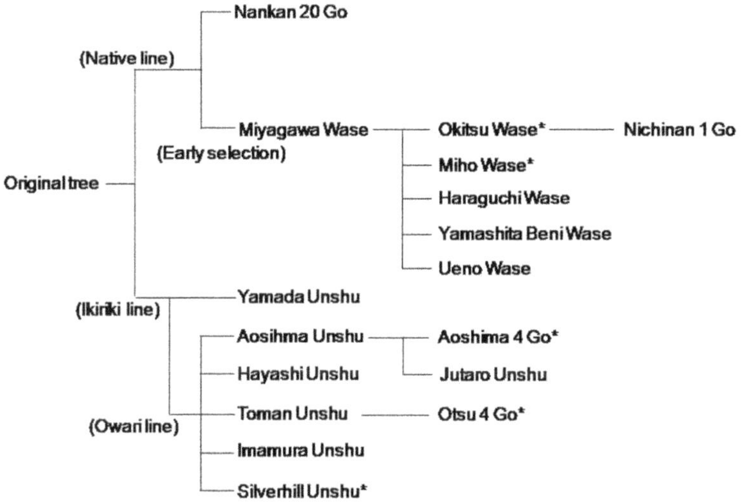

FIGURE 3.3 The diversification of the Satsuma mandarin strains.

Asterisks (*) represent the strains derived from a nucellar seedling. Other strains were selected from bud mutations (cited from 'Wadaino kankitsuhinshu, Ehime-ken Fruit Growers Co-operative Association, 1997' with some modifications).

Wase' and 'Okitsu Wase' and ultra-early maturing cultivars such as 'Ueno Wase' were selected from standard cultivars. The cultivation area exceeded 170,000 hectares in 1973, but there was an overproduction in the first half of 1975. After World War II, except for Satsuma mandarins, there were no major cultivars other than Natsudaidai, which belongs to the late maturing citruses. Therefore, in 1949, the Horticulture Research Station (the predecessor of NARO Institute of Fruit Tree and Tea Science) began breeding new citrus cultivars by crossing 'Miyagawa Wase' and 'Trovita' orange. From this breeding programme, 'Kiyomi' tangor was released in 1979. See Figures 3.3 and 3.4.

3.1.3 Recent Trends

Currently, citrus production in Japan is about 1 million tonnes, of which about 70% is Satsuma mandarin and the remaining 30% is other citrus fruits. There are many cultivars of other citrus

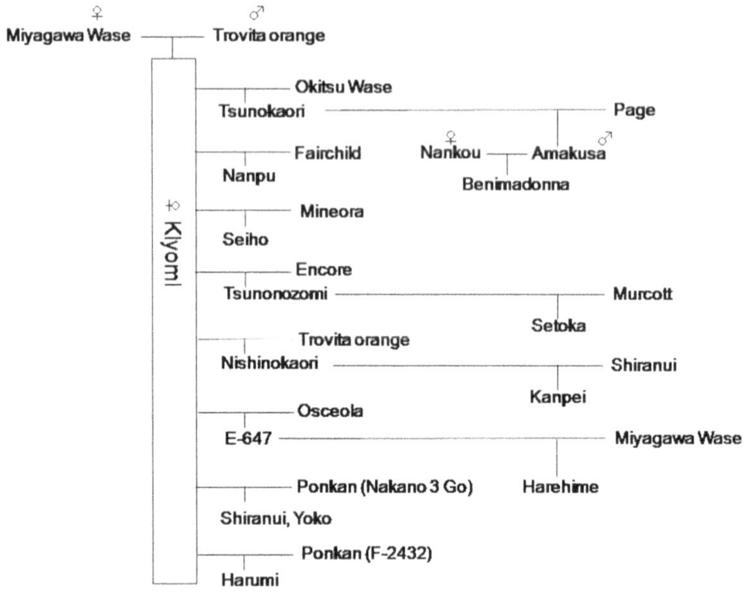

FIGURE 3.4 Pedigree of 'Kiyomi' and its descendants. (Cited from Kankitsu, NHK books, 2007 written by Nesumi with some modifications).

fruits, each of which is produced in small quantities. As mentioned above, while citrus production is increasing worldwide, it has declined sharply in Japan since the mid-1970s, and in recent years, the cultivation area and production volume of citrus have been declining at an annual rate of 1–2%. The reasons for this include overproduction, sluggish consumption, an increase in imported sweet oranges and other fruits that began in 1991, an increase in other cheap and easily eaten imported fruits such as bananas, and a decline in fruit consumption among the younger generation.

As explained later, while the number of Satsuma mandarins, Natsudaidai, Hassaku (*C. hassaku*hort. ex Tanaka), Iyo (*C. iyo* hort. ex Tanaka), Ponkan, and Navel oranges (*C. sinensis* Osbeck var. brasiliensis Tanaka) is decreasing, the number of mid-late citrus fruits such as 'Shiranuhi' (*C. unshiu* x *C. sinensis*) is gradually increasing. In addition, although the production volume is small, the cultivation of fragrant acid citrus fruits such as Yuzu and lemons [*C. limon* (L.) Burm.f.] is also increasing. New mid-to-late maturing citrus cultivars include 'Setoka', 'Reiko', 'Benibae', and 'Seinannohikari' (all from the national institute), 'Kanpei' and 'Beni Madonna' (all from the Ehime prefectural institute), and 'Nijyumaru' (from the Saga prefectural institute). However, the production volume of these new mid to late maturing citrus fruits is still not large, with only 2,200 tonnes for 'Kanpei' in 2020, for example. The supply of seedlings for new cultivars such as 'Kanpei' and 'Nijumaru' is strictly controlled to protect breeding rights, and cultivation by unlicensed producers is prohibited both domestically and overseas.

3.2 CULTURAL ENVIRONMENT

3.2.1 Citrus Production Areas

In Japan, citrus is cultivated in the Tokai (inc. Aichi, Shizuoka, Mie and southern Gifu prefectures), Kinki (inc. Wakayama prefecture, etc.), Shikoku, and Chugoku (inc. Hiroshima prefecture, etc.) regions, including the Pacific coast and the Seto Inland Sea coast, which are relatively warm even in winter due to the influence of the Kuroshio Current, as well as throughout Kyushu and the Ryukyu Islands. Looking at the production areas of Satsuma mandarins, the largest production areas are

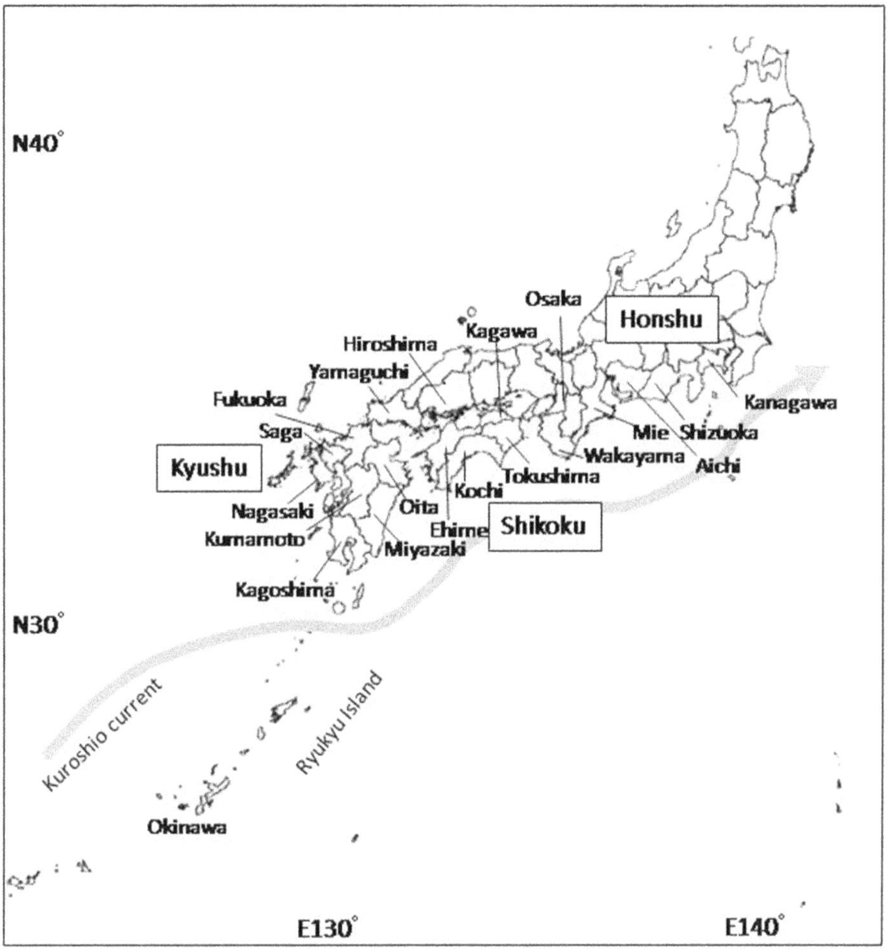

FIGURE 3.5 Major citrus production prefectures in Japan.

Wakayama, Ehime, Shizuoka, Kumamoto, Nagasaki, and Saga. For other citrus fruits (late maturing cultivars, specialty fruits), the largest production areas are Ehime, Wakayama, Kumamoto, Kochi, Oita, Hiroshima, and Tokushima. In Okinawa Prefecture, Shiikuwasha, which is famous for its high accumulation of polymethoxyflavones that are functional health ingredient, is cultivated (Nogata et al. 2006; Inafuku-Teramoto et al. 2011; Onozuka et al. 2008). In particular, the cultivation areas for these citrus fruits are in the warmer regions of Japan, with annual average temperatures ranging from 16.2°C to 18.8°C (excluding Okinawa) and annual average precipitation ranging from 1,150 to 2,626 mm. The major production areas, annual average precipitation and temperature for each production area, and distribution of cultivars other than Satsuma Mandarin shown in Figure 3.5 and Table 3.1 and Table 3.2.

3.2.2 Soil and Climatic Conditions in Citrus-Growing Areas

The main soils in Japan include brown forest soil, and soil, red-yellow soil, and lowland soil. Red-yellow soil and brown forest soil are common in citrus-growing areas. Red-yellow soils are particularly common in the Tokai region, along the Seto Inland Sea, and in northern Kyushu, where they overlap with citrus-growing areas (Figure 3.6). Granite, andesite, and volcanic ash are common as

TABLE 3.1
Production of Satsuma Mandarin in 2022 and Average Annual Temperature and Precipitation (1991~2020) in the Production Area (Statistics of Agriculture, Forestry and Fisheries, MAFF)

Prefecture	Production (t)	Cultivation area (ha)	Ave. annual temp. (°C)	Ave. annual precip. (mm)
Wakayama	152,500	6,720	16.9	1414
Ehime	109,300	5,330	16.8	1405
Shizuoka	103,000	4,880	16.9	2327
Kumamoto	75,000	3,570	17.2	2007
Nagasaki	40,400	2,510	17.4	1895
Saga	38,900	1,770	16.9	1951
Aichi	24,200	1,070	16.2	1579
Fukuoka	17,600	1,100	17.3	1687
Hiroshima	16,400	1,550	16.5	1572
Mie	15,300	950	16.3	1613
Kanagawa	12,600	1,030	16.2	1731
Osaka	12,500	681	17.1	1338
Kagawa	10,000	939	16.7	1150
Oita	9,860	628	16.8	1727
Kagoshima	9,700	776	18.8	2435
Tokushima	9,060	652	16.8	1620
Miyazaki	6,940	503	17.7	2626
Others	13,160	1,003	-	-

TABLE 3.2
Distribution and Production (t) of the Local Cultivars other than Satsuma Mandarin in 2021 (Statistics of Agriculture, Forestry and Fisheries, MAFF)

	Cultivars								
Prefecture	Shiranuhi	Natsudaidai	Hassaku	Iyo	Yuzu	Kabosu	Sudachi	Shiikuwasha	Total
Ehime	7,597	5,359	536	21,611	2,674	4	10	-	37,791
Wakayama	5,662	2,224	17,687	469	390	-	10	-	26,442
Kumamoto	10,614	6,260	151	54	381	-	-	-	17,460
Kagoshima	1,706	11,010	40	140	1,015	3	-	1	13,915
Kochi	104	-	37	-	11,769	-	22	-	11,931
Oita	858	599	139	55	1,108	5,900	-	-	8,659
Hiroshima	2,534	909	4,375	85	76		-	-	7,979
Tokushima	194	4	660	-	2,817	-	4,057	-	7,732
Okinawa	-	-	-	-	-	-	-	3,879	3,879
Saga	3,113	-	31	632	24	-	-	-	3,800
Shizuoka	409	798	220	216	51	-	-	-	1,694
Mie	534	1,000	10	20	-	-	-	-	1,563.8
Others	3,272	1,236	599	293	2,614	70	5	0	8,090
Total	36,596	29,399	24,485	23,576	22,918	5,977	4,104	3,880	150,934

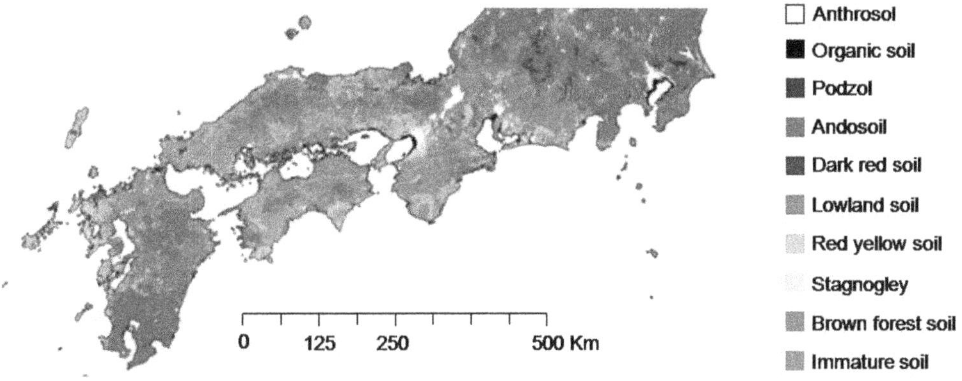

FIGURE 3.6 The soil types in southwestern Japan. (https://soil-inventory.rad.naro.go.jp/explain.html. Accessed May 20, 2024).

geological parent materials, and the geological ages range from the Paleozoic, Mesozoic, Tertiary, and Alluvial eras. Citrus is a fruit tree that is weak to moisture and strong to drought, so soils with good water permeability and clay content are considered to be suitable. Citrus is also quite tolerant to acidity, with a pH of 4.5 to 5.5 in the lower root zone and 5.5 to 6.5 in the main root zone. Therefore, citrus is adapted to the acidic soils that are widespread throughout Japan. Citrus cultivation in Japan is carried out on sloping lands due to competition for the use of limited arable land, with 44% of cultivated land having a slope of 15 degrees or more (MAFF, 2002). Mandarin orange yields and fruit quality are greatly affected by the moisture status and breathability of the soil. When comparing flat land with sloping land, the latter often produces better quality fruit, because flat land has more moisture, while sloping land has better permeability and breathability.

3.3 CULTIVARS AND ROOTSTOCKS

3.3.1 Cultivars

The production of citrus fruit has decreased remarkably (Table 3.3, Figure 3.2), mainly due to overproduction and competition with other fruits. However, the production of some cultivars with excellent fruit quality such as 'Shiranuhi' is still increasing. Production of acid citrus is also steady. Table 3.3 and Figure 3.7 show the major citruses in Japan.

Satsuma mandarin (*C. unshiu* Marcow.): It is the most produced citrus in Japan. Its parents are Kinokuni (*C. kinokuni*) and Kunembo (*C. nobilis*) (Fujii et al. 2016; Shimizu et al. 2016), and it is thought to have originated in Nagashima, Kagoshima Prefecture, more than 400 years ago. The fruit is able to escape cold damage by harvesting before the coldest months, January and February. Its tree is colder and hardy than other citrus such as sweet orange and pummelo (Kawase et al. 1982). Thus, Satsuma mandarin is widely distributed throughout the citrus production area of Japan. The fruit is large for a mandarin. It has excellent peeling properties and is seedless, making it very easy to eat. In Japan, cultivation began in earnest in the late 19th century. Originally, the ripening period was around December, but due to numerous mutations, cultivars such as the ultra-early maturing, which matures around September, and the high sugar content 'Aoshima Unshu', have arisen. This 'Aoshima Unshu', 'Miyagawa Wase', and 'Okitsu Wase' are the three major cultivars (Table 3.4, Figure 3.3). Among the ultra-early maturing, 'Nichinan 1 Go' is the leading cultivar. Satsuma mandarin is also important as a parent for cross-breeding, and 'Kiyomi', which was bred using 'Miyagawa Wase' as a parent, has produced many cultivars. The main producing areas are Wakayama, Ehime, Shizuoka,

TABLE 3.3
Production of Major Citrus in Japan, 1966–2020 (Thousand Tonnes)

Common name	Latin name	1966	1975	1990	2005	2020	Main production prefecture	Period of harvest
Satsuma mandarin	*Citrus unshiu*	1,750	3,665	1,653	1,132	766	Wakayama, Ehime, Shizuoka	Sep to Dec
Shiranuhi	Hybrid cultivar[z]	-	-	0.03	38	39	Kumamoto, Ehime, Wakayama	Feb to Mar
Natsudaidai	*C. natsudaidai*	241	372	170	62	30	Kagoshima, Kumamoto, Ehime	Mar to Apr
Hassaku	*C. hassaku*	85	158	123	51	26	Wakayama, Hiroshima, Ehime	Jan to Apr
Iyo	*C. iyo*	-	45	217	108	26	Ehime, Saga, Wakayama	Jan to Feb
Yuzu	*C. junos*	0.6	-	10	15	24	Kochi, Tokushima, Ehime	Oct to Dec
Ponkan	*C. reticulata*	5	-	28	29	18	Ehime,Kochi, Kagoshima	Dec to Jan
Kiyomi	Hybrid cultivar[y]	-	-	8	21	13	Ehime, Wakayama, Saga	Feb to Mar
Kawachi bankan	*C. kawachiensis*	-	-	3	7	13	Ehime, Kumamoto, Kochi	Mar to May
Pummelo	*C. maxima*	2	-	8	10	12	Kochi, Ehime, Kagoshima	Jan to Mar
Lemon	*C. limon*	-	-	2	5	9	Hiroshima, Ehime, Wakayama	Oct to Jan

[z] Kiyomi (*C. unshiu* × *C. sinensis*) × *C. reticulate*
[y] *C. unshiu* × *C. sinensis*

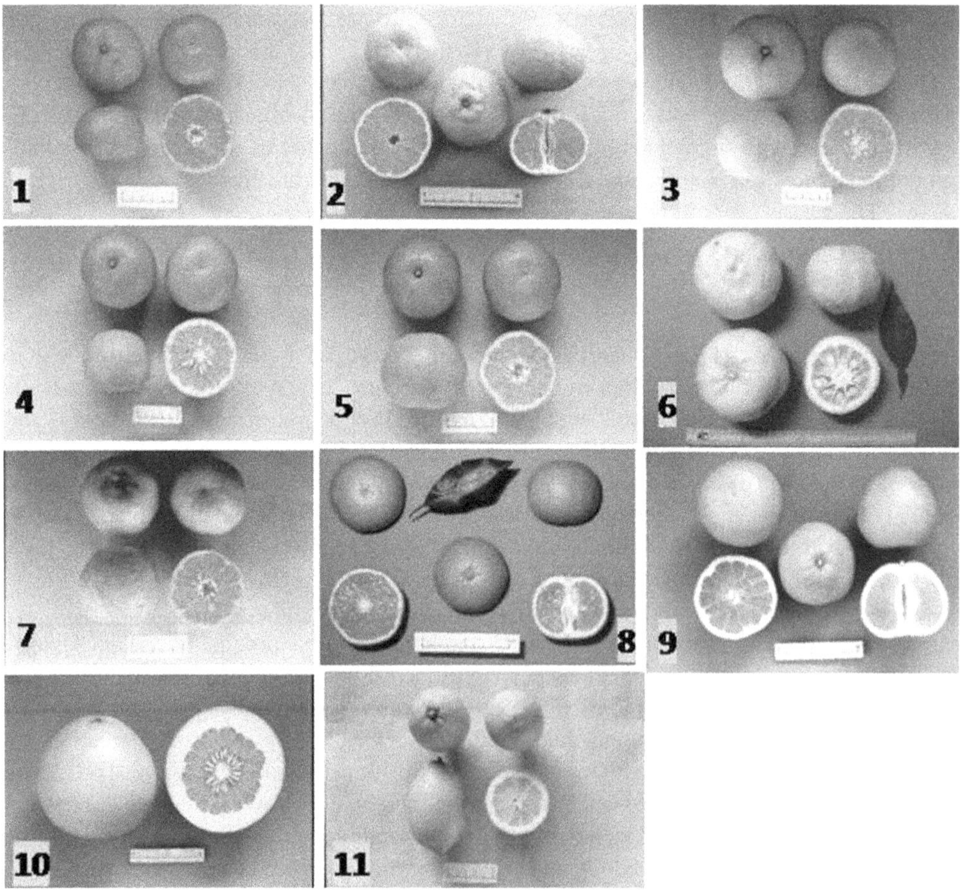

FIGURE 3.7 Photographs of fruit of major citrus in Japan. 1: Satsuma mandarin 'Okitsu Wase', 2: 'Shiranuhi', 3: Natsudaidai 'Kawano Natsidadai', 4: 'Hassaku', 5: Iyo 'Katsuyama Iyokan', 6: Yuzu, 7: Ponkan 'Yoshida Ponkan', 8: 'Kiyomi', 9: 'Kawachi Bankan', 10: Pummelo 'Banpeiyu', and 11: Lemon 'Lisbon'.

TABLE 3.4
Cultivation Area (ha) of Major Satsuma Mandarin Cultivars in Japan in 2020

Cultivar	Type	Cultivation area	Method of selection	Period of harvest
Miyagawa Wase	Early maturing	6,563	Bud sport	Oct to Nov
Okitsu Wase	Early maturing	3,671	Nucellar seedlings	Oct to Nov
Aoshima Unshu	Late maturing	3,664	Bud sport	December
Nichinan 1 Go	Ultra-early maturing	1,483	Bud sport	Sept to Oct
Hayashi Unshu	Late maturing	1,281	Bud sport	December
Nankan 20 Go	Midseason maturimg	1,243	Bud sport	November
Mukaiyama Unshu	Midseason maturimg	1,162	Bud sport	November
Otsu 4 Go	Late maturing	801	Nucellar seedlings	December
Ueno Wase	Ultra-early matuirng	770	Bud sport	Sept to Oct
Ishiji	Late maturing	753	Bud sport	December

Kumamoto, and Nagasaki, with annual production of 700,000 to 800,000 tonnes. This is more than 60% of all citrus, but compared to about 30 years ago, it has dropped to about one-fourth.

Shiraishi [(*C. unshiu* × *C. sinensis*) × *C. reticulata*)]: It was released in Japan by its parents, 'Kiyomi' (*C. unshiu* × *C. sinensis*) and 'Nakano 3 Go' Ponkan (*C. reticulata*) (Matsumoto 2001). Its common name is 'Shiranuhi'. If its fruit has a Brix of 13 degrees or higher and an acidity of 1% or less and is shipped through Japan Agricultural Cooperatives (JA), it can be sold under the registered trademark 'Dekopon'. It produces seedless fruit that are easy to peel, has a good aroma, and taste, and is highly marketable. It is popular among consumers, and among citrus fruits, it is produced in the second largest amount after Satsuma mandarin. Around 40,000 tonnes are produced annually. It is attracting attention around the world, and the fruits are marketed as 'Hallabong' and 'Sumo Citrus' in South Korea and USA, respectively.

Natsudaidai (*C. natsudaidai* Hayata): It was discovered in Yamaguchi Prefecture in the 17th century. Based on its morphology, it has been thought to be a hybrid between pummelo and mandarin, and recent genetic research has revealed that mandarin's parent is Kinokuni (*C. kinokuni*) (Shimizu et al. 2016). Large-scale cultivation began in the late 19th century and later spread throughout the country, but consumption subsequently decreased due to its strong sour taste. With the discovery of 'Kawano Natsudaidai', which was a mutant with low acidity, it became the second citrus after the Satsuma mandarin, which was produced in excess of 300,000 tonnes annually by the 1980s. However, consumption has rapidly declined due to the poor peeling properties of the fruit, its bitter taste, and a large number of seeds, and the current production amount is about 30,000 tonnes. Most of it is produced in Kagoshima, Kumamoto, and Ehime. In Kagoshima, the main cultivar is 'Beni Amanatsu' a coloured mutant of 'Kawano Natsudaidai'.

Hassaku (*C. hassaku* hort. ex Tanaka): It was discovered in Hiroshima in the 19th century. It has been thought to be a hybrid between pummelo and mandarin. Genetic analysis revealed that Mandarin's parent is Kunembo (*C. nobilis*) (Shimizu et al. 2016). Mixed planting is necessary for stable production because of its self-incompatibility. It is susceptible to the citrus tristeza virus (CTV), and the fruit of infected trees is smaller. 'Hasaku No. 55' is equipped with mild strain and can significantly reduce the damage caused by CTV (Sasaki 1967). Twenty thousand to 30,000 tonnes are produced annually. Wakayama, Hiroshima, and Ehime are the major producing areas. 'Noma Beni Hassaku' is a mutant cultivar with deep skin and flesh colour.

Iyo (*C. iyo*hort. ex Tanaka): The place of origin is Yamaguchi. According to the genome analysis, it is considered to be a hybrid between Kaikoukan and Dancy tangerine (*C. tangerina*) (Shimizu et al. 2016). Large-scale production began in Ehime in the late 19th century. 'Miyauchi Iyokan' is the most important cultivar in Iyo. It is highly productive and has contributed to the spread of Iyo. Most of the fruits are produced in Ehime. Consumption has decreased in recent years, and production has declined to about one-seventh over the past 30 years.

Yuzu (*C. junos* Sieb.): It is acid citrus and is considered to have originated in China and came to Japan by the 7th century via the Korean Peninsula. Since then, it has been heavily involved in Japanese food culture. It is the most major acid citrus in Japan, with over 20,000 tonnes harvested annually. About half is produced in Kochi, followed by Tokushima and Ehime. Among the citrus, it is highly cold-resistant (Kawase et al. 1982) and can be grown in the northern part of Japan, where regular citrus cannot be grown. It has a unique scent. Although the disadvantage is that its fruit has many seeds, seedless cultivars also exist. Superior cultivars are selected and cultivated in each region. In addition to using the fruit directly, there are also many processed products. The major processed products are fruit juices, which have high value as beverages and seasonings. It is also culturally important, such as yuzu baths (bathwater prepared with Yuzu fruits) during the winter solstice.

Ponkan (*C. reticulata* Blanco): It is native to India. In India, it is called Nagpur suntara or Santra. It is a globally important citrus that is suitable for relatively high temperatures. The fruit is large among mandarins and has a strong aroma. In Japan, cultivation is popular in Ehime, Kochi, and Kagoshima, with over 20,000 tonnes produced annually. It is broadly divided into two types based on fruit characteristics. The fruit of the 'Kosho-kei' is tall and large, while the fruit of the 'Teisho-kei' is flat and slightly smaller. The famous former cultivars are 'Yoshida Ponkan' and 'Imazu Ponkan', and the latter is 'Ota Ponkan'. It is also important as a parent for cross-breeding in Japan. 'Shiranuhi', 'Harumi', and 'Kanpei' are the progenies of Ponkan.

Kiyomi (*C. unshiu* × *C. sinensis*): It was bred in Japan by crossing 'Miyagawa Wase' with 'Trovita' orange. It bears a male sterile flower and seedless fruit. The fruit characteristics are approximately intermediate between its parents, Satsuma mandarin and sweet orange. More than 10,000 tonnes are produced annually, with most coming from Wakayama and Ehime. It is also important as a breeding material and became the parent of many cultivars such as 'Shiranuhi'.

Kawachi Bankan (*C. kawachiensis* Tanaka): It was discovered in Kumamoto about 100 years ago. The fruit is similar to a grapefruit with a lot of juice. The tree tends to be larger. It is a late maturing citrus that takes about a year from flowering to harvest, making it suitable for areas with mild winters. The cultivation of Kawachi Bankan has been steadily increasing, more than five times in 50 years. Ehime produces the most, followed by Kumamoto. Approximately 10,000 tonnes are produced annually.

Pummelo [*C. maxima* (Burm.) Merr.]: It is considered to be native to the Malay Archipelago and Indonesia and is widely distributed in Southeast Asia. The fruit is large. Generally, the skin and pulp are pale yellow, but some have red flesh. The pericarp is thick and difficult to peel. It has a bitter taste. It is self-incompatible. Although most cultivars are very seedy, there are some seedless ones. The leaves are also large and develop winged leaves. High-quality fruit is often produced in regions at lower latitudes. 'Tosa Buntan' and 'Suisho Buntan' are famous, and over 10,000 tonnes are produced annually. Most of the fruits are produced in Kochi.

Lemon [*C. limon* (L.) Burm.f.]: A major acid citrus that is better suited to cooler climates than lime (*C. aurantifolia*). It is essential for cooking and drinking. It seems to have originated in the eastern foothills of the Himalayas as a hybrid between the sour orange and the citron (Wu et al. 2018). It was introduced to Japan in the mid-19th century. The main production area is Hiroshima, which has mild winters. Due to import liberalization in 1964, domestic production became almost non-existent. Recently, however, production has increased because of strong demand for domestically produced fruit, with annual production of 5,000 to 8,000 tonnes. Approximately 50,000 tonnes are imported annually.

3.3.2 Rootstocks

In Japan, almost all citrus cultivars are grafted on trifoliate orange (*Poncirus trifoliata*). It is a superior rootstock because of its cold hardiness, tolerance to tristeza virus (CTV), and adaptability to acid soil. Trees grafted on trifoliate oranges are dwarfed and bear high-quality fruit (Iwamasa 1988). However, trifoliate orange is susceptible to tatter leaf virus (CTLV). Infected trees grafted on trifoliate orange could not be cultivated due to severe symptoms. Thus, Yuzu (*C. junos*) and Shiikuwasha (*C. depressa*) were used as rootstocks for CTLV-infected cultivars such as Ponkan (*C. reticulata*) about 40 years ago. Recently, generally, CTLV-free trees grafted on trifoliate oranges were cultivated (Takahara et al. 1980).

Dwarfing rootstock 'Hiryu' (Flying Dragon) (*P. trifoliata*) showed better performance than common trifoliate orange when vigorous scion cultivars were grafted (Kobayashi et al. 1995b; Kawabata et al. 2021). When vigorous Shiikuwasha was used as a rootstock, the productivity of weak 'Otani Iyokan' was improved (Takahara et al. 1994). In Tankan (*C. tankan*) grafted on trifoliate orange, winter defoliation hinders stable fruit production. The leaf fall during winter was lower in Carrizo citrange (*C. sinensis* × *P. trifoliata*) and Swingle citrumelo (*C. paradise* × *P. trifoliata*) than in trifoliate orange (Uchino et al. 2023).

3.3.3 Breeding Programmes

3.3.3.1 Increase of Efficiency in Breeding

Shortening the breeding cycle: Citrus seedlings grow slowly and require many years for flowering and fruiting. Yoshida (1980) reported that 43% of 10-year-old trees of seedlings did not flower under the Japanese climatic conditions. Yamada et al. (1979), Okudai et al. (1981), and Mitani et al. (2008) reported that top-grafting was a very effective method for shortening the juvenile period. By conducting top-grafting, it has become possible to shorten the period from hybridization to releasing the cultivar more than 10 years. Recently, research on breeding periods using biotechnology has been carried out. Endo et al. (2005) introduced citrus FLOWERING LOCUS T to trifoliate orange and obtained early flowering individuals. Endo et al. (2020) also demonstrated that fast-track breeding system by integrating early flowering transgenic plants with marker-assisted selection. See Figure 3.8 and Table 3.5.

Polyembryony: The polyembryonic characteristic of citrus is a constraint in cross-breeding because it prevents the production of hybrids from a polyembryonic seed parent. Ueno et al. (1967) evaluated the number of embryos per seed of 116 cultivars. Okudai et al. (1980) determined the embryo number per seed in many Satsuma mandarin cultivars. They found a few embryos in 'Imamura Unshu' and it produced a high percentage of hybrid seedlings. 'Hayaka' (Okudai et al. 1991) was a hybrid derived from 'Imamura Unshu' as a seed parent. Since polyembryony is a

FIGURE 3.8 Top-grafting for shortening breeding cycle in citrus. Left: 1st year of grafting, right: 3rd year of grafting.

TABLE 3.5
Effect of Top-Grafting on Shortening the Breeding Period

	Year		
Cultivar	Hybridization	Released	Years
	Before adopting top-grafting		
Sweet Spring	1947	1982	35
Summer Fresh	1947	1982	35
May Pummelo	1946	1985	39
Yellow Pummelo	1946	1985	39
Hayasaki	1962	1988	26
Benimadoka	1962	1993	31
	After adopting top-grafting		
Hayaka	1972	1990	18
Tsunokaori	1972	1991	19
Ariake	1973	1994	21
Amakusa	1982	1995	13
Akemi	1975	1999	24
Setoka	1984	2001	17

TABLE 3.6
Major Artificial Hybrid Cultivars of Citrus Released in Japan

Cultivar	Cross combination	Production in 2020 (tonne)	Breeder[z]	Period of harvest[y]
Shiranuhi	Kiyomi × Ponkan	38,854	1	Feb to Mar
Kiyomi	Satsuma mandari × Sweet orange	13,123	1	Feb to Mar
Harumi	Kiyomi × Ponkan	5,314	1	Jan to Feb
Setoka	Tsunonozomi × Murcott	5,003	1	Jan to Mar
Beni Madonna	Nankou × Amakusa	4,200	2	Nov to Dec
Kanpei	Nishinokaori × Shiranuhi	2,222	2	Jan to Mar
Harehime	(Kiyomi × Osceola) × Satsuma mandarin	1,560	1	Dec to Jan
Reikou	(Kiyomi × Encore) × Murcott	1,152	1	Jan to Feb
Amakusa	(Kiyomi × Okitsu Wase) × Page	414	1	Dec to Jan
Tsunokagayaki	(Kiyomi × Okitsu Wase) × Encore	395	1	Jan to Feb

[z] 1: National institute and 2: Prefectural institute.

simply inherited and dominant characteristic (Iwamasa et al. 1967), a mono-embryonic cultivar 'Kiyomi' (Nishiura et al. 1983), which is one of the major seed parents for cross-breeding in Japan could be produced from a cross between both heterozygous cultivars, Satsuma mandarin, and sweet orange. Several promising recently released cultivars were the progeny of 'Kiyomi' (Yoshida 2003) (Table 3.6, Figure 3.9).

3.3.3.2 Cross-Breeding

Although small-scale breeding had already been carried out, systematic citrus breeding in national institutes was initiated in 1937 (Nishiura 1964). The first released hybrid cultivar was 'Kiyomi' derived from the cross combination between Satsuma mandarin and sweet orange (Nishiura et al.

FIGURE 3.9 Photographs of fruit of major artificial hybrids citrus in Japan. Left: 'Setoka' and right: 'Kanpei'.

1983). Since then, various new cultivars have been released from both national and prefectural institutes. Hybrid cultivars with large production are shown in Table 3.6. The most important breeding objective is excellent fruit quality. Almost all major hybrid cultivars are mandarin-type with high fruit quality such as 'Shiranuhi', 'Setoka', and 'Kanpei' (Figures 3.7 and 3.9) (Matsumoto 2001; Matsumoto et al. 2003).

To obtain hybrid seedlings with high fruit quality efficiently, elucidation of the inheritance of quantitative traits is being carried out by statistical genetic methods (Imai et al. 2016, 2018a, 2019). The application of combinations of machine learning methods, Bayesian networks, and image analysis is also being considered to improve the efficiency of breeding (Minamikawa et al. 2022). The important breeding objectives, seedlessness, and functional components are described below.

Seedless breeding: Seedlessness is a desirable characteristic for both fresh and processed citrus markets and one of the most important breeding objectives. To develop seedless citrus at the diploid (2x=18) level, utilization of sterility is essential (Yamamoto et al. 1995). Sterility can be divided into three types: male sterility, female sterility, and self-incompatibility. Among the several types of male sterility, another abortion is the strictest male sterility in citrus and was shown to be controlled by both nuclear and cytoplasmic genes (Yamamoto et al. 1997). Segregation and heritability of male sterility in populations derived from progeny of Satsuma mandarin were reported (Goto et al. 2016). Several seedless cultivars with male sterility have been developed (Yamamoto 2014). The mechanism and inheritance of the strictest female sterility derived from 'Mukaku Kishu' were elucidated (Nesumi et al. 2001) and seedless cultivar and parental lines with this female sterility have been released (Yoshida et al. 2005b; Yamamoto 2014). Distribution and identification of self-incompatible cultivars and genes were revealed, respectively (Yamamoto et al. 2006; Kim et al. 2010, 2011, 2020; Yamamoto 2023). In addition, the production of triploid (3x=27) individuals is useful for seedless breeding. Polyploid breeding will be described later.

Functional components: Since citrus fruit is a rich source of functional components, breeding of new cultivars bearing fruit containing them is considered to be an urgent issue (Hasegawa 2013; Ogawa 2013). The flesh and peel of 'Aurastar' contain high graphene contents of 0.75 and 0.81 mg/g dry weight, respectively (Kita et al. 2021). In both 'Tamami' and 'Seinannohikari', β-cryptoxanthin content in the flesh was high [more than 2 mg/100 g fresh weight (FW)] (Yoshida et al. 2005a; Yoshioka et al. 2015). 'Kankitsu chukanbohon nou 6 gou' ('Citrus parental line norin No. 6') is rich in functional ingredients. Its fruit includes synephrine, β-cryptoxanthin, nobiletin, tangerine, and phenylpropanoids (Yoshida et al. 2005b).

3.3.3.3 Mutation Breeding

Spontaneous mutation: In particular, the occurrence of bud sports has contributed to the cultivar differentiation of Satsuma mandarin (Tanaka 1925; Iwamasa 1988). There are cultivars with

TABLE 3.7
Some Citrus Cultivars Derived from Artificial Mutation Breeding in Japan

Cultivar	Original cultivar	Variation	Kind of irradiation	Breeder[z]
		Nucellar		
Toyofuku Wase	Oura Wase (*C. unshiu*)	High tree vigour, low acid	-	1
Sasshu	F2428 Ponkan (*C. reticulata*)	High sugar	-	1
Hinoyutaka	Shiranuhi (*C. unhiu* × *C. sinensis*) × *C. reticulata*	Early maturing	-	1
Natsumi	Kara (*C. unshiu* × *C. nobilis*)	High productivity and quality	-	1
		Irradiation		
Natsumi seedless	Natsumi (*C. unshiu* × *C. nobilis*)	Seedless	Soft X-ray	1
Harushizuka	Aoshima Unshiu (*C. unshiu*)	Seedless	Heavy ion beam	1
Mie Kinan 6 Go	Hayaka (*C. unshiu* × *C. reticulata*)	Late maturing	Unknown	1

[z] 1: Prefectural Institute.

various ripening times and fruit qualities. As shown in Table 3.4, many of the major cultivars were bud sport origin. Iwamasa et al. (1984) reported that among the total of 1,648,696 trees of early maturing Satsuma examined, 35 were judged as ultra-early maturing mutants. In addition, 'Kawano Natsudaidai' and 'Miyauchi Iyokan' in Natsudaidai and Iyo, major mid and late maturing citrus, were spontaneous mutants, respectively.

Artificial mutation: In polyembryonic cultivars, a common method for obtaining mutants is to produce nucellar seedlings. 'Okitsu Wase' and 'Miho Wase' were the first nucellar origin cultivars in Japan (Figure 3.3; Iwasaki et al. 1966). Since then, several cultivars have been produced by nucellar breeding mainly at prefectural institutes. Irradiation has also been used to cause mutation. Mutants on seediness and the period of maturity were obtained (Table 3.7).

3.3.3.4 Polyploid Breeding

Since triploids are seedless, producing triploids is a useful way to promote seedless breeding in citrus efficiently. Although triploid seedlings appeared in crossing between diploids (Oiyama and Okudai 1983), interploidy breeding (2x × 4x) is efficient in producing triploids (Kaneyoshi et al. 2008). Hence, the production and selection of new tetraploid accessions have been conducted. In polyembryonic cultivars, autotetraploids (true-to-type tetraploids of a given accession) can be selected from nucellar seedlings (Kawase et al. 2005). Colchicine treatment of seeds was effective in obtaining tetraploid in polyembryonic cultivars (Yahata et al. 2004; Yasuda et al. 2022; Yamamoto et al. 2024). In mono-embryonic cultivars, a combination of colchicine treatment of isolated small buds and their micrografting is useful for the production of autotetraploids (Oiyama and Okudai 1986; Kaneyoshi et al. 2008). New tetraploids were also produced by protoplast fusion (Kobayashi et al. 1995a). Table 3.8 shows the triploid cultivars released in Japan. Among them, four cultivars were derived from the cross between diploid and tetraploid plants, and one cultivar was selected from open pollination of diploid seed parent (Yoshida et al. 2003; Tokunaga et al. 2005; Kaneyoshi et al. 2014; Nakashima et al. 2015; Yamaguchi et al. 2022). Since triploid can be produced by means of protoplast fusion between diploid and haploid (Kobayashi et al. 1997), 'White Love' was obtained by this method.

TABLE 3.8
Citrus Triplod Cultivars Released in Japan

Cultivar	Cross combination	Note	Breeder[z]
Puchimaru	Oval kumquat (*Fortunella margarita*) × tetraploid Meiwa kumquat (*F. crassifolia*)		1
White Love	Cybrid having lemon (*C. limon*) Somatic hybrid nuclear and sudachi (*C. sudachi*) cytoplasmic genome + Haploid clementine (*C. clementina*)	Somatic hybrid	1 and 2
Tokushima 3X No. 1	Tetraploid sudachi HS4 (*C. sudachi*) × sudachi Ryokuko-kei (*C. sudachi*)		3
Awasuzuka	Tetraploid sudachi 'Honda' (*C. sudachi*) × yuzu 'Yamane' (*C. junos*)		3
Yellow Bell	Open-pollinated seedling of diploid 'Michitani-line Villafranca' (*C. limon*)		3
Oita Kaken 6 Go	Tetraploid kabosu (*C. sphaelocarpa*) × kabosu 'Toyo No Midori' (*C. sphaelocarpa*)		3

[z] 1: National Institute, 2: Private company, and 3: Prefectural Institute.

3.3.3.5 Marker-Assisted Selection

Shimada et al. (2014) constructed a linkage map of Mandarin using DNA markers. Genome-wide association study (GWAS) was conducted by Minamikawa et al. (2017) and Imai et al. (2018b). DNA marker of polyembryony and CTV resistance gene was identified (Nakano et al. 2008; Ohta et al. 2011). QTL mapping was reported in male sterility, flavanone glycoside accumulation, polymethoxyflavone (PMF) accumulation, and carotenoid content (Sugiyama et al. 2011; Goto et al. 2018; Matsumoto et al. 2022; Kubo et al. 2023).

3.4 CULTURAL PRACTICES

3.4.1 Nursery System

3.4.1.1 Nursery Cultivation

Citrus growers carefully select rootstocks and scion cultivars based on various factors like soil conditions, climate, disease resistance, and market demand. In Japan, dwarfing rootstocks such as trifoliate orange (*P. trifoliata*) or its mutated strain 'Hiryu' (Flying Dragon) are widely used in citrus cultivation (Kawase et al. 1987; Yonemoto et al. 2004; Iwasaki et al. 2011b). In most cases, those rootstocks are used to graft popular scion cultivars including Satsuma mandarin and others. Two-year-old rootstocks are grafted by bark or side grafting at a height of about 100 mm from the ground surface. Generally, nursery fields are established in suitable locations with easy access to irrigation, proper drainage, and protection from extreme weather conditions. Therefore, nursery fields are often located in inland areas that were originally paddy fields, such as Tanushimaru in Fukuoka Prefecture. Nurseries require regular management practices like watering, fertilization, pest and disease control, and bud picking. It is important to monitor for pests and diseases such as citrus leafminer (*Phyllocnistis citronella*), swallowtail butterfly caterpillars (*Papilio xuthus*), aphids (*Toxoptera ciricidus*, *Aphis spiraecola*), citrus canker (*Xanthomonas citri* subsp. citri), and scab (*Elsinoë fawcetti*) disease to prevent outbreaks and maintain plant health. One to two-year-old nurseries after grafting are dug up using a special machine or shovel and then shipped. See Figure 3.10.

FIGURE 3.10 The nursery field and digging machine.

FIGURE 3.11 Planting design in flat field.

3.4.2 Planting Designs

In the flat land, the citrus trees are usually planted in rows, and the planting space is 1.5–2.5 m × 4.0–5.0 m (between trees × between rows). In the slope land, terrace field cultivation is adopted for soil erosion and ease of tree management. The tree thinning is carried out as needed when the trees grow so big that it bumps into the neighbouring. The high-density planting spacing, with about 1 m between trees, is sometimes practiced to increase the yield of young trees (Tachibana et al. 1987; Tachibana 1996). In such cases, the spacing of trees is adjusted to 2 m between trees through thinning after about 10 years. The lifespan of the commercial production is expected to be up to 30 years. See Figures 3.11 and 3.12.

FIGURE 3.12 Planting design in the slope field.

3.4.3 Cropping System

3.4.3.1 Typical Citrus Orchard in Japan

Citrus orchards are mainly established on slopes and hillsides, and most of the management units are family-owned, with about 1 ha per household (MAFF, 2020). To accommodate such slope conditions, the roads, monorails, and multi-purpose sprinkler systems have been installed. The monorail systems transport harvested fruit and fertilizers, while the multi-purpose sprinkler system provides overhead irrigation and spray pesticides and fertilizers (Morinaga et al. 2005). Recently, the cultivation area on the steep slopes, which have low work efficiency, has decreased, while the cultivation area on the flat land has increased. Air-blast sprayers are often installed on flat land to save labour. See Figures 3.13 and 3.14.

3.4.3.2 Soil Management and Fertilizer Application

The amount of fertilizers varies with cultivars, tree age, and soil properties. In the general case of mature Satsuma mandarin trees, the composition of annual fertilizer application is 200-300 kg/ha for nitrogen, 100-150 kg/ha for phosphorus, and 100-150 kg/ha for potassium. Annual precipitation in the citrus cultivation areas is relatively so large (1,500-2,500 mm) (Table 3.1) that the orchard soil pH tends to be lowered. Low soil pH causes physiological disorders such as excess absorption of manganese, which induces brown-spot injury to the leaves. Thus, the calcareous fertilizer, which is dolomite or calcium carbonate, is applied for pH amendment in winter. Organic materials including compost manure, rice straw, and pruned twigs are also applied for soil improvement. When deficiency symptoms of microelements appear in the trees, a foliar spray of the element involved is effective.

In areas with fertile soil and lots of rain, weeds grow vigorously. Weeds compete with citrus trees for fertilizer and interfere with tree management. Herbicides have often been used to control weeds, but in steep orchards, they can cause soil erosion, and fertilizer to wash away for heavy rain. Therefore, some growers have adopted sod-culture using mower.

3.4.3.3 Irrigation

There are many orchards without irrigation facilities because Japan has a high annual rainfall (Table 3.1). Recently, however, there have been frequent drought years, and irrigation equipment has been introduced to ensure stable fruit quality and tree health. Irrigation systems mainly use multi-purpose sprinklers and irrigation tubes.

Since fruit with high total soluble solids (TSS) are obtained from drought-stressed trees, growers mulch the ground with a special plastic sheet that prevents water supply to the roots (Morinaga et al.

FIGURE 3.13 Monorail systems for transportation and multi-purpose sprinkler.

FIGURE 3.14 Air-blast sprayer for control of diseases and insect pests.

2005; Iwasaki et al. 2011a; Iwasaki 2022). The sheet has the characteristics of moisture permeability and waterproofness. In such cultivations, it is preferred to introduce irrigation tubes to adjust drought stress (Yoshida et al. 2014; Shimazaki et al. 2016). See Figure 3.15.

3.4.3.4 Pruning and Thinning

The tree form of citrus is usually a 'modified open centre'. Pruning is usually conducted from February to March after the severe cold season. The main purpose of pruning is to adjust the shape and size of the tree and to penetrate solar irradiation efficiently deep into the canopy. Without pruning, tree management becomes difficult, fruit-bearing tends to occur only in the peripheral parts of the tree, and inefficient space for fruit production is formed inside the crown. See Figure 3.16.

FIGURE 3.15 Plastic sheet mulch to prevent rain entry into the soil for soil drought.

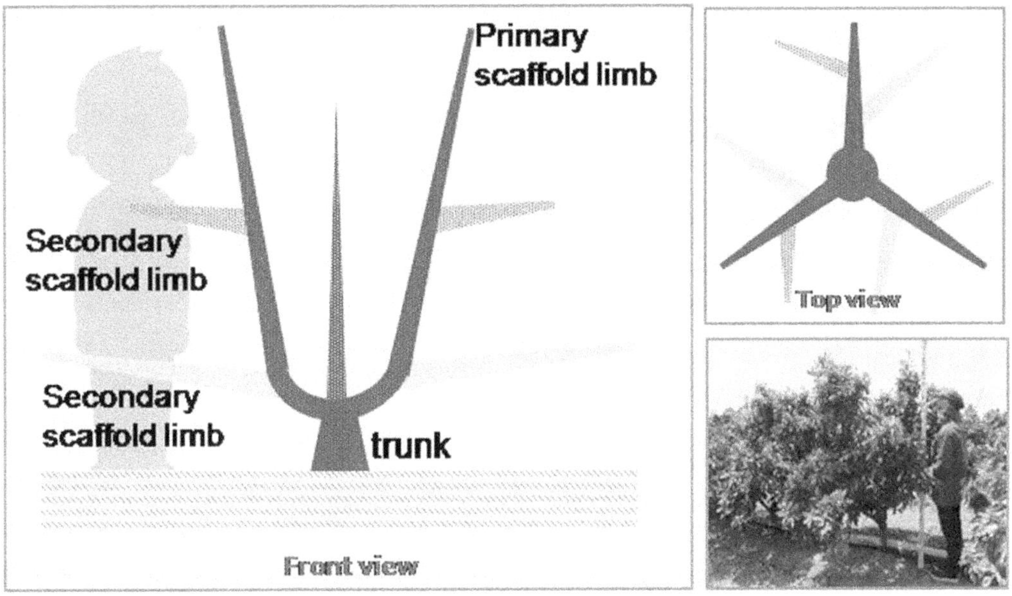

FIGURE 3.16 "Modified open-centre" as the typically tree shape in Japanese citrus cultivation.

FIGURE 3.17 Cultivation in a plastic house with the aim of off-season shipping.

Fruit thinning is practiced from July to August for quality fruit production and the prevention of alternate bearing. Hand thinning is common but ethylchlozate (ethyl 5-chloro-1H-indazole-3-acetate) and NAA (1-naphthaleneacetic acid, sodium salt) are sometimes used as a chemical fruit thinner for Satsuma mandarin and some cultivars at 10 to 40 days after full bloom (Iwahori and Oohata 1975; Hirose et al. 1984).

3.4.3.5 Plastic House Cultivation

Early maturing strains of Satsuma mandarin and other cultivars have been grown in plastic houses with the aim of shipping fruit in the off-season (Harada et al. 1985; Ikeda et al. 2012). In the case of early Satsuma mandarin, the houses are covered with vinyl sheets and usually heated from November using heavy oils and/or heat pumps. The temperature in the plastic houses is maintained within 15 to 30°C. Flowers bloom in December and fruit are harvested from June and July. Satsuma mandarin fruit produced in plastic houses is appreciated for its high quality by consumers. This is because of applied drought stress by providing limited water during the ripening period, which increases the TSS of the fruit. See Figure 3.17.

3.4.3.6 Protection From Bird and Animal Damage

Recently, damage caused by birds and animals has become a serious problem. The birds include brown-eared bulbuls, warbling white-eye, and crows, while the animals include wild boars, masked palm civets, and raccoon dogs (Suzuki et al. 1999; Takeuchi et al. 2002). There are mainly two ways to prevent bird attacks. One is to wrap the tree with a net with 30 mm mesh, and the other is to wrap the fruit in special paper bags or synthetic fibre bags. Electric wire fences are mainly used to prevent animal damage. See Figure 3.18.

FIGURE 3.18 Main birds and animals that damage citrus fruit.

3.5 PESTS AND DISEASES

3.5.1 Pests

Scale insects, yellow tea thrips, citrus leafminer, white-spotted longicorn beetle, citrus red mite, and pink citrus rust mite are the harmful insect or mite pests in the Japanese citrus industry. Although chemical control is widely practiced by citrus producers, some biological controls have been developed. Two parasitoids, *Aphytis* sp. and *Physcus fulvus*, of arrowhead scale were introduced into Japan in 1980 from China. These parasitoids were very effective in reducing the number of arrowhead scales in citrus orchards (Takagi 1983). Thus, the damage caused by the arrowhead scale has been significantly reduced. Natural enemy microorganisms are used to control the white-spotted longicorn beetle. The use of the filamentous fungus *Beauveria brongniartii* reduces the number of longicorn beetles (Kobayashi et al. 1999). *Beauveria brongniartii* preparations are sold as biopesticides.

3.5.2 Diseases

Bacterial and fungal disease: Canker, scab, and huanglongbing (HLB) or citrus greening cause huge economic loss. Canker (bacteria) is the most serious problem among them. Differentiation in the Japanese strain was reported (Shiotani et al. 2000). Varietal differences were observed in susceptibility/resistance to canker (Koizumi and Kuhara 1982). Satsuma mandarin, the most major citrus in Japan, is resistant to the disease. When susceptible cultivars such as sweet orange are cultivated, frequent spraying of fungicide is necessary. Scab (fungi) is also a harmful disease in Japan. Varietal differences were observed in susceptibility/resistance to scab (Yoshida and Shichijo 1984). The major citrus Satsuma mandarin is susceptible to this disease. HLB (bacteria) invaded the Ryukyu Islands of Japan in the 1980s. Since citrus psylla, a vector of this disease has been common there, it is easily expanded. It is currently found only in the central and southern parts of the Ryukyu Islands, but not distributed in major citrus-producing regions in Japan (Iwanami 2022).

Viral disease: Citrus tristeza virus (CTV), Satsuma dwarf virus (SDV), and citrus tatter leaf virus (CTLV) seriously decrease citrus production in Japan. CTV is the most harmful disease among them. Varietal differences were observed in susceptibility/resistance to CTV (Yoshida et al. 1983). Trifoliate orange shows immunity to CTV (Yoshida 1985). Decreases in tree vigour, productivity, and fruit size occur in the infected susceptible cultivars such as Yuzu and Hassaku. Since CTV can be naturally transmitted by brown citrus aphids, virus-free trees grown in orchards are easily reinfected. Hence, the use of mild strain such as 'M-16A' has been conducted to prevent damage caused by CTV (Ieki et al. 1997; Ieki 2003). SDV infections cause a severe decrease in fruit quality and yield in Satsuma mandarin. This virus contaminates soil and infects replanted trees (Iwanami 2023). CTLV induces a bud-union disorder in scion cultivars grafted onto trifoliate orange or citrange. Since many Ponkan and Tankan trees had this virus, it was difficult to use trifoliate orange as rootstock for them. However, this problem was solved by cultivating virus-free trees (Iwanami 2023).

Virus-free citrus plants are often produced by a semi-micrografting method developed by Takahara et al. (1986). The use of antiviral agents is also effective in producing virus-free plants (Iwanami and Ieki 1994).

3.6 RESEARCH ORGANIZATION IN JAPAN

In Japan, the institutes conducting citrus research are broadly divided into national (independent administrative agency) and prefectural research institutes. Among national research institutes, the 'Division of Citrus Research' is located in the NARO Institute of Fruit Tree and Tea Science, which has a 'Citrus Breeding and Production Group' within it. This research organization can be traced back to the Agricultural Experiment Station established in Okitsu, Shimizu, Shizuoka city founded in 1902. Here, the research is mainly conducted on citrus cultivar development, cultivation, pests, and diseases. It also works with prefectural institutes to conduct joint research on common national issues. Meanwhile, there are 47 local government offices in Japan, and the research facilities in 18 prefectures that have research organizations for citrus are listed in Table 3.9. Local government offices mainly conduct research on cultivation methods, selection of promising lines, cultivar development, pests and diseases, etc., in order to solve problems specific to their region regarding citrus

TABLE 3.9
National and Prefectural Research Institutes on Citrus Industry in Japan

Institute	Prefecture	Address
Division of Citrus Research, National Agriculture and Food Research Organization (NARO) Institute of Fruit Tree and Tea Science,	Shizuoka	Okitsu, Shimizu-ku, Shizuoka 424-0292
Fruit Tree Research Centre, Shizuoka Research Institute of Agriculture and forestry	Shizuoka	Mobata, Shimizu-ku, Shizuoka 424-0101
Department of Horticultural Research, Aichi Agricultural Research Centre	Aichi	Kaminogo-cho, Gamagori, Aichi 443-0007
Kinan Fruit Tree Science Branch, Mie Prefectural Agricultural Research Institute	Mie	Ureshinokawagita-cho, Matsusaka, Mie 515-2316
Wakayama Fruit Tree Experiment Station	Wakayama	Oki, Arida-gun, Aridagawa-cho, Wakayama 643-0022 Japan
Department of Fruit Tree Research, Agricultural Technology Research Centre, Hiroshima Prefectural Technology Research Institute	Hiroshima	Akitsu-cho, Higashihiroshima, Hiroshima 739-2402
Citrus Promotion Centre, Yamaguchi Prefectural Agriculture and Forestry General Technology Centre	Yamaguchi	Suooshima-cho, Oshima-gun, Yamaguchi 742-2805
Agricultural and Horticultural Research Division, Tokushima Agriculture, Forestry, and Fisheries Technology Support Centre	Tokushima	Ishii-cho, Myozai-gun, Tokshima 779-3233

(Continued)

TABLE 3.9 (Continued)
National and prefectural research institutes on citrus industry in Japan

Institute	Prefecture	Address
Kagawa Agricultural Experimental Station	Kagawa	Ban-cho, Takamatsu, Kagawa 760-8570
Fruit Tree Research Centre, Ehime Research Institute of Agriculture, Forestry and Fisheries	Ehime	Shimoidai-machi, Matsuyama, Ehime 791-0112
Fruit Tree Experiment Station, Kochi Agricultural Research Centre	Kochi	Asakura, Kochi, 780-8064
Fukuoka Agricultural Research Centre	Fukuoka	Yoshiki, Chikushino, Fukuoka 818-8549
Saga Prefectural Fruit Tree Experiment Station	Saga	Haruke, Ogi, Saga 845-0014
Department of Fruit Tree and Tea Research, Nagasaki Agricultural and Forestry Technical Development Centre	Nagasaki	Onibashi-machi, Omura, Nagasaki 856-0021
Fruit Tree Research Institute, Kumamoto Prefectural Agricultural Research Centre	Kumamoto	Matsubasemachi, Uki, Kumamoto 869-0524
Fruit Tree Group, Agricultural Research Division, Oita Prefectural Agriculture, Forestry and Fisheries Research Centre,	Oita	Ohara, Kunisaki-machi, Kunisaki, Oita 873-0511
Miyazaki Agricultural Research Institute	Miyazaki	Sadowara-cho, Miyazaki 880-0212
Division of Fruit and floriculture, Kagoshima Prefectural Institute for Agricultural Development	Kagoshima	Kinpo-cho, Minamisatsuma, Kagoshima 899-3401
Nago Branch, Okinawa Prefectural Agricultural Research Institute	Okinawa	Nago, Nago, Okinawa, 905-0012

production and distribution. In addition, these local research institutes work together with some local extension organizations in each prefecture to support citrus producers and the industry.

REFERENCES

Endo, T., Fujii, H., Omura, M., and Shimada, T. (2020). Fast-track breeding system to introduce CTV resistance of trifoliate orange into citrus germplasm, by integrating early flowering transgenic plants with marker-assisted selection. *BMC Plant Biol* 20: 224. https://doi.org/10.1186/s12870-020-02399

Endo, T., Shimada, T., Fujii, H., Kobayashi, Y., Araki, T., and Omura, M. (2005). Ectopic expression of an FT homolog from *Citrus* confers an early flowering phenotype on trifoliate orange (*Poncirus trifoliata* L. Raf.). *Transgenic Res* 14: 703–712. https://doi.org/10.1007/s11248-005-6632-3

Fujii, H., Ohta, S., Nonaka, K., Katayose, Y., Matsumoto, T., Endo, T., et al. (2016). Parental diagnosis of Satsuma mandarin (*Citrus unshiu* Marc.) revealed by nuclear and cytoplasmic markers. *Breed Sci* 66: 683–691. https://doi.org/10.1270/jsbbs.16060

Goto, S., Yoshioka, T., Ohta, S., Kita, M., Hamada, H., and Shimizu, T. (2016). Segregation and heritability of male sterility in populations derived from progeny of Satsuma mandarin. *PLoS ONE* 11(9): e0162408. https://doi.org/10.1371/journal.pone.0162408

Goto, S., Yoshioka, T., Ohta, S., Kita, M., Hamada, H., and Shimizu, T. (2018). QTL mapping of male sterility and transmission pattern in the progeny of Satsuma mandarin. *PLoS ONE* 13(7): e0200844. https://doi.org/10.1371/journal.pone.0200844

Harada, Y., Chien, C., and Inoue, H. (1985). Control of room temperature and growth cycles of wase Satsuma mandarin trees in a plastic greenhouse. *Bull Kagawa Agric Exp Stn* 37: 55–65.

Hasegawa, Y. (2013). Programmed citrus breeding in Japan. *Nippon Shokuhin Kagaku Kogaku Kaishi* 60: 609–613. https://doi.org/10.3136/nskkk.60.609

Hirose, K., Takahara, T., Ono, S., and Yoshinaga, K. (1984). Studies on the thinning of Kawano natsudaidai fruit by application of chemicals, 1: Effects of time of spraying (young fruit diameter) and concentration of ethychlozate (ethyl-5-chloro-1H-3-indazoly-lacetate) and 2,4-DP (2,4-dichorophenoxy-2-propionic acid triethanolamine salt) on thinning. *Bull Fruit Tree Res Stn* D: 1–13.

Ieki, H. (2003). The current situations and subject matters on research works of citrus tristeza virus in Japan. *Bull Natl Inst Fruit Tree Sci* 2: 1–7.

Ieki, H., Yamaguchi, A., Kano, T., Koizumi, M., and Iwanami, T. (1997). Control of stem pitting disease caused by citrus tisteza virus using protective mild strains in navel orange. *Ann Phytopathol Soc Jpn* 63: 170–175. https://doi.org/10.3186/jjphytopath.63.170

Ikeda, S., Shindo, T., and Tanaka, K. (2012). The utilization technology of heat pump for growing Satsuma mandarin in plastic greenhouse. *Bull Saga Pref Fruit Tree Exp Stn* 17: 9–18.

Imai, A., Kuniga, T., Yoshioka, T., Nonaka, K., Mitani, N., Fukamachi, H., et al. (2016). Evaluation of the best linear unbiased prediction method for breeding values of fruit-quality traits in citrus. *Tree Genetics and Genomes* 12: 119. https://doi.org/10.1007/s11295-016-1078-8

Imai, A., Kuniga, T., Yoshioka, T., Nonaka, K., and Mitani, N., Fukamachi, H., et al. (2018a). Predicting segregation of multiple fruit-quality traits by using accumulated phenotypic records in citrus breeding. *PLoS ONE* 13: e0202341. https://doi.org/10.1371/journal.pone.0202341

Imai, A., Kuniga, T., Yoshioka, T., Nonaka, K., Mitani, N., Fukamachi, H., et al. (2019). Single-step genomic prediction of fruit-quality traits using phenotypic records of non-genotyped relatives in citrus. *PLoS ONE* 14: e0221880. https://doi.org/10.1371/journal.pone.0221880

Imai, A., Nonaka, K., Kuniga, T., Yoshioka, T., and Hayashi, T. (2018b). Genome-wide association mapping of fruit-quality traits using genotyping-by-sequencing approach in citrus landraces, modern cultivars, and breeding lines in Japan. *Tree Genet Genome* 14: 24. https://doi.org/10.1007/s11295-018-1238-0

Inafuku-Teramoto, S., Suwa, R., Fukuzawa, Y., and Kawamitsu, Y. (2011). Polymethoxyflavones, synephrine and volatile constitution of peels of citrus fruit grown in Okinawa. *J Japan Soc Hort Sci 80*: 214–224. https://doi.org/10.2503/jjshs1.80.214

Iwahori, S., and Oohata, J.T. (1975). Chemical thinning of Satsuma mandarin (Citrus unshiu Marc.) fruit by 1-naphthaleneacetic acid: role of ethylene and cellulase. *Sci Hortic* 4:167–174. https://doi.org/10.1016/S0304-4238(76)80009-X

Iwamasa, M. (1988). Citrus cultivars in Japan. *HortScience* 23: 687–690. https://doi.org/10.21273/HORTSCI.23.4.687

Iwamasa, M., Ueno, I., and Nishiura, M. (1967). Inheritance of nucellar embryony in citrus. *Bull Hort Res Sta* B7: 1–10.

Iwamasa, M., Yamaguchi, S., and Kuriyama, T. (1984). Occurrence of very early mutants from the wase (early ripening) Satsumas and their characteristics. *Bull Fac Agr Saga Univ* 56: 99–107.

Iwanami, T. (2022). Occurrence and control of citrus greening (Huanglongbing) in Japan. *JARQ 56*: 105–120. https://doi.org/10.6090/jarq.56.105

Iwanami, T. (2023). Major citrus viruses in Japan. *JARQ* 57: 99–110. https://doi.org/10.6090/jarq.57.99

Iwanami, T., and Ieki, H. (1994). Elimination of citrus tatter leaf virus from shoots of potted citrus plants by Ribavirin. *Ann Phytopath Soc Japan* 60: 595–599. https://doi.org/10.3186/jjphytopath.60.595

Iwasaki, M. (2022). Development of 'Shielding-mulch cultivation' for high-quality fruit production of Satsuma mandarin. *Hort Res Japan* 21: 441–447. https://doi.org/10.2503/hrj.21.441

Iwasaki, M., Fukamachi, H., Imai, A., and Nonaka, K. (2011a). Effects of summer and autumn water stress on fruit Quality of medium-late maturing citrus 'Harehime'. *Hort Res Japan* 10: 191–196. https://doi.org/10.2503/hrj.10.191

Iwasaki, M., Fukamachi, H., Satoh, K., Nesumi, H., and Yoshioka, T. (2011b). Development of tree vigor prediction method at an early stage based on stem hydraulic conductance of seedling in citrus rootstocks. *J Japan Soc Hort Sci.* 80: 390–395. https://doi.org/10.2503/jjshs1.80.390

Iwasaki, T., Nishiura, M., and Okudai, N. (1966). New citrus varieties 'Okitsu-Wase' and 'Miho-Wase'. *Bull Hort Res Sta* 6: 83–93.

Kaneyoshi, J., Furuta, T., Kurao, M., and Yamaguchi, S. (2008). Induced tetraploid by colchicine treatment in some monoembryonic *Citrus* cultivars and triploidy production by using the tetraploid as seed parents. *Hort Res (Japan)* 7: 5–10. https://doi.org/10.2503/hrj.7.5

Kaneyoshi, J., Furuta, T., Shioda, T., Akasaka, S., Yanagimoto, Y., and Kurihisa, H. (2014). Appearance of triploids in natural hybrid seedlings of lemon and breeding of new cultivar 'Yellow Bell'. *Hort Res (Japan)* 13: 19–26. https://doi.org/10.2503/hrj.13.19

Kawabata, Y., Fujisawa, T., Kawakubo, Y., Sakaki, H., and Kitazono, K. (2021). In greenhouse citrus 'Hinoyutaka', a dwarfing rootstock 'Hiryu', flying dragon trifoliate orange (*Poncirus trifoliata* var.

monstrosa) is more suitable than a rootstock karatachi, trifoliate orange (*Poncirus trifoliata* (L.) Raf.). *ResBull Kumamoto PrefAgr Res Cent* 28: 66–72.

Kawase, K., Iwagaki, I., Takahara, T., Ono, S., and Hirose, K. (1987).Rootstock studies for citrus varieties in Japan. *JARQ* 20:253–259.

Kawase, K., Yahata, M., Nakagawa, S., Haraguchi, K., and Kunitake, H. (2005). Selection of autotetraploid and its morphological characteristics in meiwa kumquat (*Fortunella crassifolia* Swingle). *Hort Res (Japan)* 4: 141–146. https://doi.org/org/10.2503/hrj.4.141

Kawase, K., Yoshinaga, K., Uchida, M., and Hirose, K. (1982). Studies on the frost hardiness of citrus trees. Differences in the freezing resistance between citrus species and their seasonal variation in grafted *nursery plants and seedlings*. *Bull Fruit Tree Res Stn* D4: 25–45.

Kim, J.H., Handayani, E., Wakana, A., Sato, M., Miyamoto, M., Miyazaki, R., et al. (2020). Distribution and evolution of *Citrus* accessions with S3 and/ or S11 alleles for self-incompatibility with an emphasis on sweet orange [*Citrus sinensis* (L.) Osbeck; SfS3 or SfS3sm]. *Genet Resour Crop Evol* 67: 2101–2117. https://doi.org/10.1007/s10722-020-00964-x

Kim, J.H., Mori, T., Wakana, A., Ngo, B.X., Masuda, J., Sakai, K., et al. (2010). Production of homozygous S1 seedlings for S gene in 'Hirado Buntan' pummelo (*Citrus grandis* Osbeck) and determination of the S alleles (S9 and S10) by pollination with the S1 seedlings to *Citrus* cultivars. *J Fac Agr Kyushu Univ* 55: 239–245. https://doi.org/10.5109/18836

Kim, J.H., Mori, T., Wakana, A., Ngo, B.X., Sakai, K., and Kajiwara, K. (2011). Determination of self-incompatible *Citrus* cultivars with S1 and/or S2 alleles by pollination with homozygous S1 seedlings (S1S1 or S2S2) of 'Banpeiyu' pummelo. *J Japan Soc Hort Sci* 80: 404–413. https://doi.org/10.2503/jjshs1.80.404

Kita, M., Nesumi, H., Kuniga, T., Nakajima, N., Yoshioka, T., Outa, S., et al. (2021). New citrus cultivar 'Aurastar'. *J NARO Res Develop* 7: 1–8. https://doi.org/10.34503/naroj.2021.7_1

Kobayashi, S., Ohgawara, T., Saito, W., Nakamura, Y., and Omura, M. (1997). Production of triploid somatic hybrids in citrus. *J Japan Soc Hort Sci* 66: 453–458. https://doi.org/10.2503/jjshs.66.453

Kobayashi, S., Ohgawara, T., Saito, W., Nakamura, Y., and Shimizu, J. (1995a). Fruit characteristics and pollen fertility of citrus somatic hybrids. *J Japan Soc Hort Sci* 64: 283–289. https://doi.org/10.2503/jjshs.64.283

Kobayashi, Y., Ohno, H., Okada, M., and Shikano, E. (1995b). Effects of 'Hiryo (Flying Dragon)' rootstock on tree growth, yield and fruit quality of 'Aoshima Unshu'. *Bull Shizuoka Citrus Exp Sta* 26: 23–30.

Kobayashi, S., Onomoto, N., Negoro, M., Okada, H., Nakai, M., Ueda, M., et al. (1999). Control of white-spotted longicorn beetle, *Anoplophora malasiaca* (Thomson) on citrus by *Beauveria brongniartii*. *Proc Kansa Pl Prot* 41: 65–66.

Koizumi, M., and Kuhara, S. (1982). Evaluation of citrus plants for resistance to bacterial canker disease in relation to the lesion extension. *Bull Fruit Tree Res Stn* D4: 73–92.

Kubo, T., Matsumoto, Y., Yamashita, T., Watanabe, S., and Kotoda, N. (2023). QTL mapping of flavanone glycoside accumulation in citrus. *Tree Genet Genomes* 19:18. https://doi.org/10.1007/s11295-023-01591-0

MAFF (2002). https://www.maff.go.jp/j/council/seisaku/kazyu/h16_2/pdf/data7.pdf. Accessed 23 June 2024

MAFF (2020). Census of agriculture and forestry in Japan census results report.

Matsumoto, R. (2001). 'Shiranuhi', a late-maturing citrus cultivar. *Bull Natl Inst Fruit Tree Sci* 35: 115–120.

Matsumoto, Y., Kubo, T., Itami, Y., Islam, M.Z., Watanabe, S., and Kotoda, N. (2022). QTL mapping of polymethoxyflavone (PMF) accumulation in citrus. *Tree Genet Genomes* 18: 7. https://doi.org/10.1007/s11295-022-01539-w

Matsumoto, R., Yamamoto, M., Kuniga, T., Yoshioka, T., Mitani, N. et al (2003) New citrus cultivar 'Setoka'. *Bull Natl Inst Fruit Sci* 2:25–31.

Minamikawa, M.F., Nonaka, K., Hamada, H., Shimizu, T., and Iwata, H. (2022). Dissecting breeders' sense via explainable machine learning approach: Application to fruit peel ability and hardness in citrus. *Front Plant Sci* 13:832749. https://doi.org/10.3389/fpls.2022.832749

Minamikawa, M.F., Nonaka, K., Kaminuma, E., Kajiya-Kanegae, H., Onogi, A., Goto, S., et al. (2017). Genome-wide association study and genomic prediction in citrus: Potential of genomics-assisted breeding for fruit quality traits. *Sci Rep* 7: 4721. https://doi.org/10.1038/s41598-017-05100-x

Mitani, N., Matsumoto, R., Yoshioka, T., and Kuniga, T. (2008). Citrus hybrid seedlings reduce initial time to flower when grafted onto shiikuwasha rootstock. *Sci Hort* 116: 462–465. https://doi.org/10.1016/j.scienta.2008.03.003

Morinaga, K., Sumikawa, O., Kawamoto, O., Yoshikawa, H., Nakao, S., Shimazaki, M., Kusaba, S., and Hoshi, N. (2005). New technologies and systems for high-quality citrus fruit production, labor-saving and orchard construction in mountain areas of Japan. *J Mountain Sci* 2:59–67. https://doi.org/10.1007/s11629-005-0059-4

Nakano, M., Shimizu, T., Fujii, H., Shimada, T., Endo, T., Nesumi, H., et al. (2008). Marker enrichment and construction of haplotype-specific BAC contigs for polyembryony genomic region in *Citrus*. *Breed Sci* 58: 375–383. https://doi.org/10.1270/jsbbs.58.375

Nakashima, M., Tokunaga, T., Nii, M., Tsumura, T., Yamamoto, H., Sakaguchi, Y., et al. (2015). Breeding of a new triploid acid citrus cultivar, 'Awasuzuka', and its characteristics. *Bull Tokushima Agri Forest Fish Tech Sup Cen* 2: 9–12.

Nesumi, H., Nakano, M., and Yoshida, T. (2001). Mode of inheritance on the abnormal development of impregnated ovules derived from Mukaku-kishu. *J Japan Soc Hort Sci* 70 (suppl 2): 403.

Nishiura, M. (1964). Citrus breeding and bud selection in Japan. *Fla State Hort Soc* 19: 79–83.

Nishiura, M., Shichjo, T., Ueno, I., Iwamasa, M., Kihara, T., Yamada, Y., et al. (1983). New citrus cultivar 'Kiyomi' tangor. *Bull Fruit Tree Res Stn* B10: 1–9.

Nogata, Y., Sakamoto, K., Shiratsuchi, H., Ishii, T., Yano, M., and Ohta, H. (2006). Flavonoid composition of fruit tissues of citrus species. *Biosci Biotechnol Biochem* 70:178–192. https://doi.org/10.1271/bbb.70.178

Ogawa, K. (2013). Functional components and processing of citrus fruits. *Nippon Shokuhin Kagaku Kogaku Kaishi* 60: 603–608.

Ohta, S., Endo, T., Shimada, T., Fujii, H., Shimizu, T., Kuniga, T., et al. (2011). PCR primers for marker assisted backcrossing to introduce a CTV resistance gene from *Poncirus trifoliata* (L.) Raf. into Citrus. *J Japan Soc Hort Sci.* 80: 295–307. https://doi.org/10.2503/jjshs1.80.295

Oiyama, I., and Okudai, N. (1983). Studies on the polyploidy breeding in citrus. III Occurrence of triploids in the progenies of diploid sweet orange crossed with diploids. *Bull Fruit Tree Res Stn* D5: 1–7.

Oiyama, I., and Okudai, N. (1986). Production of colchicine-induced autotetraploid plants through micrografting in monoembryonic citrus cultivars. *Japan J Breed* 36: 371–376. https://doi.org/10.1270/jsbbs1951.36.371

Okudai, N., Matsumoto, R., Oiyama, I., Takahara, T., Asada, K., Yamamoto, M., et al. (1991). New citrus cultivar 'Hayaka'. *Bull Fruit Tree Res Stn* 21: 51–57.

Okudai, N., Oiyama, I., and Takahara, T. (1981). Studies on the improvement of zygotic seedling yield of the polyembryonic citrus. I Differences in embryo number per seed and zygotic seedling yield among its varieties and strains. *Bull Fruit Tree Res Stn* D3: 9–21.

Okudai, N., Yoshinaga, K., Takahara, T., Ishiuchi, D., and Oiyama, I. (1980). Studies on the promotion of flowering and fruiting of juvenile citrus seedlings. II. Effect of grafting. *Bull Fruit Tree Res Stn* D2: 15–28.

Onozuka, H., Nakajima, A., Matsuzaki, K., Shin R-W, Ogino, K., Saigusa, D., et al. Nobiletin, a citrus flavonoid, improves memory impairment and a pathology in a transgenic mouse model of Alzheimer's disease. *J Pharmacol Exp Ther* 326:739–744. https://doi.org/10.1124/jpet.108.140293

Sasaki, A. (1967). A mild strain of tristeza virus in Hassaku trees (*Citrus hassaku* Hort. et Y. Tanaka) in Hiroshima Prefecture. *Ann Phytopath Soc Japan* 33:162–167. https://doi.org/10.3186/jjphytopath.33.162

Shimada, T., Fujii, H., Endo, T., Urda, T., Sugiyama, A., Nakano, M., et al. (2014). Construction of a citrus framework genetic map anchored by 708 genebased markers. *Tree Genet Genomes* 10: 1001–1013. https://doi.org/10.1007/s11295-014-0738-9

Shimazaki, M., and Nesumi, H. (2016). A method for high-quality citrus production using drip fertigation and plastic sheet mulching. *JARQ* 50: 301–306. https://doi.org/10.6090/jarq.50.301

Shimizu, T., Kitajima, A., Nonaka. K., Yoshioka, T., Ohta, S., Goto, S., et al. (2016). Hybrid origins of citrus varieties inferred from DNA marker analysis of nuclear and organelle genomes. *PLoS ONE* 11: e0166969. https://doi.org/10.1371/journal.pone.0166969

Shiotani, H., Ozaki, K., and Tsuyumu, S. (2000). Pathogenic interactions between *Xanthomonas axonopodis* pv. citri and cultivars of pummelo (*Citrus grandis*). *Phytopathology* 99: 1383–1389. https://doi.org/10.1094/PHYTO.2000.90.12.1383

Sugiyama, A., Omura, M., Matsumoto, H., Shimada, T., Fujii, H., Endo, T., et al. (2011). Quantitative trait loci (QTL) analysis of carotenoid content in citrus fruit. *J Japan Soc Hort Sci* 80: 136–144. https://doi.org/10.2503/jjshs1.80.136

Suzuki, K., and Yoshinaga, K. (1999). Prevention and control of wildlife damage in fruit production. *Bull Natl Inst Fruit Tree Sci* 32: 39–64.

Tachibana, S. (1996). Effect of planting density on the quality of wase Satsuma mandarin fruits grown under different cultural treatments. *J Japan Soc Hort Sci* 65:463–470. https://doi.org/10.2503/jjshs.65.463

Tachibana, S., Morioka, S., and Nakai, S. (1987). Effect of planting density on fruit yield under different cultural treatments in Satsuma mandarin tree. *J Japan Soc Hort Sci* 56:9–15. https://doi.org/10.2503/jjshs.56.9

Takagi, K. (1983). Establishment of Aphytis sp. and *Physcus fulvus* (Hymenoptera: Aphelinidae), the imported parasitoids the arrowhead scale, *Unapsis yanoensis* (Hemiptera: Diaspidae), on citrus orchard in Japan. *Bull Fruit Tree Res Stn* D5: 93–110.

Takahara, T., Kawase, K., and Ono, S. (1980). Rootstocks for ponkan (*Citrus reticulata* Blanco.) in relation to tatter leaf virus. *Bull Fruit Tree Res Stn* D10: 35–45.

Takahara, T., Ogata, T., Kawase, K., Iwagaki, I., Muramatsu, N., Ono S, et al. (1994). Effect of rootstock on growth and fruit quality of 'Otani Iyokan' (*Citrus iyo* hort ex Tanaka). *Bull Fruit Tree Res Stn* 26: 39–60.

Takahara, T., Okudai, N., and Kuhara, S. (1986). Elimination of citrus viruses by the semi-micrografting. *Bull Fruit Tree Res Stn* D8: 13–24.

Takeuchi, M., Yamakami, T., and Wada, K. (2002). Environment-friendly prevention of citrus fruit damage from injurious birds. *Bull Agric Res Mie Pref Sci Tech Prom Center* 29: 53–60.

Tanaka, T. (1925). Further data on bud-variation in citrus. *Japan J Genet* 3: 131–143. https://doi.org/10.1266/jjg.3.131

Tanaka, T. (1926). Wild citri of the Japanese territories. *Bul Sci Fak Terk, Kjusu Imp Univ* 2: 51–58. https://doi.org/10.15017/20737

Tokunaga, T., Nii, M., Tsumura, T., and Yamao, M. (2005). Production of triploids and the breeding seedless cultivar 'Tokushima 3X No. 1' from tetraploid X diploid crosses in sudachi (*Citrus sudachi* Shirai). *Hort Res (Japan)* 4: 11–15. https://doi.org/10.2503/hrj.4.11

Uchino, K., Kukita, H., and Kumamoto, O. (2023). Effect of rootstocks on leaf fall during winter, tree growth, fruit yield and quality in young tankan 'Tarumizu 1' trees. *Res Trop Agr* 16: 6–13. https://doi.org/10.11248/nettai.16.6

Ueno, I., Iwamasa, M., and Nishiura, M. (1967). Embryo a number of various varieties of citrus and its relatives. *Bull Hort Res Sta* B7: 11–21.

Wu, G.A., Terol, J., Ibanez, V., López-García, A., Pérez-Román, E., Borredá, C., et al. (2018). Genomics of the origin and evolution of citrus. *Nature* 554: 311–317. https://doi.org/10.1038/nature25447

Yahata, M., Hashihara, Y., Kuroki, H., Kunitake, H., and Komatsu, H. (2004). Effect of colchicine and oryzalin treatments on tetraploid production of nucellar embryos in meiwa kumquat (*Fortunella crassifolia* Swingle). *Hort Res (Japan)* 3: 11–16. https://doi.org/10.2503/hrj.3.11

Yamada, Y., Iwamasa, M., Nishiura, M., and Shichijo. T. (1979). Shortening the breeding cycle in citrus. I. Effect of budding and topworking on flower induction and fruit setting of citrus seedlings. *Bull Fruit Tree Res Stn* B6: 1–30.

Yamaguchi, R., Otomo, N., and Narahara, M. (2022). Characteristics of the new triploid kabosu (*Citrus sphaerocarpa*) cultivar 'Oita Kaken No. 6'. *Bull Oita Pref Agri Forest Fish Res Cen Agri Res Div* 9: 17–23.

Yamamoto, M. (2014). Progress on studies for seedless breeding of citrus in Japan. *Adv Hort Sci* 28: 64–72. 10.1400/230099.

Yamamoto, M. (2023). Self and cross-incompatibility of *Citrus depressa* Hayata (Shiikuwasha) and its relatives. *Hort J* 92: 134–141. https://doi.org/10.2503/hortj.QH-035

Yamamoto, M., Arita, S., and Kozai, N. (2024). Simple methods for producing tetraploids in polyembryonic citrus. *JARQ* 58: 107–113. https://doi.org/10.6090/jarq.58.113

Yamamoto, M., Kubo, T., and Tominaga, S. (2006). Self-and cross-incompatibility of various *Citrus*accessions. *J Japan Soc Hort Sci* 75: 372–378. https://doi.org/10.2503/jjshs.75.372

Yamamoto, M., Matsumoto, R., Okudai, N., and Yamada, Y. (1997). Aborted anthers of *Citrus*result from gene-cytoplasmic male sterility. *Sci Hort* 70: 9–14. https://doi.org/10.1016/S0304-4238(97)00017-4

Yamamoto, M., Matsumoto, R., and Yamada, Y. (1995). Relationship between sterility and seedlessness in citrus. *J Japan Soc Hort Sci* 64: 23–29. https://doi.org/10.2503/jjshs.64.23

Yasuda, K., Nukaya, T., Sudo, M., Yahata, M., Tominaga, A., Mukai, H., et al. (2022). Effect of in vitro colchicine treatment on tetraploid induction in seeds of polyembryonic cultivars of the genera *Citrus*, *Fortunella*, and *Poncirus*. *Trop Agric Develop* 66: 51–58. https://doi.org/10.11248/jsta.66.51

Yonemoto, Y., Matsumoto, T., Furukawa, T., Asakawa, M., Okuda, H., and Takahara, T. (2004). Effects of rootstock and crop load on sap flow rate in branches of 'Shirakawa Satsuma mandarin' (*Citrus unshiu* Marc.). *Sci Hortic* 102:295–300. https://doi.org/10.1016/j.scienta.2004.02.005

Yoshida, J., and Iwasaki, M. (2014). Influence of the ratio of the irrigation zone to the root zone on drought stress of citrus trees. *Hort Res (Japan)* 13: 335–341. https://doi.org/10.2503/hrj.13.335

Yoshida, T. (1980). Relationship between length of the juvenile period and tree growth in citrus seedlings. *Bull Fruit Tree Res Stn* B7: 1–13.

Yoshida, T. (1985). Inheritance of susceptibility to citrus tristeza virus in trifoliate orange [*Poncirus trifoliata* Raf.]. *Bull Fruit Tree Res Stn* B12: 17–25.

Yoshida, T. (2003). Breeding of citrus cultivar 'Kiyomi' with high quality and monoembryony. *Breed Res* 5: 103–107.

Yoshida, T., Nesumi, H., Yoshioka, T., Ieki, H., Ito, Y., Nakano, M., et al. (2003). New kumquat cultivar 'Puchimaru'. *Bull Natl Inst Fruit Tree Sci* 2: 9–16.

Yoshida, T., Nesumi, H., Yoshioka, T., Ito, Y., Yano, M., Nakano, M., et al. (2005b). 'Kankitsu chukanbohon nou 6 gou' ('Citrus parental line norin No. 6') is useful for breeding seedless and functional component-rich cultivars. *Bull Natl Inst Fruit Tree Sci* 4: 53–59.

Yoshida, T., Nesumi, H., Yoshioka, T., Nakajima, N., and Kuniga, T. (2005a). New citrus cultivar 'Tamami. *J Japan Soc Hort Sci* 74 (suppl 1): 236.

Yoshida, T., and Shichijo, T. (1984). Test for resistance of citrus cultivars to scab, *Elsinöe Fawcett* Bitanc. & Jenkins and segregation of resistance on hybrid seedlings. *Bull Fruit Tree Res Stn* B11: 9–16.

Yoshida, T., Shichijo, T., Ueno, I., Kihata, T., Yamada, Y., Hirai, M., et al. (1983). Survey for resistance of citrus cultivars and hybrid seedlings to citrus tristeza virus (CTV). *Bull Fruit Tree Res Stn* B10: 51–68.

Yoshioka, T., Matsumoto, R., Kuniga, T., Yamamoto, M., Takahara, T., Yoshinaga. K., et al. (2015). New citrus cultivar 'Seinannohikari'. *Bull NARO Inst Fruit Tree Sci* 19: 11–22.

4 Citrus Industry of Korea
Challenges and Opportunities

Kwan Jeong Song and Kyung Uk Yi

4.1 INTRODUCTION

4.1.1 History Of Citriculture

The old literature documented that citrus had grown before the 11th century in Jeju Island, but the definite origin of citrus remains to be discovered (Yun, 2001; Yi et al., 2018). In the Joseon Dynasty (from 1392 to 1910), citrus fruit tribute was brought from Jeju Island into the central government during the early dynastic period. Citrus fruits were treated as valuable because they were offered to the royal family's shrine and special guests, and used as medical herbs, which was referred to as dried mandarin-peel. Citrus orchards belonging to the central government were established on Jeju Island to provide citrus fruits for tribute. More than 20 landrace species or cultivars have been reported in the old literature. However, as of now, only 12 species or cultivars have been preserved as genetic resources and are no longer economically grown, except "Yuzu" (*Citrus junos*). "Yuzu" has been grown out of Jeju, such as Goheung, located in a south-coastal area of mainland Korea, to produce fruit tea, dried peel, or a rootstock (Yun, 2001; Yi et al., 2018). These Korean landrace species or cultivars include "Gamza" (*C. benikoji* hort. ex Tan.), "Byungkyul" (*C. platy mamma* hort. ex Tan.), "Jinkyul" (*C. sunki* hort. ex Tan.), "Cheongkyul" (*C. nippokoreana* Tan.), "Pyunkyul" (*C. tangerina* Hort. ex Tan.), "Binkyul" (*C. leiocarpa* hort. ex Tan.), "Dongjeongkyul" (*C. erythrosa* hort. ex Tan.), "Hongkyul" (*C. tachibana* (Mak.) Tan.), "Sadoogam" (*C. pseudogulgul* hort. ex Shirai), "Dangyooza" (*C. grandis* (L.) Osb.), and "Jigak" (*C. aurantium* L.) (Figure 4.1).

Satsuma mandarin, the leading citrus species in Korea, was first introduced from Japan to Jeju Island in 1911 by French-born Father Esmile Taquet (Yun, 2001). However, the large-scale cultivation of citrus began in the middle of the 1960s with the central government-led project encouraging farmers with low-interest loans. This project aimed to increase growers' income in rural areas, which resulted in a steady increase in both the growing area and citrus production for three decades until 2000.

4.1.2 Recent Trends In The Citrus Industry

Based on an agricultural production report by Statistics Korea (2021), citrus ranks top fruit production with 635,835 M/T in 22,029 ha in 2021 (Figure 4.2). This corresponds to 30.1% of the total Korean fruit production of 2,109,490 M/T. More than 99% of citrus fruits are produced in Jeju Island, where the citrus industry producing more than 50% of the local agricultural product, is one of the key industries along with tourism, leading the local economy. Income from citrus production represented less than 10% of the GRDP of Jeju Island in 2020. However, local businesses are very sensitive to the citrus industry because more than 80% of local farmers grow citrus, and most

DOI: 10.1201/9781003584599-4

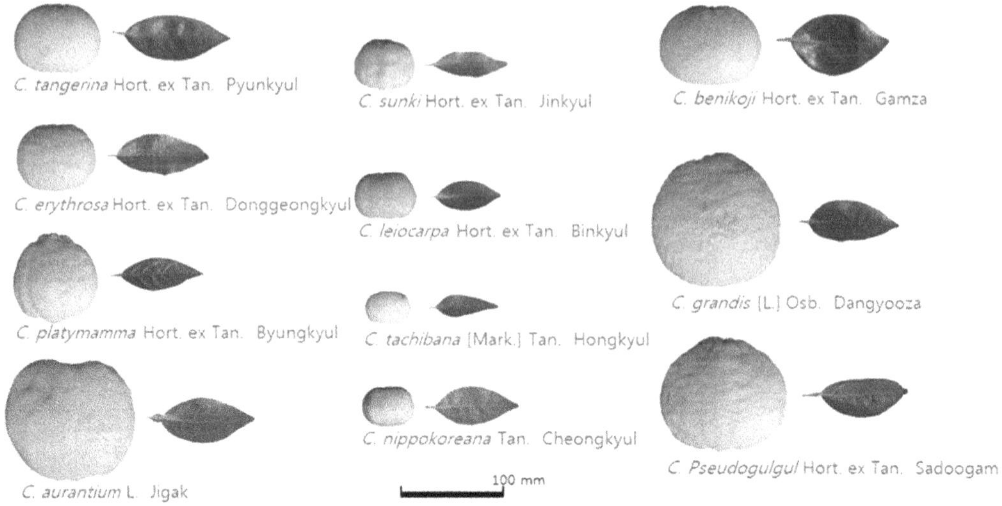

FIGURE 4.1 Shape and size of fruit and leaf of Korean citrus landraces.

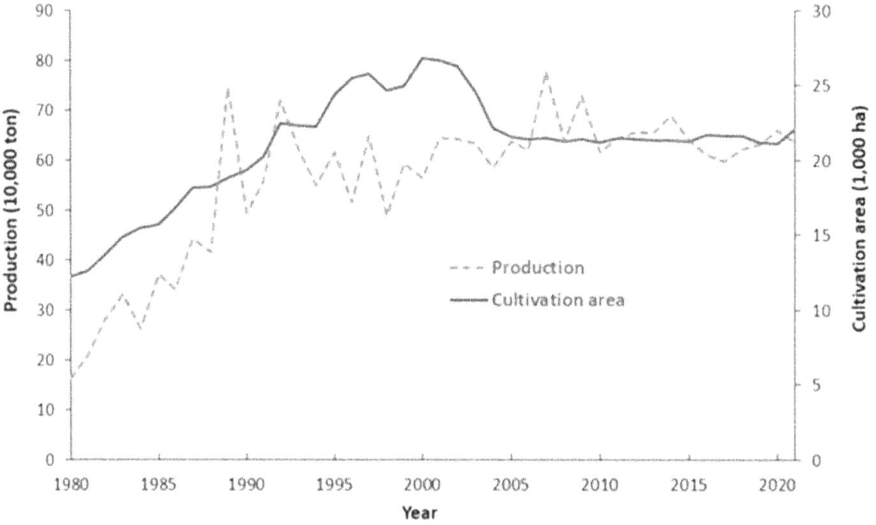

FIGURE 4.2 Changes in citrus cultivation and production in Korea.

of the income from citrus is recycled into other local businesses. Annual citrus production varies over the years, fluctuating around 600,000 M/T with a degree of alternate bearing. Also, in recent years, the citrus industry on Jeju Island has faced challenging climate change and an oversupply of diverse fruits domestically or internationally. Citrus cultivation reached a peak of 26,821 ha in 2001. After that, citrus cultivation was reduced by 20% during the following five years due to the effort of the local government to avoid overproduction. Additionally, the citrus cultivation area is gradually moving north, through the south-coastal area of the mainland, due to an increase in annual mean temperature by global warming.

The total citrus cultivation area decreased slightly as the open field cultivation area continued to shrink. On the other hand, the protected cultivation areas for late-maturing varieties, including "Shiranuhi", "Setoka", "Kanpei", and some hybrid cultivars, and off-season Satsuma mandarin growing in plastic film houses, have increased steadily in recent years. As consumer demand for

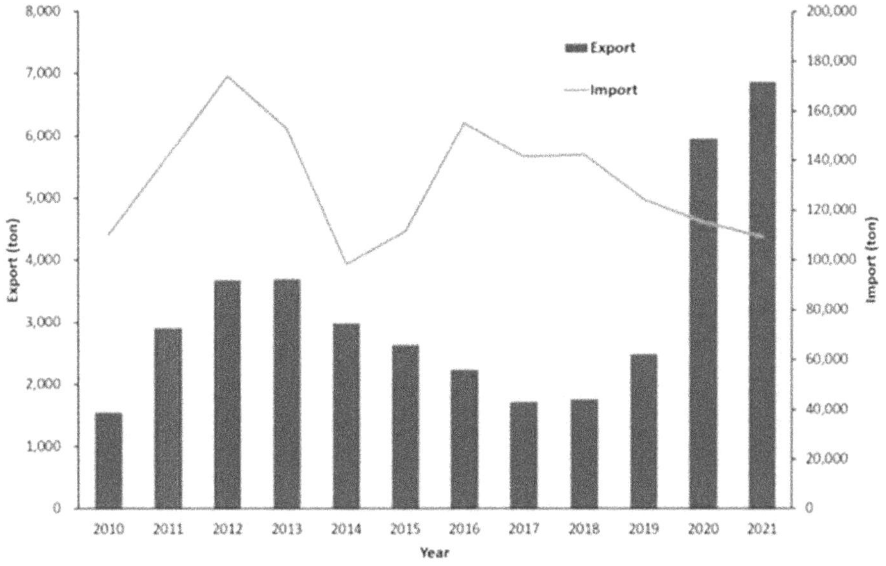

FIGURE 4.3 Changes in importing and exporting fresh citrus fruit in Korea.

high-quality, increases more growers will be expected to switch to late-maturing citrus growing in plastic film houses or to apply to intelligent farming based on sensors in the coming years to chase higher farm incomes (CREI, 2023).

Korea has pursued the expansion of exporting fresh citrus fruits. However, the amount of export has remained at a low level below 10,000 M/T per year (Figure 4.3). Only Satsuma mandarin of citrus has been exported to 20 countries worldwide. The largest export destination is Russia (about 70%), followed by Canada, the United States, Hong Kong, and Singapore in 2021. Citrus exports to Russia dropped by 50% since the Russia-Ukraine war started in 2022. On the other hand, a large amount of citrus has been imported since 2012, when the FTA between Korea and the USA went into effect. The most imported citrus fruit is orange, which is about 85%. The other imports of citrus fruit include grapefruit and lemon. Orange fruits have been chiefly imported from the USA, grapefruit from South Africa and the USA, and lemon from the USA. Recently, the amount of oranges imported from Australia has increased since 2020. Also, imports have decreased due to higher import prices and lower consumer demand in the domestic market.

4.2 CULTURAL ENVIRONMENTS

4.2.1 Geography

South Korea occupies the southern portion of the Korean Peninsula, located out from the far east of the Asian continent between 33°~43° north latitude and 125°~132° east longitude and comprises 3,358 nearby islands (Figure 4.4). The East Sea borders South Korea to the east, the Yellow Sea to the west, and Korea Strait to the south. The major production region of citrus fruits is Jeju Island, which is located between 33°11'~33°34' north latitude and 126°09'~126°58' east longitude in the most southern part of Korea. It has an oval shape of 73 km east-west, 31 km north-south, and the coastline is 258 km, which covers an area of 1,833 km^2. The extinct volcano Mount Halla is located at the center of Jeju Island, which is 1,950 m above sea level and has the highest peak in South Korea. It forms a gentle slope around the shield volcano Mount Halla from the center into the coastal area. Citrus is cultivated in coastal areas reaching 300 m above sea level, mainly in the northern and southern parts, except for the eastern and western of Jeju Island. Recently, a small amount of citrus

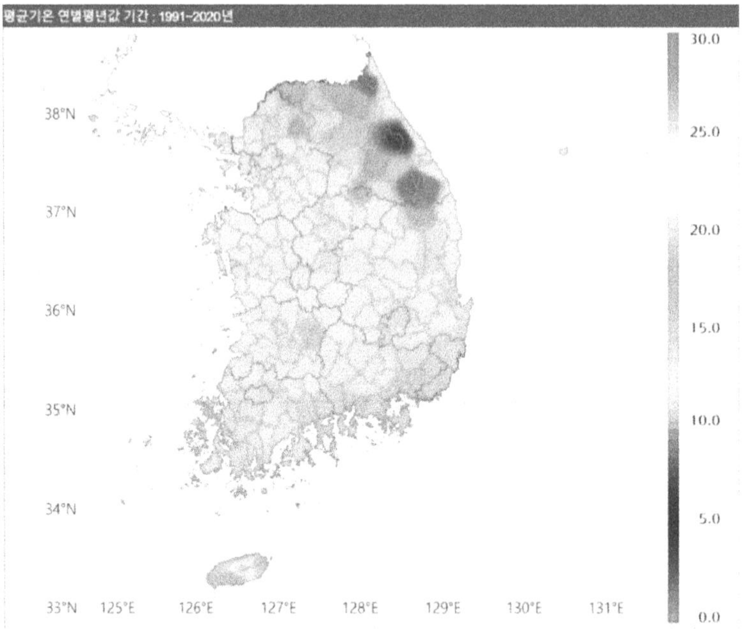

FIGURE 4.4 Location of South Korea and average air temperature for the recent 30 years (1991~2020).

cultivation has been extended up to the south-coastal area of the mainland around 35° north latitude for open field production due to an increase in annual mean temperature by global warming.

4.2.2 Climatic and Soil Conditions

Most of South Korea belongs to the temperate climatic zone, being cold to very cold depending on the latitude during winter and hot and sultry with plentiful rainfall in summer. The only coastal area of Jeju Island in the southern tip of Korea has a subtropical climate. For that reason, citrus fruits have been produced exclusively in Jeju Island except for small-scale culture in the open field or heated plastic greenhouse in the southern tip of the Korean Peninsula. Jeju Island has a humid subtropical climate with four distinct seasons. During winter, it is mild with moderate rainfall, while it is hot and humid with very high rainfall in summer (Tables 4.1 and 4.2). The annual average temperature is 15~16°C, and the average temperature of the coldest month is around 6°C (Figure 4.4) (Korea Meteorological Administration (KMA), 2023). Average annual rainfall across the island is 1,746 mm, and roughly 70% of the total rainfall occurs from April to September. Notwithstanding its location and climate, snow is not uncommon during the winter. Forest and grassland blanket the island's central area, while agricultural lands are across the coastal area.

Soils on Jeju Island have relatively thin layers and consist mainly of silt and volcanic ash. Soils have high infiltration capacity in the high rainfall areas, including Mount Halla and the eastern part of the island. In contrast, soils in the drier western part of the island have relatively lower infiltration capacity. Since Jeju Island is a volcanic island formed mostly with permeable basalt, very few perennial streams exist on the island and are found mainly on the southern and northern ridge of Mount Halla; the island is dependent entirely on groundwater. Most citrus orchards were established on volcanic ash soil, highly acidified to pH 5.0. The volcanic ash contains a high amount of humus (10%) but is most insoluble. The content of available phosphorus is very low because of strong absorption by volcanic ash.

TABLE 4.1
Annual Mean Temperature, Rainfall, and Relative Humidity of Jeju Island

Area	Average temperature (°C)	High temperature (°C)	Low temperature (°C)	Average rainfall (mm)	Annual rainfall (mm)	Relative humidity (%)
Jeju-si	16.5	19.6	13.8	126.7	1520.7	70.2
Seigwipo-si	16.9	20.2	14.0	174.1	2089.3	73.7
Gosan	15.8	18.8	13.3	101.5	1218.5	76.5
Seongsan	15.8	19.4	12.4	179.7	2156.6	73.8

TABLE 4.2
Monthly Mean Temperature, Rainfall, and Relative Humidity of Jeju Island over Ten Years (2011~2020)

	Average temperature (°C)	High temperature (°C)	Low temperature (°C)	Rainfall (mm)	Relative humidity (%)
January	6.1	9.0	3.4	52.0	66.0
February	7.0	10.3	4.0	79.2	66.2
March	10.2	13.8	6.7	104.3	66.4
April	14.3	18.2	10.8	149.8	70.1
May	18.5	22.3	15.2	153.7	74.7
June	21.5	24.6	19.1	221.7	85.5
July	25.8	28.6	23.6	194.1	87.2
August	27.2	30.2	24.9	270.8	82.7
September	23.4	26.4	20.8	227.5	77.9
October	18.9	22.2	16.0	126.0	69.9
November	14.0	17.2	10.9	99.7	69.4
December	8.2	11.2	5.3	67.7	66.7
Average	16.2	19.5	13.4	145.5	73.6
Total				1746.3	

4.3 CULTIVARS AND ROOTSTOCKS

4.3.1 Cultivars

4.3.1.1 Satsuma Mandarin

4.3.1.1.1 Yura Wase or YN-26

There are two Yura lines. One is "Yura wase", selected firstly as a bud sport from "Miyagawa Wase" in Japan, and then "YN-26", selected again from nucellar seedlings. These "Yura" lines are very early maturing cultivars commonly harvested in early or mid-October with reasonably accepted fruit quality with 120 g of fruit size, 11 °Brix of soluble solids, and 0.8% of titrable acidity.

4.3.1.1.2 Haryejosaeng

"Haryejosaeng" was selected as a nucellar seedling from "Tachima wase" artificially pollinated with *C. natsudaidai*by by the Citrus Research Institute (CRI), NIHHS, RDA in 2004. Compared to "Miyagawa Wase", the most common cultivar in Jeju Island, "Haryejosaeng" is more vigorous, and fruit acidity could be easily managed. However, the other traits, including fruit season and resistance to pests and diseases, are almost the same. Fruits mature in early- to mid-November with 90 g of fruit size, 10.6 °Brix of soluble solids, and 1.1% of titrable acidity.

4.3.1.1.3 Miyagawa Wase or Okitsu Wase

"Okita Wase" was selected as a nucellar seedling derived from "Miyagawa Wase" pollinated with *P. trifoliata*in Japan. The "Okita Wase" tree is strong and grows well compared to "Miyagawa Wase". The other traits, such as fruit quality, are almost the same. The harvest period is from mid-to-late November, with around 110 g of fruit size, 8.7~10.8 °Brix of soluble solids, and 0.85~1.25% of titrable acidity. "Okita Wase" contains a high amount of β-cryptoxanthin, which is known as a cancer prevention ingredient.

4.3.1.2 Mandarin Hybrids

4.3.1.2.1 Minihyang

"Minihyang" mandarin, a hybrid variety of "Kishu mikan" (*C. kinokuni* ex Tanaka) x "Ponkan" (*C. reticulata*) "Ponkan", was selected by the CRI, NIHHS, and RDA and applied for variety protection in 2016. The vigor of the "Minihyang" tree is moderate with 30~40g of oblate-shaped fruit. Fruits mature in mid- to late November with below 1% of titrable acidity and over 15 °Brix of soluble solids. "Minihyang" is available for open field cultivation.

4.3.1.2.2 Winter Prince

"Winter Prince" mandarin is a hybrid obtained from a cross between "Harehime" ["E647" ("Kiyomi" × "Osceola") × "Miyagawa Wase"] and "Ponkan" and registered as a variety by the CRI, NIHHS, RDA and filed an application for variety protection in 2020. The "Winter Prince" tree is vigorous and bears oblate-shaped fruits with a slightly protruding top of the fruit. The harvest period is from late November to early December. The "Winter Prince" is recommended to replace "Ehime Kashi 28 gou" with a similar harvesting period originating from Japan. Matured fruit is soft and weighs about 150g with approximately 12 °Brix of soluble solids and 0.8% of titrable acidity. Fruits typically produce no seed but will be seedy by pollination with fertile pollens.

4.3.1.2.3 Ehime Kashi 29 gou

"Ehime Kashi 28 gou" is a hybrid of "Nankou" and "Amakusa" mandarin in Japan and was registered in 2005. The cultivar is commonly called "Benimadonna" in Japan and Korea. The obovoid-shaped fruits weigh about 250 g with 12 °Brix of soluble solids and 1% of titrable acidity. The fruit is seedless and ripens in mid-December to early January. The peel is thin, but the peeling is not easy, sticky, and wetting. The tree is vigorous with a spreading tree shape and a little thorny in vigorous branches.

4.3.1.2.4 Kanpei

"Kanpei" mandarin is developed by a cross of "Nishinokaori" and "Ponkan" mandarin in Japan and is commercially available as "Redhyang" in Korea. The tree is vigorous and grows upright until the onset of fruit. The fruit is oblate-shaped with a little radial crack near the stalk. The peel is thin and reddish, and puffing is rare. The fruit ripens in mid- to late-January and weighs on average 300g. The soluble solids are about 11 Brix, and the titrable acidity is about 1%. The fruit is typically seedless, but it may contain a few polyembryony seeds.

4.3.1.2.5 Shiranuhi

"Shiranuhi" mandarin is a hybrid of "Kiyomi" x "Ponkan" mandarin, developed in Japan in 1972. The cultivar is called "Hallabong" in Korea, "Sumo" in the USA, and "Ugly fruit" in Australia, which has recently become popular worldwide. The tree is less vigorous and moderately tolerant to citrus scab and citrus canker. The fruit is 200~300g in weight and is obovoid in shape with a prominent collar at the neck of the fruit. The fruit ripens in mid-to late February. Titrable acid is about 1% and soluble solids is 13~14 °Brix. The fruit is almost seedless, although it may contain polyembryony seeds by pollination with fertile pollens.

4.3.1.2.6 Suneat

"Suneat" is a bud sport of "Shiranuhi" (M16A line) selected by a farmer in Jeju, Korea, in 2007 and registered in 2016. "Suneat" is almost the same as its parent and the other nucellar varieties in tree vigor and attitude and fruit quality, including fruit size, soluble solids, and titrable acidity. However, the peel color of "Suneat" is much darker than standard "Shiranuhi" during the early coloring period. Eventually, the difference goes away almost as the fruit ripens.

4.3.1.2.7 Setoka

"Setoka" mandarin is a hybrid originated from a cross of "Kuchinotsu No.37" ("Kiyomi" x "Encore") and "Murcott" in Japan and was registered in 2001. The fruit weighs about 200g and has an oblate shape. The rind is thin, smooth, and peelable. The fruit ripens in early March. The soluble solids is about 13 °Brix and the titrable acidity is 0.8~1.3%. "Sotoka" produces nearly seedless fruits. The tree is less vigorous and the tree shape is spreading with good fruit sets. The tree has small thorns. "Setoka" is resistant to citrus scab and citrus canker, but susceptible to CTV.

4.3.1.2.8 Natsumi

"Natsumi" is a nucellar selection of "Kara" mandarin that originated from the pollination of "Oota Ponkan" mandarin in Japan. The cultivar is a late-maturing mandarin grown in a protected plastic film house. The fruits are harvested in mid-April, weighing about 150 g with 13 °Brix of soluble solids and 1.2% of titrable acidity. The fruits are seedless or contain a couple of seeds. The tree is vigorous and its branches tend to droop. Citrus canker tolerance is similar to Satsuma mandarin.

4.3.2 Rootstocks

Trifoliate orange (*Poncirus trifoliata*) is exclusively used as rootstock for all kinds of citrus grown in Korea due to its dwarf characteristics, cold resistance, high adaptability to acidic soil, and resistance to CTV (Oh et al., 2021b). Newly introduced rootstocks such as "Flying Dragon" trifoliate orange and "Carrizo" citrange are being evaluated to find more adaptability to new cultivars and climate changes (Hong et al., 2022).

4.3.3 Genetic Improvement

4.3.3.1 Breeding Objectives

Citrus breeding activities in Korea began in 1974 and have focused only on improving scion cultivars (Yun et al., 2019; Chung et al., 2020). The main breeding objectives have been changed from cold tolerance, high fruit quality with high sweetness and low acidity, and diversification of ripening season in the first breeding era (1978~1993) into high fruit quality with high sweetness and low acidity, easy peeling, low puffiness, attractive pulp and peel coloration, and disease tolerance in the recent breeding era (1994~at present) due to severity of climatic change. Polyploidy for developing seedless triploid was added most recently.

4.3.3.2 Genetic Resources and Establishment of Core Collection

Genetic resources are used to improve the existing traits for breeding materials. Two research organizations, including CRI, NIHHS, RDA and Citrus Breeding Center (CBC), Jeju Special-governing Provincial Agricultural Research and Extension Services (Jeju ARES) have tried to scale up genetic resources collection into 519 accessions with duplication conservation in CRI and CBC

(Park et al., 2020). These genetic resources collection consists of diverse citrus species, cultivars, bud sports, and relatives, conserved with two trees grafted on trifoliate orange per accession. It is essential to increase utilization to evaluate the genetic potential of genetic resources because conservation requires high expenses and massive land area. Also, the genetic resources are collected from different regions or countries based on different accession names and phenotypes, but genotypes might be the same or very similar. Therefore, it is essential to establish core collection by evaluating the genetic potential for saving the expenses and land area. The CBC, Jeju ARES, had conducted a project to construct a core collection from 2013 to 2016 and reported establishing the preliminary core collection consisting of 164 accessions based on evaluating fruit quality, bioactive components, and disease resistance. In the future, this preliminary core collection needs to be improved based on a precise evaluation of diverse traits including qualitative and quantitative traits (Song et al., 2017).

4.3.3.3 Conventional Breeding

Sexual hybridization is very difficult in Satsuma mandarin due to high male and female sterility. Then, the improvement of Satsuma mandarin mainly relies on selection of bud sports and nucellar seedlings. From these activities in Satsuma mandarin, several cultivars such as "Haryejosaeng" originated from nucellar selection, and "Sandojosaeng" originated from bud sport have been released from the CRI, NIHHS, RDA, and the CBS, Jeju ARES. However, only "Haryejosaeng" has been a leading cultivar, replacing "Miyagawa wase" and "Okitsu wase", major cultivars of Satsuma mandarin in Korea.

To extend the citrus season in Korea, protected cultivation of late-maturing mandarin hybrids under plastic film houses has increased with cultivars originating from Japan since the mid-1990s. Recently cultivation of mandarin hybrids such as "Kanpei", "Shiranuhi", and "Setoka" has occupied about 20% of citrus cultivation in Jeju. These cultivars were released by sexual hybridization using monoembryony seed parents and pollen-fertile parents in Japan (Park et al., 2020). The CRI, NIHHS, RDA, had conducted the sexual hybridization using monoembryony seed parents such as "Kiyomi", "Harehime", and "Ehime 28 gou" and pollen-fertile parents such as "Wilking", "Fortune", and "Oota Ponkan" since 1994 (Yun et al., 2019) and the CBC, Jeju ARES, since 2011 (Chung et al., 2020), and private breeding companies such as the Hannong Bio Industry Corp., since 2013, respectively (Song et al., 2017). To shorten the juvenile stage and breeding period, a single stem training system under a plastic film house was introduced as Figure 4.5. Recently, several cultivars of mandarin hybrids such as "Winter Prince" and "Minihyang" by CRI, NIHHS, RDA, "Gaeulhyang" and "Dalkomi" by CBC, Jeju ARES, and "Dertamna" by a private breeding company have been released.

4.3.3.4 DNA Marker Development and Biotechnology

It takes a long time to evaluate the fruit traits of seedlings due to long juvenility. It is also difficult to predict the phenotype of fruit traits by chromosome recombination due to the complexity and uncertainty of quantitative trait inheritance. It is tough to evaluate disease tolerance due to the inaccuracy of field-based screening and the inefficiency of facility-based screening. Therefore, it is essential to develop DNA markers closely linked to fruit traits and disease tolerance to increase the efficiency of conventional breeding (Kim et al., 2016). The development of DNA markers needs analysis of the correlation between genetic information and phenotyping. Phenotyping is laborious and tedious and needs to increase accuracy through repeat evaluation, which becomes obstruct in developing DNA markers. Korean citrus breeding programs have constructed huge cross populations since 1978, but these segregant populations have been demolished soon after a short evaluation of fruit phenotypes to increase the efficiency of breeding fields and labor. Recently, some segregant populations for genetic analysis have been constructed and conserved at Jeju National University. Based on these populations, preliminary linkage maps were constructed using "Fina Sodea" clementine and Byungkyul, a Korean landrace, based on RAPD and SSR markers (Chang et al., 2018). In the near

FIGURE 4.5 Single stem training of seedlings applied in the citrus breeding system of Korea.

future, these works will be further advanced after most seedlings will bear fruits. Polyembryony is a common trait in citrus and the identification of zygotic embryo-derived individuals is essential when polyembryony seed parents are used for cross breeding. Seventeen SSR markers for distinguishing polyembryony and polyembryony were reported from genotyping 101 citrus genetic resources (Woo et al., 2019). Red pulp coloration is one of the significant aims of Korean citrus breeding programs and many cross hybridizations have been conducted to select a novel blood mandarin type. Then, molecular markers were developed to screen the progeny candidates of the blood mandarin genotype early and correctly (Oh et al., 2021a). In addition, SSR markers were developed to identify citrus cultivars bred in Korea for protecting the breeders' rights (Woo et al., 2020).

Cultivar improvement is achieved by introducing beneficial traits lacking in existing cultivars by sexual hybridization or mutation. In the case of sexual hybridization, from closely related cultivars or species to lacking those. However, artificial crossing is not possible in cases of sexual incompatibility and somatic hybridization or protoplast fusion has become an alternative. The CRI, NIHHS, RDA, has conducted somatic hybridization to produce somatic hybrids being allotetraploids. These somatic hybrids are getting into fruiting stages and will be used for triploid breeding by sexual crosses with diverse diploidy genotypes in the future (Kim et al., 2020). Also, the obtainment of spontaneous polyploids and production of artificial autotetraploids by colchicine treatment have been explored and sexual crosses using these autotetraploids will be carried out in the near future (Song et al., 2011).

4.4 CULTURAL PRACTICES

4.4.1 Nursery System

The nursery tree production system has not been modernized yet in Korea. Seeds of trifoliate orange are collected in late October and sowed in outdoor fields in early March. After growing seedlings for one year, the healthy and vigorous seedlings are transplanted in mid- or late March with spaces of 70 cm between rows and 20 cm between trees. One year later, seedlings are transplanted again in March with spaces of 70 cm between rows and 30 cm between trees and scion budwoods are veneer-grafted on rootstocks after cutting the tops at a height of 5~10 cm in mid-April to May. The nursery trees are dug out and transplanted in orchards (Figure 4.6).

FIGURE 4.6 Conventional citrus nursery system in Korea.

This conventional nursery system can be improved with establishing virus-free certification and sheltered pot-grown nursery systems in advanced countries such as the USA and Spain (Aubert and Vullin, 1998; Vapnek, 2009). The CRI has provided scion trees originating from virus-free foundation trees to private nurseries with an insect-proof plastic house since 2022. Based on these activities, voluntary virus-free certification was established in 2023. At this moment, scion trees should be tested periodically and nursery trees should be tested against citrus tristeza virus (CTV), Satsuma dwarf virus (SDV), citrus mosaic sadwavirus (CiMV), citrus tatter leaf virus (CTLV), and citrus exocortis viroid (Cevd) and adequately labeled or tagged before being sold.

4.4.2 Cropping System

Due to the climatic conditions of Jeju Island, Satsuma mandarin grafted on trifoliate orange rootstock is the primary citrus species (84.8% by production in 2021), which is cultivated in the open field (Table 4.3), and most cultivars consist of "Miyagawa wase" and "Okitsu wase" which were originated from Japan. The very early or early maturing type is much preferred over mid or late-maturing types due to the low-temperature approach from late December. In general, Satsuma mandarin season ends in late February due to short storage (Figure 4.7). Protected cropping of Satsuma mandarin without heating has been applied since the early 2000s to extend fruit season into March to April. The early cropping system heated under a plastic film house targets an off-season supply of Satsuma mandarin from May into September. This heated cropping has been applied since the early 1990s to replace low-profitable banana culture using a heated cropping facility. These two types of protected cropping are based on two Satsuma mandarin cultivars, "Miyagawa Wase" and "Okitsu Wase", which are early maturing types.

Mandarin hybrids are late-maturing type, which ripen after January in Jeju and then require protection from cold weather conditions (Figure 4.7). This cropping has been applied since the late

TABLE 4.3
Area and Production of Citrus by Species and Cultivation Type on Jeju Island in 2021

Species	Cultivation type	Area (ha)	Production (M/T)
Satsuma mandarin	Open field	14,607	467,293
	Heated plastic film house	373	27,009
	Unheated plastic film house	916	25,833
Mandarin hybrid	Open field	624	4,293
	Unheated plastic film house	3,458	88,690
Total		19,978	613,118

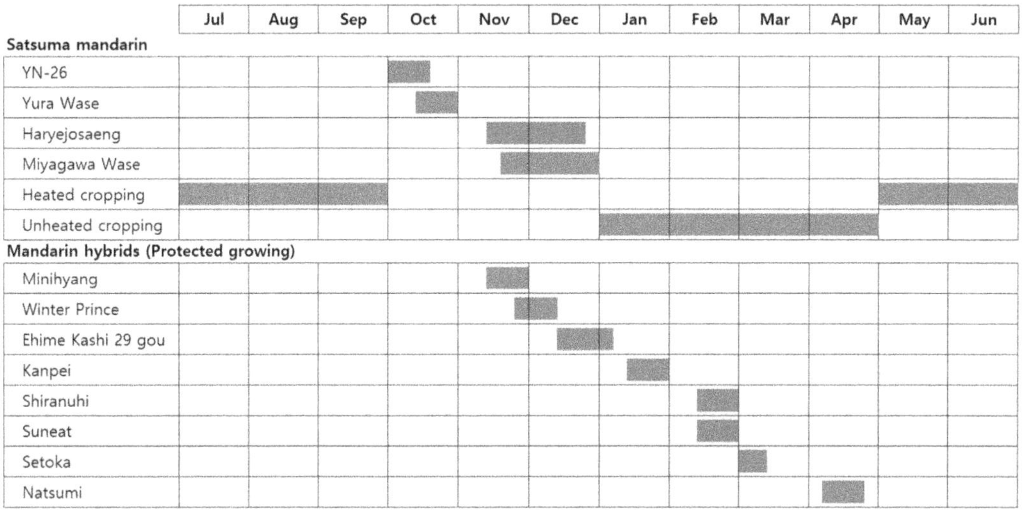

FIGURE 4.7 Maturing season for Satsuma mandarin and mandarin hybrid.

1990s to meet consumers' preference for higher sweetness compared to Satsuma mandarin. These mandarin hybrids include "Ehime 28 gou", "Kanpei", "Shiranuhi", "Setoka", and "Natsumi", the season of which is from mid-January to April for targeting the early off-season of Satsuma mandarin. A small amount of late-maturing mandarin hybrids such as "Ehime 28 gou" and "Shirahuhi" have been produced in open fields without or with fruit protection by paper bagging. The fruits produced in open fields are slightly smaller, but are similar in soluble solids and lower in titrable acidity compared to those that come from protected plastic film houses.

4.4.3 Open Field Cultivation (General and Soil Covering)

Most citrus orchards in Jeju Island were established in three decades from the late 1960s to 1980s, with an ultra-high density of 3,000 trees per ha (1.8 x 1.8 m) (Moon and Kim, 2006). The planting space must be adjusted with the final tree density of 750 trees per ha (3.6 x 3.6 m) by removing half of the trees every 10 years to avoid overcrowding of the tree canopy. However, most farmers evaded this recommendation due to fear of a reduction in yield. In recent years, many farmers began to adjust planting space based on the recognition that close spacing causes difficulties in orchard management, such as pruning, fertilizer application, control of pests and diseases, and fruit harvesting, and deterioration of fruit quality by obstructing sunlight penetration. Most recently, the planting space has been adjusted to be 4.0 x 3.0 m for very early maturing cultivars or 5.0 x 4.0 m for early maturing cultivars as the final tree density in newly establishing orchards. The initial planting space is 4.0 x 1.5 m for very early maturing cultivars or 5.0 x 2.0 m for early maturing cultivars. This planting density aims to promote automatic mechanization due to labor shortage.

Most citrus trees are trained basically to build tree shape of open-center natural form with the frame of three main branches with two or three secondary branches each, a total of six to nine secondary branches per tree. Pruning is performed manually from March to April every year. Fruit thinning is generally performed 3 to 4 times by hand from August to September as physiological fruit drop ends in July, which will improve the uniformity of fruit size, alternative bearing, and efficiency of fruit harvesting and sorting operations. Also, flower thinning by cutting out the bearing mother branch with many flower buds before full blooming is recommended for adjusting fruit load in case too much abundance of flowering is predicted.

TABLE 4.4
Recommended Amount of Nutrient Supply in Open Field Orchard of Satsuma Mandarin

	Volcanic ash soil			Non-volcanic ash soil		
Tree age	N	P	K	N	P	K
10	148	140	123	132	126	107
20	230	200	190	210	180	170

Partial sod culture or integrated vegetation management is common because it has many advantages, such as reducing weed competition for resources, preventing the erosion of soil by wind or water, reducing chemical inputs, and improving fruit quality and yield. Natural vegetation with herbs such as southern crabgrass (*Digitaria adscendense*) or artificial vegetation with sowing seeds of exotic plants such as clover, alfalfa, or rattail fescue *(Vulpia myuros*) could be optional depending on the preference of growers for soil covering resources. Mulching is also applied by using diverse straw or crushed twigs originated from the trees thinned out or windbreaks. Recently, porous films have been used for mulching materials to control soil moisture levels, targeting the increase of soluble solids in fruits.

Nutrient management is based mainly on soil type and tree age. The recommended amount, depending on soil type and tree age, is indicated in Table 4.4. For the volcanic ash soil, 148~230 kg nitrogen (N), 140~200 kg phosphate (P), and 123~190 kg potassium (K) are recommended per ha, while for the non-volcanic ash soil, 132~210 kg N, 126~180 kg P, and 107~170 kg K are recommended. It is recommended that this amount be distributed three times at different rates (50:20:30 in N, 100:0:0 in P, and 30:40:30 in K). Recommended application timing is early March, late May or early June, and mid to late October. The application of compost is recommended with 20 M/T per ha every year in both soil types.

The open fields citrus cultivation area has been steadily decreasing as the number of farmers switching to facility cultivation (plastic film house) increases due to the shrinkage of income from open field cultivation, and it is expected to continue to decrease. On the other hand, the citrus cultivation area of very early maturing type increased from 1,532 hectares in 2010 to 2,657 hectares in 2021 by 173% (Figure 4.8). This is related to fruit price stabilization through dispersion of massive supply with a new introduction of high-quality citrus varieties such as "Yura Wase", which is matured in early October with 11.5 °Brix of soluble solids and 0.9% of titrable acidity.

Early maturing Satsuma mandarin, representing most citrus fruits produced in Korea, is generally characterized by relatively high acidity and low soluble solid contents. The fruit quality could be improved by shifting the soluble solid contents and lowering the acidity. Moderate drought stress during the fruit developmental stage has been recognized to increase the accumulation of soluble solid contents through promoting the unloading of transported carbohydrates (Yakushiji et al., 1998; Han et al., 2014). Mulching of the soil surface with the porous film is practiced in order to induce drought stress (Figure 4.9). Porous film prevents the penetration of the rainwater underground, but permits the air circulation and evaporation of the soil moisture through the pores of the film, resulting in the hastened drying of soil moisture in the root zone. However, the effect of soil mulching is different depending on mulching material, mulching timing, soil moisture level before soil mulching, mulching tightness, and soil type. The soil mulching practice requires installing irrigation equipment under the covering to control soil moisture level based on monitoring fruit acidity, lowering which is not easy at the late maturation stage.

In addition, the installation of umbrella pillar supporters holding fruiting branches is recommended to protect from fruit loss due to branch breaking and fruit scratching due to branch

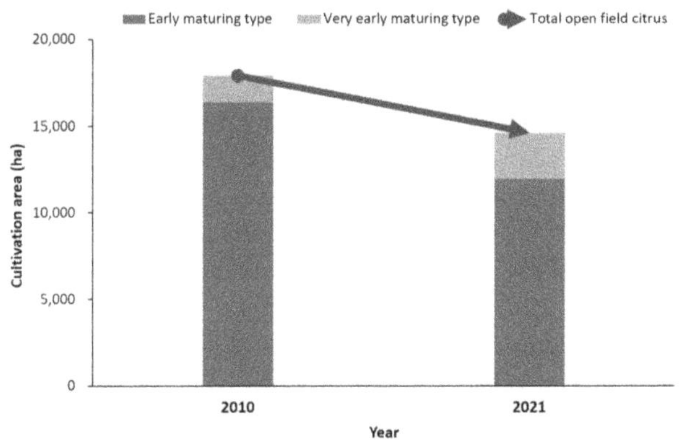

FIGURE 4.8 Change in open field citrus cultivation as per maturity type.

FIGURE 4.9 Umbrella pillar supporters holding fruiting branches.

drooping (Figure 4.9). It helps to improve the sunlight-interception of fruiting parts. Many growers practice foliar spray of CaCO3 to improve fruit quality, three to four times at the interval of 15 days from the beginning of coloration at the stylar end of fruit. It is recognized to improve peel coloration and extend the storage life of fruit by reducing peel puffiness.

4.4.4 Protected Cultivation

In order to extend the citrus season and improve fruit quality, citrus trees are widely cultivated under protected conditions covered by plastic film in Jeju, Korea (Moon and Kim, 2006). Two types of protected production consist of Satsuma mandarin and late-maturing mandarin hybrids. Also, the protected production of Satsuma mandarin is classified with or without heating, which determines the harvest time. These protected cropping methods allow for adjusting flowering time and improving fruit quality by controlling soil moisture level and air temperature in the plastic film house.

The heating time will depend on the intentional harvest time in cultivating the heated plastic film house. In order to advance the harvest time to late April, the plastic film house is heated before or in November, while heating starts in or late January to harvest in September. The starting time of heating affects flush types to be used for flowering and fruit-bearing, and therefore the method of pruning. Cultivating Satsuma mandarin in an unheated plastic film house is not to advance fruit maturity but to delay harvest time. The plastic film house protects fruits on trees from snow, cold, or birds during winter. Moreover, the additional advantage is that it enables the control of soil moisture levels during fruit maturation stages and the decrease of peel puffiness in fruits.

In Korea, citrus species such as late-maturing mandarin hybrids, kumquat, and lemon, other than Satsuma mandarin, are grown fundamentally in plastic film houses. Several mandarin hybrids, including "Shiranuhi", "Kanpei", and "Setoka", have expanded occupancy percentage by replacing kumquat and old late-maturing mandarin hybrids such as "Kiyomi". The cultivated area of hybrids, which increased from 2,112 ha in 2015 to 4,082 ha in 2021 by 193%, continues to increase due to the continuous government support projects to strengthen the competitiveness of the citrus industry and the conversion of crops from Satsuma mandarin grown open field. "Shiranuhi" was the most cultivated hybrid, with 4,082 ha in 2021. Some growers attempt to grow it even in the southern part of the Korean Peninsula. However, this cultivar has been switched to other hybrids as fruit prices decline due to oversupply, trees gradually become less vigorous, fruits become much more acidic, and yield deteriorates with the aging of trees. The cultivation area of "Kanpei" has increased dramatically from 145 ha in 2015 to 897 ha in 2021. However, the increment of the area is expected to slow down due to the high frequency of fruit splitting and alternate bearing. The yield per unit area and consumer preference of "Setoka" are higher than those of other hybrids. For these reasons, the cultivation area of "Setoka" will likely continue increasing, from 260 ha in 2015 to 930 ha in 2021 by 350%.

4.5 HARVEST, PACKING, AND MARKETING CHANNELS

4.5.1 Harvest and Storage

The harvest time is determined based on the degree of pee coloration and the soluble solids content of the fruit. Fruits with low and soluble solids content or high acidity will be harvested after ripening in the tree. Before harvest, it is recommended to apply anti-rotting materials or fungicides for storage. Selective harvest is typical in citrus fields. Fruits on the sunny side outskirts of the tree are harvested first. If the surface of the fruit is wet from rainwater or morning dew, harvesting should wait until it is dry. It is crucial to avoid any physical damage to the fruit during harvesting, and then wearing gloves, handling scissors with care, and picking the fruit by double cutting the fruiting branch are commonly practiced. Harvest for the open field cultivated citrus is recommended to be terminated not late on the 20th of December to prevent low-temperature damage caused by hail or snow.

A short pre-drying of harvested citrus fruits just before storing lowers the moisture content in the peel. This pretreatment preserves freshness and increases storage life. The treatment aims for a 3% reduction of fruit weight for 1~2 weeks in shady and dry conditions.

Citrus fruits are kept in storage warehouses before being sorted and shipped to improve fruit quality or adjust the amount of supply to fresh fruit markets. The storage conditions for Satsuma mandarin are 3~5°C and 85% humidity. The storage warehouse should be thoroughly ventilated. Every 15 to 20 days, fruit quality should be inspected, and rotten fruits should be removed. Low-temperature damage of the stored fruits occurs on the peel around 1°C and the pulp freezes around -2.5~-3°C.

4.5.2 Packing

There are lots of citrus packing houses on Jeju Island. Most of the packing houses are operated on a small scale, handling about 3,000 M/T of fresh fruits per year, and are operated mainly through distributors or corporations. There are 11 Agricultural Products Processing Centers that operate on a medium scale handling over 10,000 M/T per year by the National Agricultural Cooperative Federation (NACF) and Jeju Citrus NACF. All 11 citrus APCs are equipped with a non-destructive fruit quality evaluation facility. The processing steps from washing to packing are fully automated and controlled by a programmed system (Figure 4.10). The washing and drying line ensures that only clean fruits are processed. Fresh citrus fruits are sorted and graded based on maturity, ripeness, firmness, size, weight, and soluble solids. In the case of Satsuma mandarin from the open field, the standard of fruit size is classified into five grades (2S, S, M, L, and 2L). Too small size below 48 mm or 52 g and too large size above 71 mm or 136 g are treated as non-commodity. Visual inspection also helps to screen damaged fruits. It reduces postharvest losses during marketing.

4.5.3 Marketing Channels

Most fresh citrus fruits are consumed through domestic markets and a small amount is exported (Figure 4.11) (KAMIS, 2022). The supply of fruits to markets is initiated by growers associations or local private collectors operating the packing house with facilities for sorting and packing. To ship citrus fruits from the packing house to the consumer market, a shipment report with a quality certificate must be submitted to the Jeju Citrus Federation following the Jeju Special Self-governing Province Ordinance. In the wholesale market, fruit price is determined through auction depending on fruit quality and amount. In the direct market channel from local collector to large-scale distributor,

FIGURE 4.10 Citrus packing house operated with non-destructive fruit quality evaluation.

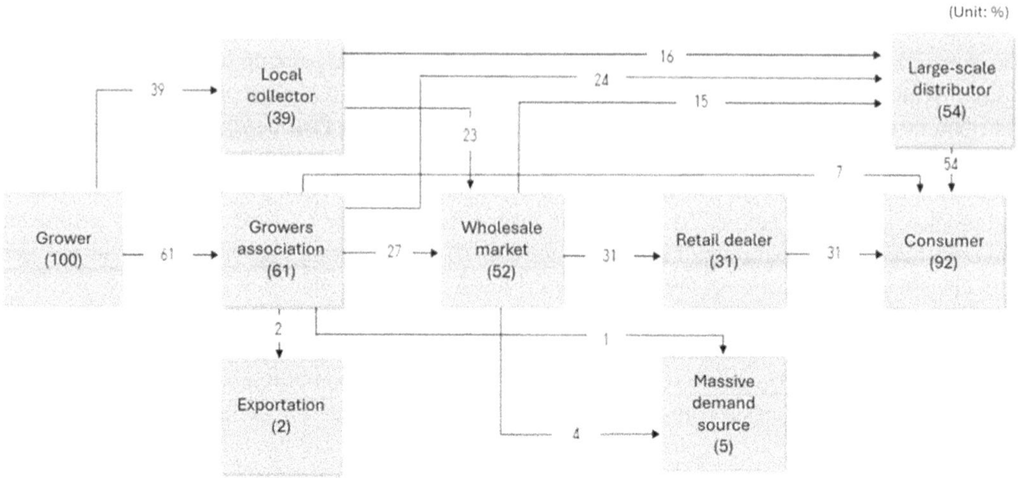

FIGURE 4.11 Marketing channels of citrus in Korea and market share of marketing channels (2022).

the fruit price will be determined through negotiation. Recently, the direct market channel between growers and consumers by social network service (SNS) and door-to-door delivery service has been established and market share has rapidly expanded.

4.6 PROCESSING

Various processed products using citrus fruits have been developed, which has become another source of income for citrus farmers. Sixteen to eighteen thousand citrus fruits, about 11% of total citrus products, are processed annually. The primary processed product is citrus juices that are extracted, concentrated, or sac of citrus fruits. Other processed products on the market are canned citrus, citrus brew wine, citrus-dried chip, chocolate, and jelly. Secondary processed citrus extracts are also being used as the raw material for medicine and cosmetics.

4.7 PESTS AND DISEASES

4.7.1 Pests

More than 70 pests have been reported, but only the major pests, including mites, aphids, and leaf miners, need to be controlled. Sometimes, thrips, scabs, bugs, or small green leafhoppers attack citrus (Kang et al., 2016).

Citrus red mite (*Panonychus citri*) preys on citrus leaves, stems, and fruits in the fall and spring. It causes a silvery stippling of the leaves as they aspirate the contents of the plant cells in the epidermis and mesophyll. Citrus red mites turn leaves and green immature fruit noticeably pale. If uncontrolled, citrus red mites may cause complete leaf drops.

The spiraea aphid (*Aphis citricola*) is the dominant aphis. It begins in May, but peaks in the summer. The affected leaves are usually dry or become deformed. When the occurrence is high, the secretions can cause sooty mold.

Citrus leafminer (*Phyllocnistis citrella*), its larvae, burrows into the new leaves, creating a curved tunnel. It occurs from May to June on the spring flushes, and begins to cause damage, but it is mild and usually causes severe damage in the summer and autumn flushes.

Western flower thrips (*Frankliniella occdentalis*) inflict harm on leaves and flowers. They leave many silvery-white spots on the leaves and, in severe cases, gray to light brown blotches. They often occur when hot and dry weather continues and in citrus grown in the greenhouse from spring to summer.

The types of citrus scales are reported on citrus trees in Jeju, Korea. Among them, cottony cushion scale (*Icerya purchase* Maskell), citrus mealybug (*Planococcus cryptus* Hempel), and arrowhead scale (*Unaspis yanonensis* Kuwana) are mostly found on citrus trees. Instar scales infest twigs and leaves along the veins. Adult scales are found on branches and trunks located in a gloomy place. Citrus scales extract plant sap from the citrus tree, thus reducing tree vigor. If infestation of citrus scales is severe, defoliation and fruit drop could occur. The scales produce honeydew, where sooty mold grows on top of it.

4.7.2 Diseases

The major diseases that occur on citrus in Jeju, Korea are citrus canker, melanose, and scab, among which melanose causes the most damage (Kang et al., 2016).

Citrus canker is caused by the bacterium *Xanthomonas citri* subsp. *citri*. Citrus canker affect most citrus species' leaves, twigs, young stems, and fruits. It develops the typical symptoms of distinctive corky necrotic lesions, sometimes possessing chlorotic or water-soaked halos, which are rough to the touch. The pathogen penetrates through the stomata of the leaves or open wounds and affects the tree's vigor and growth. The spotty appearance of the infected fruit makes it unmarketable. In highly favorable conditions, citrus canker can cause defoliation, shoot dieback, and fruit drop. Control of citrus canker relies on integrated management measures consisting of windbreaks, removal of infected leaves, twigs, and fruits, periodically spraying 1~3 times with copper products or antibiotics from the start of floral abscission to August, and control of leaf miner to protect susceptible cultivars.

Citrus melanose, caused by *Diaporthe ctri*, is a fungal disease. It occurs intensively from mid-June to mid-August and damages external leaves, twigs, and fruits of citrus trees. Dark brown to black spots are produced, causing a decline in fruit quality, which lowers the market value of fruits. Citrus melanose can be prevented or managed by pruning dead twigs and avoiding excess moisture conditions in plastic film houses. Applying fungicides before rainfall and annually 4~6 times before the onset of the monsoon season is recommended.

Citrus scab is a fungal disease caused by *Elsinoe fawcettii* and appears on young leaves, twigs, stems, and fruits. It produces raised spongy-like pustules in pinkish color, which turn grayish then brown. Citrus scab affects the appearance and quality of the fruit and lowers its market value. It is recommended that infected leaves and twigs be removed, thus reducing the density of a source of infection. The pathogen often develops on sprouts during the spring in late April under damp conditions. Fungicides must be applied during the season, especially before rainfall.

4.8 STATUS OF CITRUS RESEARCH AND EXTENSION

Three organizations, including the central government-based Citrus Research Center (CRC), NIHHS, RDA, the local government-based Agricultural Research & Extension Services (ARES), and the College of Applied Life Sciences of Jeju National University (JNU), have supported the sustainable development of Korean citrus industry with implementing independent or joint-collaborative programs for the cultivar improvement and the development of innovative cultivation and post-harvest management techniques. These three organizations have been working by dividing their roles under a close cooperative relationship. The JNU has mainly focused on basic research and technology development, the CRI on cultivars improvement and applied technology development research, and ARES on disseminating developed technologies.

The representative research achievements by the JNU included the construction of genetic linkage maps in mandarins, the development of DNA markers for identifying cultivars and sexual hybrids, the karyotyping of citrus, the development of triploid hybrids and tetraploids, the resistance of citrus canker, the pathogenicity of melanoses and citrus scab, the life cycles of several insects such as citrus thrips and integrated pest management against their control. The CRI has maintained the collection of citrus genetic resources, has released several elite cultivars by selecting sexual hybrids or nucellar mutants, has conducted research on improving citrus cultivation methods and reducing production costs, on efficient management of citrus pests and diseases, and the development of post-management technology and marketability enhancement technology. In addition, the CRI is also working on developing processed products containing functional ingredients to use citrus by-products and increase added value, as well as developing scaffolds for artificial skin and medical antibacterial paper using citrus fruits. The Jeju Special Self-Governing Province ARES has strived to replace the new cultivars and disseminate improved cultivation technologies through operating technical education programs and demonstration fields. Recently, the CRI has shifted breeding programs to combine biotic and abiotic stress tolerance with fruit quality, and the ARES has initiated the breeding programs for fruit quality improvement to speed up the release of elite new cultivars.

4.9 CHALLENGES OF THE KOREAN CITRUS INDUSTRY

The citrus industry ranks next to apples in the fruit industry in Korea and has contributed significantly to the development of the national economy. However, it is facing a range of challenges for future sustainable development. These challenges include climate change, labor shortages, the rapid rise in material prices, changes in consumption trends, and oversupply of diverse fruits domestically or internationally (Song et al., 2022). The recent climate change pattern in Korea has represented an increase in annual mean air temperature and an increase in rainy days and rainfall amounts in the maturation stage of Satsuma mandarin. This climate change pattern affects the flowering and fruit quality of Satsuma mandarin grown in open fields. To cope with climate change in citrus growing, early replacement of old trees with new elite cultivars and soil moisture management with soil covering and irrigation facilities have been strongly recommended. Labor shortages and rising material prices lead to an increase in production costs. The foreign seasonal worker system was recently initiated to resolve the labor shortages. Automation technology, such as soil moisture control based on sensing systems and unmanned chemical spray facilities has been applied to orchards. However, if fruit consumption does not increase, all strategies are useless. Fruit consumption has gradually decreased due to the decrease in the young generation population and the increase in one- to two-person households. Moreover, as the supply of various domestic and imported fruits has increased, competition is intensifying. Then, fruit quality improvement has become a critical factor in increasing competitiveness. These consumption patterns make changes in packing size and style. Sales for a small package of fresh citrus with 1~5 kg and a tiny package of piece type with peeled citrus have continuously increased. It will be crucial to sustain the development of the citrus industry to overcome these challenges in Korea. Therefore, comprehensive and cooperative approaches to cultivar improvement, promotion of new cultivar fruits, application of intelligent farm technology, and fruit quality improvement have been applied.

4.10 CONCLUSION

The citrus industry has contributed significantly to national economic development and improvement of public health since establishing a modernized growing of Satsuma mandarin. Continuous improvement of production technology through innovative research has led to the rapid expansion of growing areas and production amount by the 2010s. However, afterward, the adjustment of production amount through reducing the growing area was strategically achieved due to the intensifying competition with other fruits. Especially the protected growing system under plastic film house for the unseasonal production of Satsuma mandarin and late-maturing mandarin hybrids such as "Kanpei", "Shiranuhi", and "Setoka" and the soil covering cultivation system has contributed significantly to achieving all year around supply and keeping the competitiveness. Nevertheless, the citrus industry in Korea has recently faced several challenges, such as climate change, labor shortage, rising material prices, and changes in consumption patterns. Therefore, the comprehensive and cooperative approaches to cultivar improvement, promotion of new cultivar fruits, application of intelligent farm technology, and fruit quality improvement must be applied continuously in the future.

REFERENCES

Aubert, B., and Vullin, G. (1998). *Citrus nurseries and planting techniques*. CIRAD. Montpellier, France.

Chang, Y., Kim, H.B., Oh, E.U., Yi, K., Song, K.J. (2018). Construction of genetic linkage maps of 'Fina Sodea' clementine (*Citrus clementina*) and Byungkyul (*C. platymamma*), a Korean landrace, based on RAPD and SSR markers. *Hortic Environ Biotechnol* 59:263–274.

Chung, K.H., Nam, E.Y., Kwon, J.H., Hur, Y.Y., Kwon, S.I., et al. (2020). The history, current status, and future prospects of fruit breeding in Korea. *Korean J Breeding Sci Special Issue*: 144–160.

Han, S.H., Kang, H., and Chae, C.W. (2014). Effect of Tyvex mulching and trickle irrigation on fruit quality in Satsuma mandarin (*Citrus unshiu* Mark.). *Korean J Hort Sci Technol* 32:18–25.

Hong, H.R., Oh, E.U., Han, S.G., Yun, S.H., Kim, H.B., and Song, K.J. (2022). Characterization of soluble sugar content, related enzyme activity and gene expression in the fruits of 'Minihyang' mandarin on different rootstocks. *Horticulturae* 8:47.

Kang, J.H., Kang, S.B., Koh, S.U., Kim, D.S., Moon, Y.I., Park, J.H., Song, K.J., Yun, S.H., Han, S.H., Han, S.G., and Choi, Y.H. (2016). Citrus. In Y.J. Yim (ed.) *Fruit science specialized*. Hyangmunsa, Korea, pp. 242–283.

Kim, H.B., Kim, J.J., Oh, C.J., Yun, S.H., and Song, K.J. (2016). Current status and prospects of molecular marker development for systematic breeding program in citrus. *J Plant Biotechnol* 43:261–271.

Kim, M.J., Yun, S.H., Park, S.M., Jin, S.B., and Song, K.J. (2020). Characterization of allotetraploids derived from protoplast fusion between navel orange and kumquat. *In Vitro Cell Dev Biol-Plant* 56:634–643.

Korea Agricultural Marketing Information Service (KAMIS) (2022). *Comprehensive analysis of domestic marketing trends in the year of 2022*. Korea Agro-Fisheries & Food Trade Corporation, Korea.

Korea Meteorological Administration (KMA) (2023). *Average yearly air temperature during the last 30 years (1991–2020)*. KMA Weather DATA Service. https://data.kma.go.kr/climate/average30Years/selectAverage30YearsList.do?pgmNo=113/. Accessed 13 Aug. 2023.

Korea Rural Economic Institute (KREI). (2023). *Agricultural outlook 2023 Korea*. KREI, Sejong, South Korea.

Moon, D.K., and Kim, K.S. (2006). Growing practices of citrus and other evergreen fruit trees. In I.G. Mok, Y.U. Shin, S.J. Kim (eds) *Fruit growing in Korea*. NIHHS, RDA, Korea.

Oh, C.J., Woo, J.K., Yi, K., Park, Y.C., Lee, H.Y., Kim, M., Park, S.M., Yun, S.H., Lee, Y., Kim, H.J., Song, K.J., and Kim, H.B. (2021a). Development of molecular markers for genotyping of *Ruby*, a locus controlling anthocyanin pigment production in *Citrus* with its functional anaylsis. *Sci Hortic* 289:110457.

Oh, E.U., Kim, H.B., and Song, K.J. (2021b). Changes of fruit quality characterisitics affected by rootstocks in citrus. *J Jeju Stud* 56(2):223–239.

Park, Y.C., Hong, S.Y., Kang, S.H., Lee, C.H., Oh, M.H., and Yang, C.J. (2020). *Citrus genetic resources*. Jeju ARES, Korea.

Song, K.J., Kim, H.N., Han, S.H., Ko, S.B., and Kim, B.S. (2022). *Developing policies to strengthen the competitiveness of Jeju citrus industry in response to free trade trends*. Jeju Nat Univ, Korea.

Song, K.J., Kim, S.B., Park, J.H., Oh, E.U., Yi, K., Kim, D.W., Kang, J.H., Kim, J.S., Oh, J.H., and Gmitter, F.G. (2011). Frequency and growth characteristics of polyploids occurred spontaneously in some mandarin hybrids. *Korean J Hort Sci Technol* 29:617–622.

Song K.J., Park, Y.C., and Kim, H.B. (2017). *Establishment of core collection and polyploid breeding platform in citrus. A report of golden seed project*. Jeju Nat Univ, Korea.

Statistics Korea. (2021). *Crop production surveys*. Korean Statistical Information Service (KOISI). https://kosis.kr/statHtml/statHtml.do?orgId=101&tblId=DT_1ET0292&conn_path=I3/. Accessed 13 July 2023

Vapnek, J. (2009). *Legistaltive establishing a health certification programme for citrus*. FAO legal papers online #81.

Woo, J.K., Park, Y.C., Lee, J.W., Yun, S.H., Kim, M., Park, S.M., Lee, Y., Song, K.J., and Kim, H.B. (2019). Evaluation of polyembryony for genetic resources and efficacy of simple sequence repeat markers for the identification of nucellar and zygotic embryo-derived individuals in citrus. *Appl Biol Chem* 62:30.

Woo, J.K., Yun, S.H., Yi, K., Park, Y.C., Lee, H.Y., Kim, M., Lee, Y., Song, K.J., and Kim, H.B. (2020). Identification of citrus varieties bred in Korea using microsaterllite markers. *Hortic Sci Technol* 38:374–384

Yakushiji, H., Morinaga, K., and Nonami, H. (1998). Sugar accumulation and partitioning in Satsuma mandarin tree tissue and fruit in response to drought stress. *J Amer Soc Hort Sci* 123:719–726.

Yi, K.U., Kim, H.B., and Song, K.J. (2018). Karyotype diversity of Korean landrace mandarins by CMA banding pattern and rDNA loci. *Sci Hortic* 228:26–32.

Yun, S.H. (2001). *Classification of genus citrus and its related genera using RAPD*. Master diss Jeju Nat Univ, Korea.

Yun, S.H., Kim, M.J., and Park, S.M. (2019). *New citrus cultivar*. Citrus Research Institute of NIHHS, RDA, Korea.

5 Citrus Industry of Thailand

Daruni Naphrom, Chantalak Tiyayon, and Phuangphet Hemrattrakun

5.1 CITRUS PRODUCTION IN THAILAND

The major citrus species in term of commercial fruit trees in Thailand are mandarin (*Citrus reticulata Blanco),* pomelo (*Citrus grandis*), and lime (*Citrus latifolia* and *Citrus aurantiifolia*). The mandarin planting and production areas in 2023 were 20,787 and 20,112 ha, respectively with 7,908 households, producing around 282,599 tonnes/year (Office of Agricultural Economics, 2023b). In the last decade, after the antibiotic treatment was applied effectively to inhibit greening disease in the vascular bundle of mandarin trees, the growers turn back to renew mandarin orchard, therefore the planting area, production area, and yield in Thailand have been increased continuously (Figure 5.1 and 5.2). Currently, the northern part of Thailand is a major production area, where taking 96% of the country. Fang, Chaiprakarn, and Mae Ei Districts, Chiang Mai province are main production areas. There are small (<1.6 ha), medium (> 1.6 ha), and large (> 16 ha) scale farmers (Sruamsiri et al., 2009).

5.2 CITRUS CULTURAL PRACTICES

5.2.1 MANDARIN

'Sai Nam Phung' and 'Khiaw Waan' (*Citrus reticulata* Blanco.) are almost all mandarin cultivars in the market. Botanic characteristics of those cultivars are different. Dense and erect canopy form appear in 'Sai Nam Phung', while 'Khiaw Waan' has a gauzy and weeping canopy form. Smaller and darker green leaves are found in 'Sai Nam Phung', but it has bigger flowers than 'Khiaw Waan'. 'Sai Nam Phung' has more fruit weight than 'Khiaw Waan' and the fruit peel and segment membrane are thin and soft. The pulp is juicy, acidity, and sweet. Therefore, 'Sai Nam Phung' is more admired than 'Khiaw Waan'.

The northern part of Thailand has a sub-tropical climate where is typically characterized by three distinct seasons: hot season, rainy season, and cool season. Hot season usually begins in March and lasts until May, temperatures can soar, reaching highs of around 40–43°C in some areas. This season is marked by clear skies and minimal rainfall. Humidity levels can also be high during this period, making it feel even hotter. Furthermore, hot spots of forest fire have been occurring frequently. The rainy season typically starts in June and lasts until October, although the exact timing can vary from year to year. The precipitation is 1,365 mm on average. This period sees frequent rainfall, often in the form of heavy showers and thunderstorms. Under this climatic condition, the citrus growers should beware about Asian citrus psyllid outbreak. The cool season begins in November and lasts until February, temperatures are milder and more comfortable during this period, with daytime temperatures averaging around 28–35°C. Night time can be cooler around 10–15°C, especially in

DOI: 10.1201/9781003584599-5

FIGURE 5.1 The planting and production area of mandarin in Thailand. (Office of Agricultural Economics, 2023.)

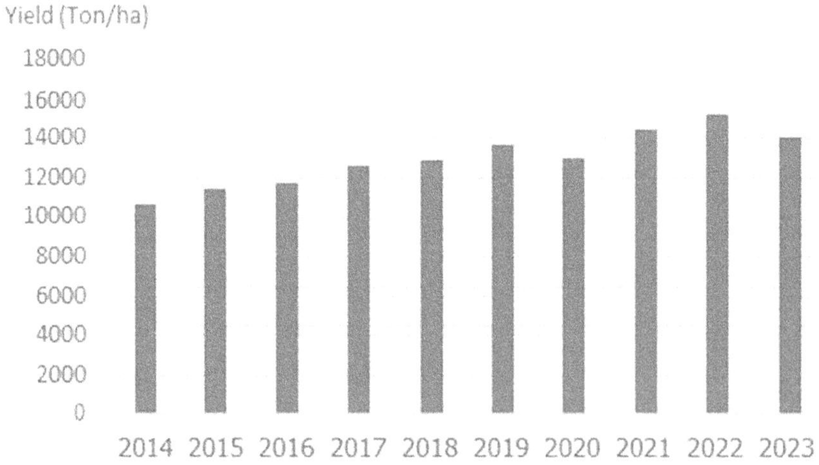

FIGURE 5.2 Yield of mandarin during 2014–2023. (Office of Agricultural Economics, 2023.)

higher elevations, with temperatures can drop to around 5–10°C. On the other hand, the southern part of Thailand has a tropical climate characterized by high temperatures and humidity throughout the year. There are two main seasons: the wet season (May to November) and the dry season (December to April). During the wet season, heavy rainfall is often accompanied by thunderstorms and occasional flooding, especially in low-lying areas. The precipitation is 3,026 mm on average. The wet season is also associated with the southwest monsoon, which brings moisture-laden air from the Indian Ocean, resulting in abundant rainfall. The daytime temperatures are averaging around 32–35°C, while the night-time is around 24–26°C. In contrast, the dry season sees less rainfall and lower humidity levels. The daytime temperatures are averaging around 35–40°C, while the night time is around 25–30°C. The mandarin fruits in the northern part of Thailand produce more yellow colour than in central and southern parts, because there is quite a difference between day and night temperature. 'Sai Nam Phung' grown in the southern part, namely 'Sho Kun' has a green colour peel instead of a yellow colour (Figure 5.3). In general, peel colour development is enhanced by low temperatures and inhibited by high temperatures (Koshita, 2015).

FIGURE 5.3 'Sai Nam Phung' (Left) and 'Shokun' (Right).

TABLE 5.1
Mandarin Production Seasons in Northern Thailand

Production seasons	Flower initiation	Set young fruit	Fruit harvesting
On-season	Late February	June–July	December–January
1st Off-season	Late March	July–August	January–February
2nd Off-season	Late April	August–September	February–March
3rd Off-season	September	December–January	June–July
4th Off-season	October	February–March	July–September

Soil characteristics in northern Thailand vary depending on factors such as topography, climate, parent material, and human activities. The northern area contains four types of landscapes including mountain, hill land, piedmont, and valley. Soils are distributed over three slope positions, i.e. summit, backslope, and footslope. Soil variability is high, including oxisols, ultisols, alfisols, inceptisols, mollisols, and entisols, found either in consociation or association (Neyret et al., 2018). In the mountainous areas, at higher elevations, mountain soils tend to be shallow and rocky, with lower fertility and organic matter content. In contrast, lower slopes may have deeper soils with better water retention and nutrient levels.

The mandarin trees in northern Thailand respond very well to the environmental condition. They can produce flowers and set fruit all year round. Five crops can be harvested from the healthy trees (Table 5.1). Anyhow it is not recommended to produce many crops per year, because high applications of fertilizer, insecticide and fungicide are necessary and lead to high cost. Furthermore, the life span of the mandarin trees will be shorter due to lack of pruning time to promote leaf flushing and photosynthate accumulation.

Mandarin can be grown in sandy soil, clay, and sandy loam, but should be well drained. It can be grown both upland and lowland. Mandarin planting in lowland which the soil is clay, high ground water, and close to natural water are planted in furrow system, with spacing between rows 6–8 m and between plants 3–4 m. Mandarin usually grown from air layering plants. However, lately grown mandarin are grafted plants. The size of stock plants that are ready to plant to the field are that the circumference of the stem base is at least 1.5 cm and height from base to the top at least 60 cm. In the upland area, planting holes were digging size 50 cm x 50 cm x 50 cm or at least larger than the bag of mandarin plant. Ferment or manure is mixed with the soil, approximately 10 kg, 0.5 kg rock phosphate, and 100 g 15-15-15 fertilizer are added to the soil. Planting the mandarin plant by breaking the soil in the bag until it separates from the roots. Cut out the crooked root and trim the shoot and some leaves so there is balance between shoot and root. Provide support pole and watering (Ornwimol, 2004).

Cultural practices after planting are important because it promotes growth and development of mandarin plants and provide high yield. The practices include irrigation, fertilization, weed management, canopy management, pruning, and pests and diseases management.

***Irrigation*:** First planted mandarin plants need water every day, 3–4 dippers a day. After two weeks, the water could be given every other day. After the mandarin plants mature, watering is still important. However, it depends on growth stages and environment. A period before flowering requires a small amount of water for food preservation. After the fruit set, water requirement is higher until fruit maturation. Water refrain two weeks before harvest enhances sweetness. If the water is given so that the soil is moist all the time, it could postpone fruit ripening approximately 20 days.

***Fertilization*:** In the first year, giving high nitrogen fertilizer, for example, 20-10-10, 25-7-7, or 15-15-15 + 46-0-0 (1:1) 0.5–1.0 kg per plant, separate to once in 4–6 months. In addition, the organic fertilizer is given 10–20 kg per plant, apply once at the end of rainy season. After the mandarin plants are four years old, they will start to be productive. Fertilizer 12-24-12 is given before flowering stage 1 kg per plant and apply foliar fertilizer to supply minor nutrients and supplements. At the fruit set stage, calcium, magnesium, iron, zinc, copper, boron, and manganese may be given by foliar application. Close to the harvest stage, 13-13-21 fertilizer are given 1–2 kg per plant. After harvest, 25-7-7 or 15-15-15 + 46-0-0 (1:1) 1–3 kg per plant, foliar fertilizer with minor nutrients and supplements, and organic fertilizer 20–50 kg per plant are given to the mandarin. The amount of fertilizer in fruit growth and development period depends on crop load. At least nitrogen 144 g/tree, phosphorus 39 g/tree and potassium 209 g/tree have been recommended for mandarin crop load at 100 kg/tree. The losses of N, P, K fertilizers in soil were reported as 30, 60 and 30%, respectively in the northern part of Thailand. Amount of soil nutrition suitable for fruit tree in Thailand is shown in Table 5.2 (Supakamnerd, 2005).

***Weed management*:** Weeds often found in the mandarin orchard are cogon grass, torpedo grass, and para grass. Mandarin growers usually use shoulder mounted lawn mower. Pesticides are not usually used close to mandarin plants because it is harmful.

Propagation: The cultivars which are used to make rootstock for budding and grafting are 'Cleopatra', 'Troyer', and 'Carrizo'. Planting rootstock from seeds may be done in nursery tray then transplant later, or plant in 15x30 cm plastic bag. The planting media consist of 20% sand, 15% rice husk, 60% coconut coir, and 5% rice husk ashes, ferrous sulfate and marl 500 g each and gypsum 1 kg per cubic metre.

***Branch supporting*:** Branch supporting is done for two reasons: prevent tearing of branches especially during fruit setting period and elevate level of fruits from the ground, which will reduce loss from disease and insects. Mandarin plants bear approximately 800–1,200 fruit per plant, branch supporting is therefore important for weight acception. In the upland area, the support is done by putting bamboo in four corners and tying bamboo horizontally to take mandarin branches. In the

TABLE 5.2
Amount of Soil Nutrition Suitable for Fruit Tree in Thailand

Soil nutrition	Suitable range
Organic matter (%)	2.0–3.0
Available P (mg/kg)	35–60
Potassium (mg/kg)	100–120
Calcium (mg/kg)	800–1,500
Magnesium (mg/kg)	250–400
Ferrous (mg/kg)	60–70
Manganese (mg/kg)	20–60
Copper (mg/kg)	3–5
Zinc (mg/kg)	3–15
Boron (mg/kg)	1–6
Sulphur (mg/kg)	25–150

Source: Supakamnerd, 2005.

lowland area, the top of bamboo is grooved to hold mandarin branches. This groove is placed under the branches and the bamboo poles are positioned vertically. Each plant uses 40–100 bamboo poles. Support is done when the mandarin fruit are approximately three months from flowering and will remove the support after harvesting. The bamboo can be reused, but the decayed bamboo will not be reused. The bamboo life is approximately one to two years.

***Windbreak*:** Wind causes great loss to mandarin plants. Wind speed at 25–30 km/hr causes wound and bruise to mandarin fruit. Wind speed at 45–60 km/hr can cause all fruits drop. In the windy area, windbreak trees are required to reduce loss and help with water evaporation from soil and evapotranspiration from mandarin plants. Windbreak should be planted at least 8 m from mandarin plants. Plants which are used as windbreak are eucalyptus (*Eucalyptus globulus* Labill.), brown salwood (*Acacia mangium* Willd.), horsetail tree (*Casuarina junghuhniana* Miq.), and some types of bamboo. Planting windbreak also needs to consider shading, competition for nutrients, and toxicity that may have caused from windbreak trees to mandarin plants.

***Canopy management and pruning*:** Things to consider in canopy management and pruning are light intensity and light acception. Light intensity has a direct role in photosynthesis. At 25 percent, light intensity is light compensation point. Light intensity higher than this point varies directly to photosynthesis. Light acception limit growth and development of plants. If we divide area of light acception into three parts from outside to inside canopy, the first and the second layers receive adequate light, while the innermost do not get enough light, which makes low yield quality. In the past, the innermost part was pruned to remove parts which did not get enough light. However, this does not solve the problem of yield quality and resulted in larger canopy plants.

Pruning is done every year starting when the plants are bearing fruits. The central leader is kept to support the canopy and keep branches in layers. Branches that need to be removed are water sprouts, narrow angle branches, overlapping branches, crooked branches, branches with diseases, and branches lower than 45 cm from the ground. Branches which are too high to harvest the fruit are also removed. Small branches are pruned by pruning shears and large branches are pruned by a saw or electric saw, pruned at the base of the branch next to the main branch. The cut wounds are covered with red lime or fungicide to prevent moulds. If the pruning is done after harvest, fertilizer is applied to the plants to provide nutrients for new growth.

***Water control for off-season production*:** Mandarin in middle part of Thailand usually flower in February to March and harvest in November to December. In northern Thailand plants flowering in April to May and harvest in January to February. Forcing flowering for off-season production is done by counting from the date that would like to be harvest back approximately ten months, then refrain water for 15–20 days. Mandarin leaves will show drought symptom by leaf curling. After water refrain until mandarin plants show drought symptom, the water is fully given to the plants. The 8-24-24 or 12-24-12 fertilizer may be applied at this time. The plants will start flowering in approximately seven days.

During flowering to fruit maturation watering regularly is required. From flowering to fruit set takes time approximately 30 days, which pesticide application is required to prevent damages. After the fruit set approximately one month, the fruit would be thumb size, 15-15-15 fertilizer is applied to the plants. Fruit quality is improved by 13-13-21 fertilizer application when the fruits are five months old. The water is refrained 10 days before harvest to enhance flesh and taste qualities.

Harvesting: Mandarin can be harvested eight to nine months from flower blooming. At first fruit maturation period, the sugar concentration in fruits will rapidly increase and sugar increase rate will decline before fruit ripening. Total soluble solids (TSS) in mandarin fruits is approximately 13 percent and total acidity (TA) 1.0-2.5 percent. The ratio between TSS:TA of mature mandarin fruit is 5:8. If the ratio is higher than 1:2, those fruits are too ripe for the market, but may be sent to the processing factory.

Mandarin fruit harvesting index is calculated by the time from flowering because the appearance is difficult to distinguish the mature fruits. Harvesting quality fruit requires pruning shears to cut the pedicel. A study ladder is helpful to harvest fruits on higher position. Hydraulic basket attached to the tractor is sometime used for harvest fruits at higher position, but it comes with higher cost and it can be used only on a flat surface. Harvesting mandarin in the rainy season requires fungi prevention. In addition, bruises from transportation are to be concerned.

***Pests and diseases management*:** Citrus greening disease (Huanglongbing) is a serious bacterial disease transmitted by the Asian citrus psyllid. Agrochemical substances: fertilizers, pesticides, and antibiotic substances are crucially used in mandarin plantation to prevent disease and pests. Pesticides such as imidacloprid, abamectin, emamectin benzoate, fenbutatin oxide, hexythiazox, pyridaben, profenofos, thiamethoxam, and buprenorphine have been used. The antibiotic substance: ampicillin, 250–2,000 ml/tree has been injected into the trunk four times a year to inhibit greening bacterial disease (Farmer interview).

Besides disease and pests, the farmers are often faced with thunderstorms in summer and hail, which damage yield significantly. Lacking labour for harvesting becomes a serious problem in mandarin plantation. Small-scale farmers spent around 17,400 USD/ha, 70% occupied by those agrochemical substances and another 30% is a labour cost. The mandarin production in the northern part of Thailand earns about 31,250 USD/ha (Farmer interview). The market is divided into two channels: 80% for wholesalers and 20% for farmer self-distributors. Since 2020, the online market both B2B and B2C has fast extended due to the pandemic of COVID-19.

5.2.2 Pomelo

Pomelo is one of the economic citrus fruits in Thailand. It has been distributed throughout Thailand. The major production provinces from north to south are Chiang Rai, Chiang Mai, Lampang, Phichit, Chaiyaphum, Chai Nat, Kanchanaburi, Samut Songkhram, Nakhon Pathom, Prachin Buri, Surat Thani, and Nakhon Si Thammarat. The production area is about 15,113 ha. They can produce 172,518 tonnes/year. The produces are almost for domestic consumption, anyhow they are also exported to international markets such as China, Vietnam, and Hongkong. In 2023, Thailand exports about 30,598 tonnes with value as 27.8 million USD (Office of Agricultural Economics, 2024b). Pomelo trees thrive in tropical to sub-tropical climates with warm temperatures and well-draining, fertile soil. They prefer full sun exposure for optimal growth and fruit production. Pomelo trees are evergreen and can grow quite large, reaching heights of up to 15 metres or more in optimal conditions. They have a rounded or spreading canopy with dense foliage. The leaves are glossy, ovate to elliptical in shape, and typically dark green in colour. They are arranged alternately on the branches and have slightly serrated edges. Pomelo trees produce fragrant, white, or pale pink flowers that are usually solitary or clustered in small groups. The flowers have five petals and numerous stamens, characteristic of plants in the Rutaceae family to which citrus trees belong. The pomelo fruit is the largest citrus fruit. It has a thick, spongy rind (peel) that can vary in colour from green to yellow to pinkish orange, depending on the variety and ripeness. The flesh inside is segmented, juicy, and typically ranges from pale yellow to pinkish white in colour. Pomelo fruits can contain seeds, although some seedless varieties have been developed. Pomelo trees have a fibrous root system that helps anchor the tree and absorb water and nutrients from the soil. They may also develop adventitious roots near the base of the trunk.

In Thailand, there are many commercial pomelo cultivars such as 'Thong Dee', 'Khao Yai', 'Khao Tangkwa', 'Khao Nam Phung', 'Thakhoi', and 'Tubtim Siam'. Pomelo fruit characteristics are shown in Table 5.3. The 'Thongdee' pomelo is registered to be a Geographical Indications (GI) for Nakhon Chai Si District, Nakhon Pathom province. The flavour of Thong Dee pomelos is often described as exceptionally sweet and aromatic. They have a well-balanced taste that is neither too sour nor bitter, making them enjoyable to eat fresh or in various dishes. Normally the pomelo tree can produce yield all year. On-season production is around July–September (Figure 5.4). Fruit harvesting is around 6–7.5 months after the fruit set.

Pomelo can be grown in almost all types of soil, silt, sand, sandy loam, or clay, which have already gone through improvement. Soil, which works well for pomelo is fertile, contains high organic matter, pH 5.5–6.0, and well drained.

***Planting*:** The size of the holes to plant pomelo should have width, length, and depth 75–100 cm, spacing between rows and between plants 8–10 m. Top soil and lower soil are separated in different piles and leave them under the sun for one to two months to destroy fungi and germs. After that, top soil is brought to mix with 10–25 kg compost or manure and 0.5–1.0 kg rock phosphate each hole. The hole is lined with decayed leaves, filled with mixed top soil, then topped with lower soil. A mound of 50 cm radius with 30 cm height is made to prevent water logging, which would lead to root and stem rot diseases. After watering until the soil is set, the pomelo can be planted. After planting, two supporting poles are placed diagonally. If the sun is strong, banana leaves, coconut leaves, or betel nut leaves are used for shading on the east and west sides until the pomelo plants are settled. Planting is usually done during rainy season in May to August, because rain safe labour

TABLE 5.3
Characteristics of Pomelo Cultivars in Thailand

Cultivars	Fruit shape	Fruit weight (g)	Flesh colour	TSS (%)	Taste
'Khao Tangkwa'	Flat round	900–1,600	Yellowish White	10–12	Sour sweet
'Thongdee'	Flat round	940–1,200	Light pink	10–12	Sweet
'Tubtim Siam'	Round with neck	1,100–1,800	Dark pink to red	9–14	Sweet
'Thakhoi'	Round	700–1,800	Pinkish white	9.5–12	Sour sweet with unique smell
'Khao Nam Phung'	Tall round	1,600–1,800	Honey white	10–12	Sour sweet

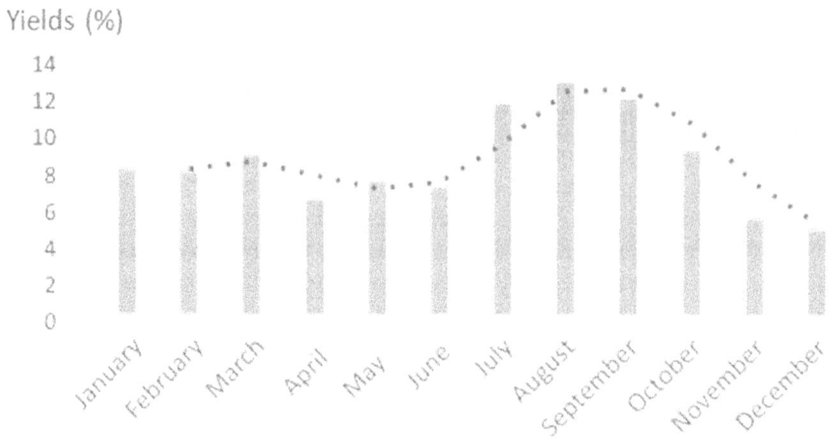

FIGURE 5.4 Pomelo yield all year round. (Office of Agricultural Economics, 2018.)

and cost in irrigation. In addition, the plants settle rapidly and have good growth from getting adequate water.

In low land areas with high ground water, furrow systems with raised planting beds are used. The planting bed is 6 m wide; the furrows are 1.0–1.5 m wide on top and 50–70 cm wide at the bottom with 1 m depth. The length of planting beds and furrows depends on orchard size.

Planting can be done in two ways: straight and slant planting. Straight planting will give one stem pomelo plant. The base of the stem is not crowded. The light can penetrate through the ground and the wind can go through, which will reduce spreading of pests and diseases. Straight planting is usually done on highland pomelo. Slant planting is done by planting pomelo 45–70°. Planting with this method provides many branches at the base of the stem. The plants have high yield. However, when the plants are getting old, the base of the stem will be too crowded, which is prone to have pests and diseases.

***Irrigation*:** Pomelo requires high moisture, especially newly planted plants. During the dry season or period without rain watering must be done every day. Pomelo orchard should have a water source, so there is water when the plants need, for example, during fruit development or for off-season production. Lack of water during fruit development causes small fruit, fruit drop, low yield, and the plants may die if the drought is severe. Orchard upland water the plants by scoop up water and pour to each plant. If the water seeps into the ground rapidly, the ridge is made around the trunk to store water, and ground cover materials, i.e. straw, rice husk, and dried grass are used. Lowland orchard uses water pump in small size boat to pump and spray to pomelo plants. However, spraying must be careful not to let water splash to leaves because it may bring microorganism from the soil to infect the plants.

***Fertilization*:** Fertilizer in pomelo production can be divided into two groups: organic and chemical fertilizers. Organic fertilizers usually have very low nutrients, but help improve soil structure. Types of organic fertilizers are manure, compost, ground bones from slaughter house, bat manure, debris from factory, for example, bagasse, and plant stubbles. Chemical fertilizer is also important for pomelo plants. Newly growth pomelo requires high nitrogen level. Fertilizer used in this period should have 2:1:1 ratio like 20:10:10. Fertilizers for plants in the fruiting stage should have 1:1:1 or 2:2:3 ratio, for example, 15-15-15 or 14-14-21.

Newly planted pomelo do not yet have a complete root system. Fertilizer will be given four months after planting or after the plants are well established. Manure can be given to the plants every year, because it helps in loosening the soil and improving aeration. Chemical fertilizer is given at half of the plant age, for example, if the plant is two years old, 1 kg of fertilizer is given (separate into many applications). When the plants are bearing fruit, high phosphorus fertilizer is used, for example, 12-24-12, 13-27-13, and 9-24-24. Sometimes foliar fertilizer 15- 30-15, 10-52-17, or 0-52-34 are used. When the fruits are 5-6 months old, changing the colour, the plants require high potassium and medium nitrogen and phosphorus. The fertilizer used in this stage are 14 -14-21, 13-13-21, and foliar supplement 10-20-30 helps improve fruit quality. After harvest, fertilizer should be applied to nourish the plants. Fertilizer used in this stage should have 2:1:1 ratio, for example, 21 -10-10, and apply foliar supplement 30-10-10.

Fertilizer application for newly planted pomelo should be applied 15–30 cm from the trunk, because the root system is not spread well. Mature pomelo plants have a complete root system, the fertilizer can be given around the canopy.

***Weed management*:** In the past, growers did not let any weeds grow in pomelo orchard. Currently, weed management is done instead of weed elimination. Grasses as a ground cover help retain the soil, reduce the heat from the sun, and build moisture in the orchard by absorbing water from the ground and evaporate through the leaves. Pomelo plants in an orchard with grasses have less dried leaf margin symptom than the orchard without grasses. However, large size weeds compete for water and nutrients with pomelo plants. Lights could not get through, which accumulates pests and diseases. Weed elimination can be done by pulling or cutting. Avoid cutting too close to the stem and branches, which will cause damage to the plants. Weed debris can be used as a ground cover around the base of the stem to keep moisture and provide nutrients after decay. Chemical herbicide is also used but it needs to be applied carefully by not allowing it to touch the pomelo plants.

***Propagation*:** Pomelo grown in Thailand, usually propagated by air layering. However, in the high-land area where tap root system is desired, budding and grafting are used. The problem found from using air layering plants is that if the mother plants have tristeza disease, the plants used for propagation will have the disease too. So, choosing mother plants is very important.

***Pruning*:** Pomelo requires regular pruning from planting to retain canopy in desired shape. Leaving the plants to grow naturally may cause uneven canopy, prone to overthrow or tearing. When the plants are firstly planted, they should be controlled to have only one main trunk and keep branches 30–45 cm above the ground. After one-year, dried branches, branches with diseases, water sprout, weak branches, and unhealthy branches should be pruned. When the plants bear fruit, criss-crossed branches and branches that compete with the trunk should be pruned. This will help light penetration and reduce the chance of pest and disease infection. After pruning, fungicide or red lime should be painted over the wound.

***Water control for off-season production*:** Pomelo flowering naturally has a long period of flowering, which makes fruit mature unevenly and makes orchard management difficult. Forcing flowering is done by refrain water for 10–30 days until the leaves are wilting, then fully giving water. Additional chemical fertilizer 12-24-12, 13-27-13, or 9-24-24 is applied to stimulate flowering. After watering 15–60 days, pomelo plants will start flushing and flowering. This method can force pomelo to flower. However, it will cause plant health to decline faster than natural flowering.

Fruit thinning: If the pomelo plants set so many fruits, they should be thinned. If all fruits are let to grow, the plant health will be declined, short-lived, have small fruit, and low fruit quality. Fruits that should be thinned are small, malformed, or blemish fruits. Plants age 2–3 years should keep 2–4 fruit, 10–15 fruit for 4 years old plants, 20–30 fruits for 5 years old plants, 30–50 fruits for

6 years old plants, and 100–120 fruits for more than 10 years old plants. Thinning is done at two stages: betel nut size and fist size.

Harvesting: Pomelo is harvested 8–9 months after flowering. In the middle of Thailand, pomelo flower during November to March and mature in August to September. Pomelo plants in the middle part of Thailand start to be less productive after 10 years old because high ground water limits root growth. Pomelo plants grown in other parts of the country with good care can be productive for 15–20 years. Mature pomelo fruit can be observed by the brown dot opposite to the stem side. The oil glands around this dot are separated apart. If we use the nail to scratch the skin, mature fruit would have a light smell. Healthy mature fruits are heavy. Harvesting pomelo can be done many ways, hand pick or using tools. The tools used to harvest pomelos are pruning sheer, pruning sheer with long handle which can hold the pedicel after cutting, and long handle with hook at the end which have rope to pull the blade on the hook (this requires another person to hold the sac to receive the cut fruit).

Pests and diseases management: Pomelos in Thailand also face on Huanglongbing diseases that can affect their growth and yield. It causes yellowing of leaves, reduced fruit size, and bitter-tasting fruits. Management involves controlling the psyllid vector and removing infected trees. However, it seems less severe than mandarin. Antibiotic treatment is also used in some areas. Other diseases and pests found in pomelo plantation are citrus black spot, citrus canker, powdery mildew, citrus leaf miner, citrus aphids, citrus root weevils, and fruit flies. To effectively manage pests and diseases in pomelo orchards, integrated pest management (IPM) strategies are often employed, combining cultural, proper orchard sanitation, biological and chemical control methods while also focusing on orchard hygiene and disease-resistant varieties when available.

5.2.3 Lime

Lime production in Thailand is a significant agricultural and industrial activity. Thailand is one of the major lime producers in Southeast Asia, with a climate and soil conditions conducive to lime cultivation. There are several types of limes, including Persian limes (*Citrus latifolia*) (seedless lime; Bearss lime and Tahiti lime) and Key limes (*Citrus aurantiifolia*). These limes vary in size, flavour, and usage. The lime planting and production areas in 2023 were 16,720 and 16,372 ha, respectively with 52,636 households, and producing around 166,775 tonnes/year (Office of Agricultural Economics, 2023a). Thailand exports a significant portion of its lime production. Key export markets include countries in Asia, Europe, and North America. Thai limes are valued for their quality and taste in international markets. Lime production plays a crucial role in Thailand's economy, providing income for farmers, supporting Agro-industries, and contributing to the country's export revenue.

The lime cultivation is widespread across various regions in Thailand. The top five provinces are in the central part, e.g., Phetchaburi, Ratchaburi, Samut Sakhon, Kamphaeng Phet, and Phichit. Lime cultivation in Thailand typically involves careful soil preparation, irrigation management, and pest control. Farmers often use modern farming techniques and equipment to improve yield and quality. Lime can be produced all year round. Anyhow there are low yield during January–May (Figure 5.5). In addition to fresh limes, Thailand also processes limes into various products such as lime juice, lime oil, and lime extract. These value-added products contribute to the country's lime industry. Lime production in Thailand faces challenges such as climate change and market fluctuations. In 2023–2024, the yield decreases about 60% due to extreme high temperature and water scarcity. The price is increasing to 10–15 Baht/fruit during summer season.

One factor that contributes to successful lime plantation is site selection. Site selection should consider soil, water, temperature, moisture, wind, and transportation. Lime grows well in well-drained soil, top soil at least one-metre, fertile, high organic matter, pH 5.6–6.0. Lime has high

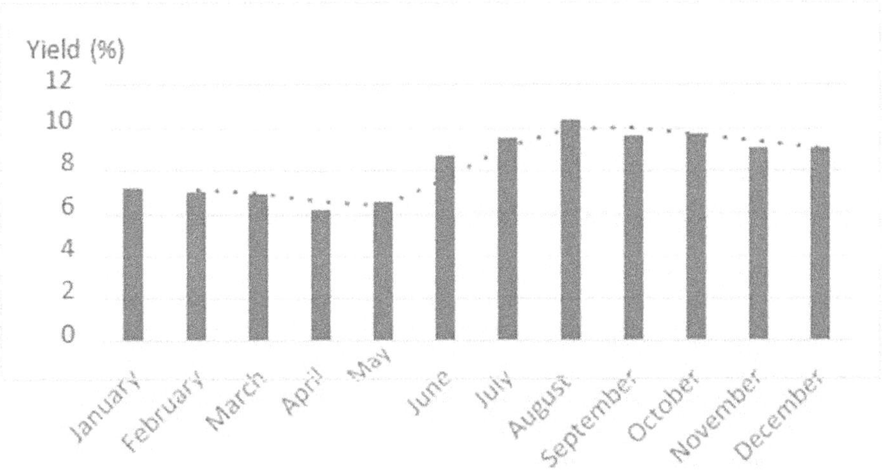

FIGURE 5.5 Lime yield all year round. (Office of Agricultural Economics, 2023.)

water requirement and requires water from planting through fruit set. Shortage of water causes slow growth and low fruit quality. Very low temperature can cause death to lime plants. This is not a problem in Thailand. However, temperature influences growth, for example, the plants have high transpiration in high temperature. Moisture results in thickness of lime peel. Low moisture causes think peel and high moisture causes thin peel. Planting intercropping such as banana could help maintain moisture in the field. Wind can cause direct damage to lime plants. It causes flowers to fall and cause blemish to fruit peel. Wind break should be planted in the area where there is strong wind regularly. Lastly, the location of the orchard should not be too far from the market.

***Planting*:** Lime usually planted from air layering plant. The plant from air layering should be grown in a plastic bag for a while that it established before transplant to the field. When planting, place the air layering bulb under the ground approximately 5 cm. Planting can be done two ways: straight planting and slant planting. Straight planting is generally used for lime planting. It provides a single stem which makes the base of the trunk open and allows airflow, reducing chances of pest and disease spreading. Slant planting is planted at 45°. This promotes branching at the base of the stem. The plant will grow rapidly at the beginning and have more space to bear fruit. However, when the plants are getting older, the supporting system is required. Lime planting is done in the rainy season because rain helps save time and cost for irrigation, and also reduces loss from death after planting.

***Irrigation*:** Lime plants need water almost all year round. When limes are firstly planted, they should be watered every day. Take care not to give water too much that it is slushy, otherwise the plants will be prone to diseases. Watering can be done by pumping and water through the hose or giving water through the water channel. During there is no rain, the water needs to be pumped and give to the plants 3–4 times a month, especially during very hot weather in March to April the water needs to be given more often.

***Fertilization*:** Fertilizers for lime are both organic and chemical fertilizers. Chemical fertilizer is very important for lime planting, especially after the plants start fruiting, because there are fruits on the plants almost all year. Nutrients necessary for lime growth are major and minor nutrients. Major nutrients are needed in large amount. These are nitrogen, phosphorus, and potassium. There are 13 minor nutrients, but minor nutrients usually deficient in lime growing are zinc, manganese, magnesium, and iron.

Soils in each area are different. Taking a soil sample for nutrient analysis helps determine nutrient management. There is a general principal of lime fertilization as follows. Before planting, the soil should be improved to loam soil by adding organic fertilizer. After planting lime 2–3 months, there is no need for fertilizer. However, after four months, which lime established and starts new growth, high nitrogen fertilizer should be given to stimulate vegetative growth, for example, 20-10-10, 20-11-11, or 15-15-15 at one handful per plant, separated into every three months. When lime plants are two years old, the fertilizer is increased to 300–500 g per plant each time. When the plants are three years old, 1 kg of fertilizer should be given. Chemical fertilizer should be given two times a year and manure should be applied once in the dry season.

If lime plants are healthy, they will bear fruit on the third year. Fertilizer, which helps stimulate flowers, must be given to the plants approximately two months before flowering. This is 12-24-12, apply 1 kg per plant per year. The second fertilizer application is done after the fruit set by applying 15-15-15 fertilizer 2 kg per plant per year. To help with fruit expansion, 13-13-21 fertilizer is used. The third fertilizer application is done after harvest and pruning to nourish the plants and get ready for flowering the next season. Fertilizer given in this period is 15-15-15 or 16- 16-16 200–300 g per plant.

Manure or compost application for fruit bearing lime plants depends on soil moisture. If the soil has low moisture, cow manure mixed with bat manure should be used. If the soil has high moisture, old bat manure should be used. In addition, ground bones, which contain calcium phosphate, should be added to reduce soil acidity. The manure should be given 0.5 kg per plant and increase 200–300 g each time.

***Weed management*:** Weeds in lime orchards compete for nutrients and are a place to live of pest and diseases. Large orchards usually use a grass cutter or the cutter attached to the tractor. However, machine cutting requires caution not to cause wound on the trunk or cause impact to the roots, otherwise pest and disease will infect the plants easily. Weed cutting should be fully eliminated or leave it short, depending on the orchard condition. Weed debris can be used as a ground cover to retain soil moisture and when they decay, they can be soil amendments.

***Propagation*:** The propagation which is usually done for lime is air layering. Plants from air layering bear fruits early, plants are not too tall, and the plants have the same characteristics as mother plants. However, plants from air layering do not have a tap root, which may be prone to falling and they tend to not live long. Rooting hormone may be used when performing air layering to promote rooting. Normally, air layering should be done in the rainy season during May to August. Lime will be rooting after air layering 20–30 days, shorter with rooting hormones. Air layering plants should be cut from the mother plants when the roots turn from white to brownish colour with many later roots. They should be cut in the evening because the leaves stop transpiration. If the new plants have too many leaves, they should be thinned to reduce water loss. New plants should be planted in plastic bags in media soil: compost or decayed leaves 2:1. After the new plants establish, they are transplanted in the field.

***Pruning*:** Lime pruning is required regularly from planting to keep the canopy in desired shape. In the first couple of years, only dried branches and branches that are too long are pruned. When the lime plants are three years old, one metre under the canopy from the ground should not have any branches. After that, pruning should be done once a year, dead and diseased branches, water sprouts, untidy branches, and crisscross branches should be pruned.

***Water control for off-season production*:** Lime has been demanded all year round, especially March to April, there are very limited produce in the market. The price during this period could be ten times higher than during when there are a lot of produces. To harvest lime in March to April, the flowering time needs to be counted back 4.0–4.5 months for fruit development and 12–13 days for flowering, which is October to November. If the lime cultivar takes a longer time for fruit development, the flower needs to be forced earlier. There are many methods of flower forcing, water refrain, smoking, and urea application.

Water refrain is one method to stimulate flowering in lime. Lime has two habits of reproduction. When there is drought, the leaves fall, followed by flowering. Another habit is that the flower will follow the new set of leaves. The water refrain limits the availability of water for vegetative growth, diverting towards reproductive functions like flower bud development. It affects the balance of plant hormones, particularly abscisic acid (ABA) and cytokinins. ABA levels increase under water stress conditions, while cytokinin levels decrease. This hormonal shift can stimulate flowering processes in citrus trees. In commercial orchards, water refrain-induced flowering may be strategically used to manipulate flowering timings, aligning them with market demands or optimizing pollination and fruit set. However, this approach requires precise monitoring of water stress levels, timing, and duration to ensure optimal results while minimizing potential risks to tree health.

Generally, lime flowers in the dry season in February to April. However, if the grower desires to produce off-season lime during this dry season, the plants have to flower in September to October, which is the rainy season, making it very difficult to refrain water. For the lime, which are grown in furrow system, the water level in the furrow should be reduced to make the planting bed dry for 15 days in mid-August. If there is rain, the time to reduce water level must be extended. In the beginning of September, 9-24-24 fertilizer should be dissolved in water and applied to the lime. After that one-week, foliar fertilizer 0-52-34 is sprayed to the plants. At this time, pruning should be done so the nutrients will not be used to the unimportant parts. Lime will start flowering in October and start to be able to harvest after January, which is the time lime is in high demand in the market.

Smoking is done to make mature leaves fall. The fire is made with fresh grass, fresh leaves, or wet straw to generate only smoke, not the flame. Lime leaves will turn wilt and fall later. After leaf fall, water and fertilizer are applied to the plants to stimulate new growth and flowering.

Urea application is done by using 5 percent by weight urea concentration spraying to the plants. This will make the leaves fall. Then chemical fertilizer 12-24-12 or 9-24-24 is applied 1–2 kg per plant around the trunk. Lime will flower after these 10–15 days. 20-14-14 is applied 200–300 g per plant 3–4 times once a month to help with fruit development. The water must be applied during fruit development and it also prevents flowering during the fruit development period.

Harvesting. Lime should be harvested when it is nearly ripe. This can be observed by the stem side, which is slightly yellow, the skin is smooth and thin, the colour of the skin is lighter green than fruits which are not yet mature. For the lime plants that are not very tall, they can be hand picked. For the plants that are tall, a knife attached to the bamboo pole is used to harvest lime fruits. Three years old lime plant can have 150–200 fruit. Six to seven years old plants can bear 800–1,000 fruit if the plants are well cared.

***Pest and disease management*:** It is crucial for lime production in Thailand, given its significant impact on the quality and quantity of the harvest. Some major pests and diseases that affect lime cultivation are

1. Huanglongbing,
2. citrus canker caused by the bacterium *Xanthomonas citri*,
3. citrus leaf miner: these larvae mine the leaves of citrus trees, causing them to curl and distort, which can reduce photosynthesis and overall tree vigour,
4. citrus tristeza virus (CTV): this virus is transmitted by aphids and can lead to decline syndromes and tree death, particularly in trees grafted on susceptible rootstocks,

5. fruit fly: the fruit fly is a significant pest in Thailand, laying eggs in citrus fruits. The larvae then feed inside, causing the fruit to drop prematurely and decay, and
6. melanose: this fungal disease causes brown, scabby spots on leaves, twigs, and fruits, affecting the aesthetic value of the fruit and sometimes its internal quality.

Effective management strategies in Thailand often combine cultural practices, biological control, and the careful use of pesticides. Disease-free plant materials and resistant rootstocks are considered to prevent the lime trees. Additionally, integrated pest management (IPM) programmes are advocated to minimize pesticide use and reduce the development of resistance in pests.

5.3 STATUS OF CITRUS INDUSTRY

5.3.1 Harvesting Window Of Major Cultivars Of Citrus In Thailand

The harvesting window for major citrus cultivars in Thailand varies depending on the specific type of citrus. The chart below shows fruit season in Thailand both "on season" and "off season". On season means fruits are more flavorful and cheaper, while out of season fruits are less flavorful and more expensive. See Table 5.4.

5.3.2 Marketing Channels of Mandarins

The marketing network of citrus in Thailand is a complex and multifaceted system that facilitates the distribution of citrus fruits from producers to consumers, both domestically and internationally. As a vital component of Thailand's agricultural sector, the marketing network encompasses various channels and stakeholders involved in the production, distribution, and sale of citrus fruits. From traditional wholesale markets to modern online platforms, the marketing network of citrus in Thailand has evolved to meet the changing demands of consumers and the global marketplace. In summary, the distribution system of mandarins in Thailand is divided into five primary categories as follows.

TABLE 5.4
Harvesting Season of Major Cultivar of Citrus in Thailand

Fruit	Jan	Feb	Mar	Apr	May	Jun	Jul	Aug	Sep	Oct	Nov	Dec
Mandarin												
'Sai Nam Phung'												
'Khiaw Waan'												
Pomelo												
'Thong Dee'	whole year											
'Khao Yai'	whole year											
'Khao Tangkwa'	whole year											
'Khao Nam Phung'	whole year											
'Thakhoi'	whole year											
'Tubtim Siam'	whole year											
Lime												
Persian limes												
Key limes												

Source: Radanachaless and Thumdee, 1999.
Remark: off season on season.

5.3.3 Wholesaling

Wholesale markets are large-scale distribution centres where mandarins are bought and sold in bulk quantities. These markets serve as central hubs where farmers and wholesalers sell their produce to retailers and distributors. Wholesale markets are essential for efficiently handling large volumes of mandarins and distributing them to various retail outlets across the country. The essential wholesale market for mandarins in Thailand can be classified into local wholesalers operating within the primary production regions and wholesale markets located in other provinces (Muenmanee, 2016).

Wholesalers based locally gather fruits directly from nearby production sources in districts or provinces like Fang and Chai Prakan in Chiang Mai Province. They then distribute these fruits to local traders or consumers within a timeframe of approximately 2–3 days. Typically, wholesalers acquire pre-prepared fruits that have already been washed, graded, and coated before selling them to retailers and consumers who need mandarin fruit for resale (Muenmanee, 2016). Wholesale markets are also spread across different regions, covering areas such as the lower northern region (Phitsanulok), upper central region (Nakhon Sawan), northeastern region (Khon Kaen), western region (Ratchaburi), and upper southern region (Nakhon Si Thammarat). In Bangkok, wholesale markets like Pak Klong Talad Market and Mahanak Market serve as major trading hubs, alongside surrounding provinces' markets like Simummuang Market, Tha Tien Market, and Nakhon Pathom Market (Intakin et al., 2019). The procurement of mandarins can be classified into two categories: firstly, mandarins that undergo grading, coating, and packaging, and secondly, mixed or combined grades that may not meet grading standards, potentially due to lacking or uneven coating. Mandarins acquired in the latter manner are typically packed in plastic baskets, predominantly consisting of a mix of medium and small-sized fruits. Thailand's major wholesale citrus markets, like Talad Thai in Pathum Thani Province, are concentrated in the central region. These markets feature designated sections specifically for trading mandarins and supply them to retailers across Thailand, both directly at the market and from various regions (Muenmanee, 2016).

5.3.4 Retailing

The mandarin retail market in Thailand comprises three main types of outlets. Firstly, fresh market retailers offer whole mandarin fruits alongside other seasonal produce, with mandarins being the predominant offering during the season. Secondly, specialized mandarin stalls are scattered throughout the province, particularly along routes leading to destinations like Chiang Mai International Airport, exclusively focusing on mandarin sales. These stalls operate mainly during the mandarin season and may also sell other fruits to cater to different consumer preferences. Lastly, retail centres within department stores directly sell mandarin fruits procured from farmers, either as standalone stores or within superstores (Muenmanee, 2016).

5.3.5 Direct Sales

Direct sales from farmers' orchards are not yet widespread due to the remote locations of most orchards, which are far from communities or tourist destinations, reducing the popularity of this distribution channel. However, orchards located near main roads close to communities or tourist attractions have the opportunity to market their produce directly (Intakin et al., 2019).

5.3.6 Online Sales

In Thailand, the online sale of fruits is gaining momentum, driven by the growing preference of modern consumers for the convenience of ordering fresh produce online. This trend is supported by the increasing availability of transportation services tailored to online markets, making it easier for farmers to reach customers. As a result, many farmers are now leveraging online channels to promote and sell their fruits directly to consumers, contributing to the expansion of the online fruit market in Thailand. Many local farmers have their websites or social media pages where they promote and sell their fruits directly to consumers, offering a more personalized shopping experience. Many local farmers have their websites or social media pages where they promote and sell their fruits directly to consumers, offering a more personalized shopping experience.

Additionally, live online sales of fruit in the Thailand market have become increasingly popular, especially through platforms like Facebook Live and TikTok Live. These live online sales events create a sense of urgency and excitement among viewers, driving impulse purchases and increasing sales for sellers. Moreover, the interactive nature of these platforms allows for direct communication between sellers and buyers, fostering trust and transparency in the online fruit market.

5.3.7 Agricultural Tourism

In key mandarin-producing areas like Fang District in Chiang Mai, there's notable growth in agricultural tourism, with citrus orchards becoming popular destinations. These orchards market themselves through agritourism, offering tickets for visitors to explore and engage in activities while enjoying picturesque surroundings. Fruit picking, a favoured activity, allows tourists to personally harvest fruits, adding depth to their experience beyond just sightseeing. Orchards may have varying entry policies, some charging fixed fees for fruit picking while others base it on the harvested fruit's weight. Additionally, they offer consumer engagement activities like educational sessions and field trips, where visitors can learn about pre and post -harvest management. Typically, larger orchards or those strategically located become significant tourist draws, especially during the peak winter tourism season (Muenmanee, 2016; Intakin et al., 2019).

The marketing of mandarin follows a straightforward channel in Thailand. According to the production source of mandarin in Thailand, primarily relies on middlemen, most entrepreneurs depend on these intermediaries to bring produce to the market. This approach is favoured by producers because they believe that utilizing middlemen is an effective method for distributing products efficiently and for swiftly exchanging information, particularly regarding price fluctuations and quantities of the produce in both wholesale and retail markets. Additionally, many farmers do not operate their own retail stores (Sithirangsan, 2001). Meanwhile, mandarin distribution can be enhanced by utilizing distributor groups in channel 2, which offer better access to retailers and consumers than wholesalers. These distributors, scattered across various regions, coordinate product delivery based on order information at each production phase, providing flexibility in distribution. Meanwhile, mandarin can also be distributed through contractual agreements with companies, stores, and supermarkets in channel 3. These channels require production sources to adhere to specified processes and obtain standard certifications to distribute products to consumers. Each delivery of oranges is carefully planned and executed according to the terms of the agreement, ensuring efficient distribution to end-users. Lastly, channel 4, mandarin growers distribute their products directly to consumers through different means, such as opening orchards for visitors and distributing citrus fruits within the plantation area. They connect with consumers through educational sessions, field trips to orange plantations, and chances to join in harvesting activities. Consumers have the option to purchase oranges on the spot for immediate consumption, as well as through online sales (Muenmanee, 2016). See Figure 5.6.

5.3.8 Marketing Channels of Pomelo

Pomelo trading in Thailand involves both domestic and international distribution channels. Local farmers and middlemen sell pomelos across the country, with the Mae Klong market serving as a key trading hub. Additionally, pomelos are exported to various countries, including the United

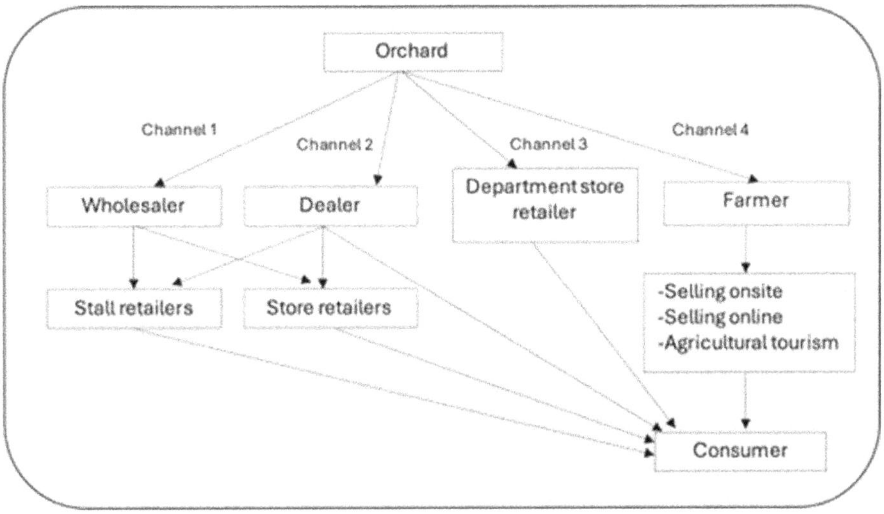

FIGURE 5.6 Marketing channel of mandarins. (Adapted from Muenmanee (2016).)

States, United Kingdom, Australia, China, Hong Kong, Taiwan, Japan, and Singapore. The fruit is highly sought after for its quality and flavour. The Agriculture Office of Samut Songkhram Province plays a prominent role in organizing events such as the 'Samut Songkhram Big White Pomelo Day' to promote and showcase the region's pomelo produce. The price of pomelo production increases every year due to various factors both internal and external related to cultivation and orchard maintenance. Factors such as rising fertilizer prices and transportation costs contribute to the upward trend in prices each year. In Samut Songkhram Province, pomelo production prices are determined throughout the year based on the weight of the fruit and the care provided by the farmers.
Retail prices of Khao Yai pomelo in 2020

- Large-sized: 50–65 baht per kilogram
- Medium-sized: 30–49 baht per kilogram
- Small-size: 15–29 baht per kilogram

See Figure 5.7.

5.3.9 Marketing Channels of Lime

The sale of lime mostly relies on intermediaries, with certain farmers selling their harvest to intermediaries at designated locations or having intermediaries visit their farms for direct purchase. However, farmers do not collaborate to sell collectively to intermediaries, resulting in them avoiding transportation costs. This lack of collective action also diminishes their ability to negotiate and leads to lower selling prices. Limes are visually sorted into three grades by farmers and merchants, with further detailed sorting by size into six grades at the merchant level. The majority of limes produced are used for domestic consumption, with a smaller portion sent to factories for processing into products like herbal cosmetics and cleaning agents. Thailand used to export limes for millions of baht annually, but due to increased domestic demand and industrial use, all production is now consumed within the country, leading to a halt in lime exports in recent years (Rasmeethamwong, 2005).

Despite the uncertain and inconsistent quantity of lime production each year, leading to challenges in aligning production with marketing, limitations stemming from environmental factors still play a significant role.

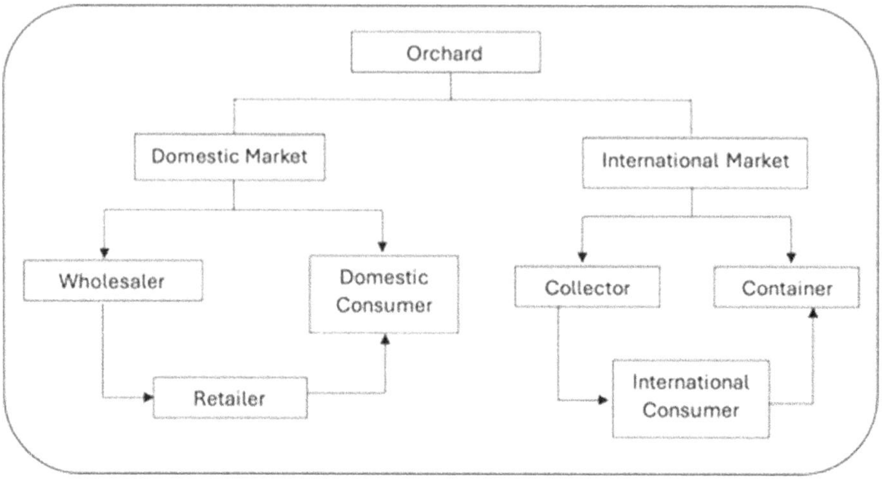

FIGURE 5.7 Marketing channel of pomelo. (From Samut Songkhram, Provincial Agriculture and Cooperatives Office, 2020.)

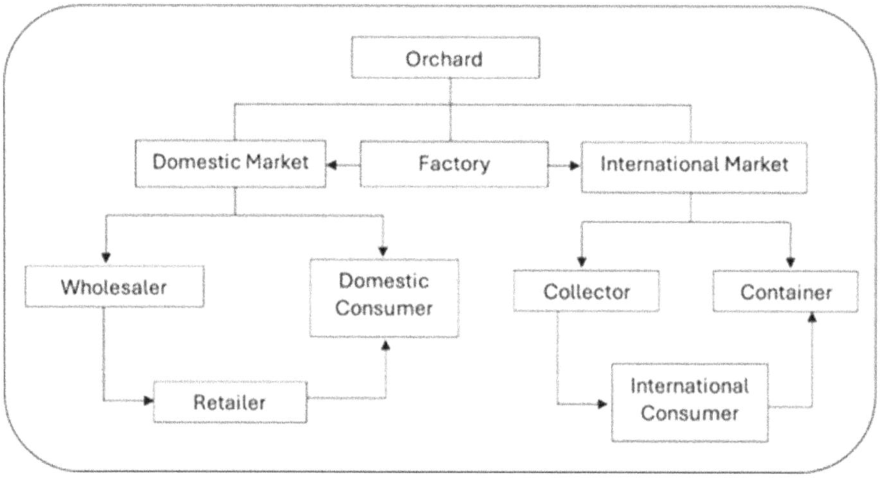

FIGURE 5.8 Marketing channel of lime. (From writer.)

While advancements in technology have enabled more effective flowering and fruiting outside of the usual season, adoption of these practices among farmers remains limited. Consequently, the production volume experiences fluctuations and is not evenly distributed throughout the year (Pongsomboon, 2003). See Figure 5.8.

5.4 STATUS OF EXPORT AND IMPORT OF CITRUS

5.4.1 Mandarin

In 2023, there was a reduction in the area of 'Khiaw Waan' trees standing compared to the previous year. This decline can be attributed to farmers cutting down older mandarin trees, resulting in lower production. Over the past 2–3 years, the prices of 'Khiaw Waan' sold by farmers have consistently dropped, while the cost of maintenance remains high. Consequently, many farmers have opted to cultivate alternative crops that offer better returns and require less intensive care, such as rubber, durian, limes, bananas, animal feed corn, and various vegetables. However, areas that saw new mandarin plantations in 2020 are beginning to yield fruit this year, notably the Bang Mot oranges (Bangkok), Nan golden oranges (Nan), and Mae Sin oranges (Sukhothai). Moreover, yield per unit of land has declined due to unfavourable weather conditions. Hot and dry weather during the mandarin season led to flower shedding and inadequate rainfall for growth. Additionally, Chiang Mai province, a key production area in the northern region, was impacted by a storm and hail in April 2023, resulting in damage to mandarin fruits and an overall reduction in the national production (Office of Agricultural Economic, 2023).

The vast majority of mandarin yield is destined for domestic consumption, distributed through various channels including central markets, cooperative networks, entrepreneurial ventures, and department stores, with an emerging presence in online sales channels. Export volumes remain relatively minimal, with most trade partners located within ASEAN countries, notably Myanmar, Vietnam, and Cambodia (Office of Agricultural Economic, 2023). According to the export data from the Office of Agricultural Economic (2023), it was found that the export of 'Khiaw Waan' mandarin, in the year 2023 amounted to 41,817.02 tonnes, valued at 1,258.40 million baht. This showed a decrease from the same period of the previous year, which had exports of 43,632.85 tonnes, valued at 1,159.88 million baht, representing a 4.16% decrease in quantity but a 8.49% increase in value. The top three key trading partners include Myanmar, Vietnam, and Cambodia. Meanwhile, the export quantity in the year 2023 showed the lower of 12.30% and 2.83% decrease in import quantity and value, respectively. See Tables 5.5 to 5.7.

The government's support policy for mandarin production and marketing focuses on several key initiatives. Firstly, it encourages the adoption of large-scale farming practices, including organic methods, in key production areas. Secondly, it supports the expansion of marketing channels, both online and offline, facilitating connections between producers and target markets through cooperative networks, entrepreneurial ventures, and retail outlets. Lastly, the policy promotes value addition and income generation at the local level by assisting in the development of Geographical Indication (GI) products, such as Bang Mot oranges (Bangkok), Nan golden oranges (Nan), and Mae Sin oranges (Sukhothai) (Office of Agricultural Economic, 2023).

5.4.2 Pomelo

Pomelo is an economically potential fruit tree for export due to its popularity as a consuming fruit with high nutritional value. Therefore, it is in demand in both domestic and international markets, especially for pomelos certified with Good Agricultural Practices (GAP) standards. This certification adds value to exported products and creates consumer confidence in the safety and quality of the produce.

TABLE 5.5
The Annual Import and Export Quantity and Value of Mandarin During 2020–2024

	2020		2021		2022		2023		2024	
	Quantity	Value	Quantity	Value	Quantity	Value	Quantity	Value	Quantity	Value
Import	118,619.77	3,564.78	114,121.70	4,075.40	96,659.75	4,291.09	84,767.15	4,169.58	22,181.72	938.81
Export	39,151.16	891.04	47,318.94	1,138.26	43,632.85	1,159.88	41,817.02	1,258.40	5,611.26	176.74

Unit: Quantity (tonne), Value (million baht).
Source: Office of Agricultural Economic (2024).

TABLE 5.6
The Monthly Import and Export Quantity and Value of Mandarin During 2021–2022

	2021				2022			
	Import		Export		Import		Export	
Month	Quantity	Value	Quantity	Value	Quantity	Value	Quantity	Value
Jan	22,943.67	645.16	2,338.63	60.57	19,537.12	832.61	3,865.91	89.45
Feb	12,546.95	335.86	2,844.93	63.76	7,447.22	240.50	3,778.94	100.52
Mar	17,492.40	468.67	6,044.74	153.24	16,817.12	525.57	5,695.60	142.68
Apr	11,414.38	283.83	5,154.82	112.58	10,821.89	415.78	4,340.23	100.37
May	5,080.03	164.46	5,856.92	121.03	2,649.96	144.98	3,897.49	103.91
Jun	3,158.37	113.89	3,651.59	92.44	2,993.92	163.53	2,366.38	65.73
Jul	2,831.64	165.32	3,314.68	84.57	3,289.76	217.28	3,163.53	85.33
Aug	5,419.81	348.08	4,079.85	98.30	4,770.11	326.15	3,252.90	88.91
Sep	4,779.82	277.96	3,308.69	76.27	6,802.77	427.81	3,374.73	102.38
Oct	4,777.94	253.19	3,425.50	86.85	5,424.28	287.97	2,947.72	85.86
Nov	10,024.35	508.09	3,644.34	89.76	7,477.95	318.00	3,622.19	99.82
Dec	13,652.34	510.89	3,654.25	98.90	8,627.67	390.91	3,327.24	94.90
Total	114,121.70	4,075.40	47,318.94	1,138.26	96,659.75	4,291.09	43,632.85	1,159.88

Unit: Quantity (tonne), Value (million baht).
Source: Office of Agricultural Economic (2023).

TABLE 5.7
The Annual Quantity and Value of Mandarin Exports from Thailand During 2018–2023

List	2018	2019	2020	2021	2022	2023
1. Domestic use (tonne)	191,023	212,937	203,261	256,025	282,073	288,536
2. Export (tonne)	1,001	806	1,476	3,329	6,393	3,369
2.1. 'Khiaw Waan' (fresh)						
• Quantity (tonne)	409	136	522	75	3,033	1,381
• Value (million baht)	7.39	2.53	9.72	2	63	40
2.2. Mandarin (fresh)						
• Quantity (tonne)	481	659	464	1,681	2,332	1,413
• Value (million baht)	7.22	10.35	19.17	41	49	41
2.3. Other Citrus (fresh)						
• Quantity (tonne)	111	11	490	1,573	1,028	575
• Value (million baht)	5.52	0.17	12.27	98	36	22
3. Export price (baht/tonne)						
3.1. 'Khiaw Waan' (fresh)	18,062	18,685	18,616	26,045	20,632	28,796
3.2. Mandarin (fresh)	15,008	15,701	41,317	24,282	21,127	28,735
3.3. Other Citr us (fresh)	49,874	15,189	25,023	62,448	34,783	38,165
4. Important trading partners						
4.1. 'Khiaw Waan' (fresh)	Vietnam, China			Myanmar, Vietnam, Cambodia		
4.2. Mandarin (fresh)	Myanmar, Hongkong, Laos			Myanmar		
4.3. Other citrus (fresh)	Myanmar, Malaysia, Laos			Myanmar		
5. Important competitors	China, Vietnam, Philippines			China, Vietnam, Philippines		

Source: Office of Agricultural Economics (2023a).

Currently, Thailand has expanded its pomelo export market to 30 countries, including key markets such as China, Malaysia, and the United States. In 2022, Thailand exported 1.68 million tonnes of pomelos worth 45.55 million baht, showing a growing demand for Thai pomelos internationally (Office of Agricultural Economic, 2023). Furthermore, a significant development in July 2023, Thailand shipping its first pomelo export to Washington DC, USA, comprising four varieties: 'Thong Dee', 'Khao Yai', 'Khao Nam Phung' and 'Khao Tangkwa'. This development opens up more export opportunities, particularly to Japan, where the high demand and health-conscious population present a lucrative market. By displaying Functional Foods labels that highlight the nutritional value, strengthen the immune system, and reduce disease risk, Thai agricultural products can increase their value and build confidence in safety and quality for target consumers. This highlights the promising future of pomelos due to high demand in the international market (Gilkey, 2024). Moreover, pomelos are well-suited for export because they have a shelf life of approximately 3 weeks to 1 month with consistent quality, enhanced taste, and thick peel that allows for long-distance transportation, making shipping by sea cost-effective. Key export markets for Thai pomelos include China, Hong Kong, Canada (Chitakornkijsil, 2010).

In one annual cycle, pomelo production in various regions of Thailand peaks from August to October, leading to oversupply and lower selling prices. Off-season pomelo production is a strategy to promote better distribution of harvest seasons for farmers, providing a way to control and manage market supply. This allows for the scheduling of pomelo production entry into the market according to demand, particularly during periods of high market prices such as the Chinese New Year (January–February), the New Year, and the Hungry Ghost Festival (August–September). It also serves as a means to plan and control pomelo production quality to meet the requirements of different markets in the future (Pongsomboon, 2017). See Tables 5.8 and 5.9.

5.4.3 Lime

Limes are an economically important crop that provides significant income to the country each year. This leads to widespread cultivation by Thai farmers. Limes can be used in a variety of dishes and can also be made into beverages or used as ingredients in various drinks. Additionally, they have medicinal properties due to their high vitamin content, often used in herbal medicine and cosmetics. This increases the demand for limes as a raw material in industries every year, leading to a corresponding increase in market demand. Especially during the dry season from March to April, lime prices are typically higher than usual due to lower production entering the market.

Popular lime cultivars in Thailand include 'Pan', 'Nang', 'Kai', 'Puang', and 'Tahiti'. However, cultivar 'Pan' are particularly favoured, known for their year-round productivity and various strains such as 'Pan Rumpai', 'Pan Sai', and 'Pan Phichit'. On the other hand, 'Tahiti' limes are not as popular among Thais, hence they are predominantly exported. According to data from the Ministry of Trade Policy and Strategy Office (2020), lime production in 2020, amounted to 483,930 tonnes,

TABLE 5.8
The Annual Import and Export Quantity and Value of Pomelo During 2020–2024

	2020		2021		2022		2023		2024	
	Quantity	Value	Quantity	Value	Quantity	Value	Quantity	Value	Quantity	Value
Import	2,604.66	44.70	7,717.41	130.58	2,052.15	34.84	1.21	0.10	0.11	0.02
Export	20,306.78	671.00	30,598.85	1,061.84	33,531.68	1,475.43	25,283.07	1,203.18	2,520.21	127.96

Unit: Quantity (tonne), Value (million baht).
Source: Office of Agricultural Economic (2024).

TABLE 5.9
The Monthly Import and Export Quantity and Value of Pomelo During 2021–2022

	2021				2022			
	Import		Export		Import		Export	
Month	Quantity	Value	Quantity	Value	Quantity	Value	Quantity	Value
Jan	196.58	3.47	946.81	32.47	500.63	8.49	1,781.94	76.50
Feb	677.99	11.53	1,311.94	32.38	423.46	7.20	1,928.08	87.36
Mar	2,316.75	39.38	4,882.40	132.78	152.10	2.59	3,535.26	120.12
Apr	541.28	9.13	2,855.32	88.12	376.15	6.56	4,229.58	117.69
May	905.34	15.39	3,454.15	120.38	84.20	1.46	1,871.74	60.77
Jun	878.69	14.94	2,681.89	88.73	270.83	4.54	3,086.40	100.44
Jul	486.52	8.27	3,787.50	131.55	244.80	4.00	5,077.16	221.79
Aug	801.89	12.91	4,442.72	172.90	N/A	N/A	5,734.30	300.78
Sep	365.63	6.24	3,313.05	126.90	N/A	N/A	2,350.15	122.33
Oct	189.27	3.25	745.80	41.35	N/A	N/A	1,470.95	96.01
Nov	150.15	2.55	1,065.49	53.50	N/A	N/A	1,308.58	72.92
Dec	207.31	3.52	1,111.79	40.77	N/A	N/A	1,157.55	98.71
Total	7,717.41	130.58	30,598.85	1,061.84	2,052.15	34.84	33,531.68	1,475.43

Unit: Quantity (tonne), Value (million baht).
Source: Office of Agricultural Economic (2024).

an increase of 32.09% from 2019, which had 366,362 tonnes, largely due to favourable weather conditions. The top five provinces with the highest lime production are Phetchaburi, Ratchaburi, Samut Sakhon, Kanchanaburi, and Phichit. Regarding exports, data from the Ministry of Commerce (2021) shows that lime exports in 2021 from January to August amounted to 1,413 tonnes, valued at 24.55 million baht, a decrease of 75.91% in quantity and 87.3% in value compared to the same period in 2020, which saw exports of 5,866 tonnes valued at 194.29 million baht. The top five important trading partners include Laos, Malaysia, Hong Kong, Myanmar, and Bahrain.

Mr. Anok Prasom, aged 65, a resident of Ban Chae Khan, Thung Phueng Subdistrict, Chae Hom District, Lampang Province, is the owner of the Kanjana lime orchard. He cultivates the Tahiti lime variety, also known as seedless limes, which are large, juicy, seedless, fragrant, and have a round and mild sour taste. He exports and sells his limes both domestically and internationally. He has over 3,500 lime trees on approximately 20 rai of land and is expanding his cultivation area by about 2,000 more trees. Due to the decrease in lime production caused by the COVID-19 pandemic, many lime trees have been abandoned by farmers, leading to a decrease in production while demand remains high. This year, Mr. Anok's orchard is expected to produce around 25–30 tonnes of limes, which may not be enough as he also needs to export to countries like the Middle East, Brunei, and Qatar, as well as Taiwan and China. These customers specifically request the Tahiti lime variety. Hence, there is still a high demand for limes in the market (MRG online, 2023).

5.5 STATUS OF CITRUS PROCESSING

5.5.1 MANDARIN

Food processing and preservation offer additional avenues for enhancing product value and meeting the increasing demand for innovative mandarin-based products among consumers. This trend has spurred technological advancements and business expansions within the processing and value-added sectors. A variety of processed mandarin products are now available, including candied peel, powdered beverages, jams, ice creams, juices, preserved peel, purees, flavoured dairy products, pre-packaged jellies, scented skin creams, and mosquito repellents (KMUTT Library, 2024). Mandarin also finds utility in the production of dietary supplements, soaps, and as decorative elements to elevate the visual appeal of dishes. In Thai traditional medicine, 'Khiaw Waan' mandarin peel and seeds are recognized as herbal remedies and are incorporated into contemporary pharmacological formulations (Medthai, 2020).

5.5.2 Pomelo

Apart from fresh consumption, pomelo can also be processed in various ways, from their immature stage to fully ripe. In cases of excessive fruit set, farmers often discard the excess fruit to ensure larger yields. The immature pomelo can be used to make candied pomelo, which fetch a good price but require three months of soaking. Fully mature pomelo can be processed into various products. The peel of the pomelo fruit can be candied or preserved in sugar syrup, which is relatively simple and possesses medicinal qualities, rendering it a product with market potential. Dried pomelo peel can be used for cancer treatment, but must come from organic sources. Leftover peels from consumed fruit are used, the peels are soaked in clear lime water, cut into small pieces, boiled with salt and sugar until sticky, then formed into candy for sale. Additionally, mandarin peels are utilized to create mosquito repellent, whereby they are dried, ground into powder, mixed with lemongrass leaves and other materials, and then fashioned into incense sticks that effectively repel mosquitoes, posing no harm to humans and are marketed as eco-friendly repellents (Samut Songkhram Provincial Agriculture and Cooperatives Office, 2020).

5.5.3 Lime

The period from June to November is typically when lime production peaks in lime-growing regions each year. However, the high volume of production during this time often exceeds market demand, leading to lower prices. Therefore, farmers are encouraged to explore alternative avenues to add value to their lime products by establishing and operating lime processing industries to create a range of products. This initiative aims to enhance income generation and offers a variety of processed lime products as lime juice, concentrate lime juice, lime syrup, lime cordial, lime squash, frozen lime juice, lime powder, lime jam, lime peel jam, lime jelly, lime marmalade, pickled lime, candied lime, candied lime peel, picked lime, etc. (Ruamrak, 1996).

5.6 RESEARCH

Researches in citrus have been done in many aspects from cultural practices, fertilizer and water management, propagation, fruit quality improvement, postharvest management, breeding, and molecular biology.

Fertilizer is one important factor in citrus growing. Phunchaisri and Tunsuwan (2001) studied 600, 800, and 1,000 meq/L nitrogen application to lime plants. It was found that titratable acidity (TA) and total soluble solids (TSS) of lime fruits received 1,000 meq/L nitrogen were higher than those of fruits in the other two treatments. However, fruit size, fruit weight, juice content, and pH were not different between treatments. Kaewtubtim (2021) studied effects of nitrogen fertilizer rate on quality of Puko pomelo by applying 1, 2, and 3 kg per plant of urea. Different rates of urea fertilizer did not show significant differences in the number of inflorescences per branch, fruit set percentage, fruit weight, fruit width, fruit height, TSS, TA, vitamin C, and chlorophyll. The plants received urea fertilizer at 2 and 3 kg per plant produced fruits with higher levels of sucrose, glucose, fructose, total sugar, and lycopene than those received 1 kg fertilizer.

A review by Issariyaporn (2007) on 'Shokun' mandarin (*Citrus reticulata* Blanco cv. Shokun) explained that N, P, K, Ca, Mg, and B affected fruit quality. If the fruit received insufficient N, K, Mg, it decreases fruit size. Higher levels of N and K could delay maturity. N and Ca support the postharvest fruit quality. Higher levels of P and K results in thinner peel. Lower levels of N and K tend to result in lower TSS and total acid whereas lower level of P results in higher total acid. The citrus plants deficiency in B produces fruit with a lack of fruit juice and might be related to physiological disorder called granulation. The basic considerations for plant nutrient management are compensation the removal of mineral element in the harvested fruits and fertilizer application according to soil and plant analysis results. Fertilizer and nutrient management in pomelo were researched under nutrients research project for quality pomelo production written extensively by Maneepong (2013).

There has not been research in irrigation or water requirement in citrus. However, automatic sprinkler system was installed in pomelo orchards in Ban Thaen district, Chaiyaphum province. It received a very good satisfaction level from the growers (Laopa and Puttamat, 2023).

Heavy metal in soil is one problem in fruit growing. Cadmium contamination in pomelo fruit and soil were studied in Pichit province (Wachirawongsakorn, 2022). The research was conducted in three cultivars of pomelo, 'Khao Tangkwa', 'Thongdee', and 'Thakoi', ten orchards each cultivar. Pollution index and bioaccumulation of cadmium were examined. Cadmium contamination in the soil of all orchards did not exceed the soil quality standard of the Department of Land Development.

Disease management is one important practice of citrus growing. In lime, actinobacteria obtained from Naresuan University were evaluated for their antagonistic activities against pathogenic microorganisms in citrus. There was an isolate which showed potent antagonistic activity against microorganisms cause of canker in kaffir lime, lime, and pomelo (Pimpanuwat, 2021). Huanglongbing or citrus greening is another major citrus disease. Akarapisan et al. (2008) surveyed the occurrence in pomelo in northern Thailand. This disease also found in mandarin, lime, and kaffir lime. Since the symptom of Huanglongbing can be mistaken with zinc deficiency, a specific primer for polymerase chain reaction is used to detect the disease. Sornsaket et al. (2024) studied effects of silver nanoparticles on fruit quality of mandarin in mandarin plants with greening disease by comparing with ampicillin application. It was found that plants received 300 ppm silver nanoparticles had lower fruit drop than plants received 15,000 ppm ampicillin treatment. However, fruit size and chemical qualities of fruits were not different.

Off-season production is a target for many growers due to the better price of crops. Panyakaew and Naphrom (2010) applied ethephon to mandarin 'Sai Nam Phung' at bloom time to make flower drops, then applied paclobutrazol to induce flowering.

Research on rootstock for commercial citrus has been going on for a long time. Thawornrat (2002) studied Som Jeed (*Citrus japonica* Thumb.), pomelo (*Citrus grandis* (L.) Osbeck), kaffir lime (*Citrus hystrix* DC.), Troyer Citrange [*Citrus sinensis* (L.) Osbeck x *Poncirus trifoliate* (L.) Raf.], Swingle Citrumelo [*Citrus paradise* x *Poncirus trifoliate* (L.) Raf.] and neck orange (*Citrus reticulata* Blanco) rootstocks on development of the neck orange. The results showed that Troyer Citrange was the best rootstock for neck orange. Recent research by Chuchert et al. (2024) compared growth of neck orange (*Citrus reticulata* Blanco) on five *Citrus* sp. rootstocks, local pomelo (*C. grandis* (L.), 'Thongdee' pomelo (*C. maxima* (Burm.f.), citron (*C. medica*), kaffir lime (*C. hystrix*), and bitter orange (*C.* × *aurantium*). The results revealed that bitter orange was the best rootstock. Rodsuwannoi (2007) studied rootstocks for 'Shokun' mandarin, 1-year pomelo (*Citrus grandis* (L.) Osbeck) and Manao kwai (*Citrus medica* Linn. Var. Linetta.), and 5-year Troyer Citrange and Swingle Citrumelo were used. It was found that Manao kwai was a better rootstock than pomelo and Troyer Citrange was a better rootstock than Swingle Citrumelo.

Multiple shoot induction in vitro was done in 'Thongdee' pomelo. There were five shoots obtained from shoot explant in MS medium with IBA supplement and 5.6 shoots obtained from node explant in MS medium with BA supplement (Obsuwan et al., 2016).

Season influences on fruit quality. Lime harvested in the dry season had larger fruit, higher fruit weight and juice weight, thicker rind, higher TSS and TSS:TA ratio than fruit in rainy season, whereas vitamin C content and TA were lower (Buathong, 2021).

Postharvest research has been conducted to extend shelf life of citrus. In 'Sai Nam Phung' mandarin, Kumpoun (2009) studied waxing materials and plastic bag (polyethylene and polypropylene) and storage temperature of 15, 25, 30, and 35°C. The research found that at lower temperature, the fruit can be kept longer than at higher temperature. At 15°C, mandarin fruit in all treatments were still in good condition for 50 days. However, waxed fruit developed abnormal odour after 4–5 weeks. Carnauba wax was used to coat lime. It was found that lime coated with wax emulsion had lower weight loss than control fruit. In addition, at 13 days the fruits had yellowish green colour, while control fruit already had unacceptable brown colour skin (Jutamongkon et al., 2022). Previous research by Klodpeng et al. (2008) indicated that washing lime fruit with ozone could extend shelf life (more than 27 days) and also provide better postharvest qualities.

Breeding has been conducted by a government agency which can manage to work continuously, for example, Agricultural Research and Development Centre, Pichit province is working on pomelo and lime breeding. Molecular biology technique is used to assist with breeding. Chuenwarin et al. (2022) used ISSR markers to study genetic relationship of 55 accessions of lime and acid citrus germplasms. The researchers could categorize studied citrus into four groups, Mexican lime and Tahiti lime, citron and lemon, kaffir lime, and finger lime.

5.7 CHALLENGES

The citrus industry in Thailand is confronted with a range of challenges that significantly impact both the yield and quality of production. Key issues include the prevalence of citrus greening disease and other pests, the susceptibility of crops to changing climatic conditions, soil health problems, the ageing of orchards and farmers, difficulties in market access and price stability. Addressing these challenges effectively requires a multifaceted approach, involving improved agricultural practices, enhanced researches, includes breeding to get good quality fruit, high yield, and disease resistance, better market strategies, and policies that support technological adoption and sustainability. Off-season production to distribute citrus crops all year round is also something to put the efforts to. The technology to preserve quality of fruit postharvest is a topic that is still important. By tackling these problems comprehensively, Thailand can secure the future of its citrus industry, ensuring it remains viable and competitive in the global market.

5.8 CONCLUSION

The citrus industry in Thailand is primarily focused on mandarin, pomelo, and lime cultivation. Mandarin production in the northern region accounts for a large portion of the country's production. 'Sai Nam Phung' and 'Khiaw Waan', are prevalent cultivars in the market, each with distinct botanical characteristics. Pomelo and lime are widely cultivated throughout the country, with major production areas spanning provinces from north to south. The pomelo fruit is primarily grown for domestic consumption but i s also exported to international markets such as China, Vietnam, and Hong Kong. Various commercial pomelo cultivars are grown in Thailand, with the 'Thongdee' pomelo being a prominent Geographical Indications (GI) product known for its sweetness and aroma. Persian and Key limes are major lime production in Thailand. The climate in Thailand varies between the northern and southern regions, impacting the growth and development of citrus. Soil characteristics also play a significant role in citrus cultivation, with different landscapes and soil types affecting plant growth. Cultural practices and management techniques, including irrigation, fertilization, weed control, canopy management, and pruning, are essential for successful citrus production. Proper irrigation schedules, fertilization practices, and weed management help maintain plant health and yield quality. Branch supporting, windbreaks, and canopy management techniques further contribute to successful citrus cultivation. Pest and disease management, including control measures for citrus greening disease and other threats, are crucial for ensuring healthy and productive citrus trees. Labour costs, ageing farmers, market channels, and the impact of external factors like weather and the COVID-19 pandemic also play key roles in the citrus production.

The citrus industry in Thailand encompasses various aspects including the harvesting window of major cultivars, the complex marketing network, and the status of export and import. The marketing network for citrus involves wholesalers, retailers, direct sales, online sales, and agricultural tourism, with different channels catering to diverse consumer preferences. Mandarin distribution is primarily domestic, with emerging online sales channels and minimal exports to ASEAN countries. Pomelos, popular for export due to their nutritional value, see increasing demand internationally, with Thailand expanding its market to 30 countries. Limes, economically significant for Thailand, face challenges of oversupply during peak production seasons but offer opportunities for value addition through processing into various products like juices, jams, and pickled limes. Government support policies focus on promoting large-scale farming, expanding marketing channels, and enhancing value addition through initiatives like Geographical Indication (GI) products. Processing of citrus fruits into value-added products like candied peel, powdered beverages, and jams further adds to their market appeal and economic significance.

Researches in citrus encompass a broad spectrum of topics, including cultural practices, nutrient balance, propagation, irrigation, disease and pest control, postharvest management, breeding, and molecular biology. Disease and pest management remains a priority, with research focusing on biological control methods and molecular detection techniques for diseases like citrus greening. Off-season production strategies, including the use of plant growth regulators, aim to capitalize on market demand and improv e economic returns for growers. Rootstock selection continues to be a focal point for citrus research, with ongoing evaluations to identify rootstocks that promote optimal growth and fruit quality. In vitro propagation techniques offer additional avenues for citrus cultivation, enhancing the efficiency of multiple shoot induction and propagation. Lastly, seasonality significantly influences fruit quality attributes, with variations observed in parameters such as fruit size, weight, juice content, and biochemical composition. Understanding these seasonal dynamics is essential for informed decision-making in citrus production. Overall, ongoing research efforts contribute to the sustainable advancement of citrus cultivation practices, addressing challenges and optimizing outcomes for growers and consumers alike.

REFERENCES

Akarapisan, A., Piwkhao, K., Chanbang, Y., Naphrom, D., and Santasup, C. (2008). Occurrence of huanglongbing disease of pomelo (*Citrus grandis*) in Northern Thailand. In *International Research Conference on Huanglongbing Proceedings, Orlando, Florida*, 381–385.

Buathong, S., Imsabai, W., and Chuenwarin, P. (2021). A study on harvesting seasons and different fruit sizes on ‘Pan Baan Phaeo’ lime fruit qualities. *King Mongkut’s Agriculture Journal* 39(3): 190–197.

Chitakornkijsil, P. (2010). Potential of production and export of pomelo. *NIDA Development Journal* 30(3): 148–166.

Chuchert, S., Sakdiset, N., Sanputawong, S., and Choopeng, S. (2024). The comparison on growth of neck orange (*Citrus reticulata* Blanco) grafted on five *Citrus* sp. rootstock. *Rajamangala University of Technology Srivijaya Research Journal* 16(1): 255–265.

Chuenwarin, P., Thongleow, S., and Auvuchanon, A. (2022). Genetic relationships analysis of lime and acid *Citrus* by ISSR markers. *The Agricultural Research Journal* 40(2): 188–199.

Gilkey, M. (2024). Pomelo diplomacy: How a favorite Thai fruit advances U.S trade tile. American Foreign Service Association. *The Foreign Services Journal*. https://afsa.org/pomelo-diplomacy

Intakin, A., Sawliu, K., and Chaovanapoonpho, Y. (2019). Marketing and purchasing behavior of consumers and wholesalers for 'Sai Nam Phung' Mandarin in Fang district, Chiang Mai province. *Khon Kaen Agricultural Journal* 47(1): 883–888.

Issariyaporn, D. (2007). Plant nutrient elements and fruit quality of 'Shokun' mandarin (*Citrus reticulata* Blanco cv. Shokun). *Journal of Yala Rajabhat University* 2(1): 56–71.

Jutamongkon, R., Valung, P., and Thongpradit, S. (2022). Extending the shelf life of lime and banana by coating carnauba wax emulsion. *Rajamangala University of Technology Srivijaya Research Journal* 15(3): 677–686.

Kaewtubtim, M. (2021). Effects of nitrogen fertilizer rate on quality of 'Puko' pomelo (*Citrus grandis* (L.) Osbeck) cv. Puko. *Prince of Naradhiwas University Journal* 13(2): 342–353.

Klodpeng, K., Pankasemsuk, T., and Nikornpun, M. (2008). Effects of washing agents on shelf-life and postharvest qualities of lime. *Journal of Agriculture* 24(2): 125–132.

KMUTT Library. (2024). Value adding in agricultural product and processing. www.lib.kmutt.ac.th/ processing-value-adding/. Accessed 25 April 2024.

Koshita, Y. (2015). Effect of temperature on fruit color development. In *Abiotic stress biology in horticultural plants*, ed. Y. Kanayama and A. Kochetov, 47–58. Springer, Japan. doi: 10.1007/978-4-431-55251-2_4

Kumpoun, W. (2009). Storage of 'Sai Nam Phung' tangerine fruit at different temperature. *Journal of Agriculture* 25(1): 79–84.

Laopa, S., and Puttamat, W. (2023). Design and construction of automatic sprinkler system control training package: A case study of pomelo orchards Ban Thaen sub district, Ban Thaen district, Chaiyaphum province. *Industrial Technology Journal* 8(2): 137–148.

Maneepong, S. (2013). *Nutrient management for quality Pomelo production*, 131. The Thailand Research Fund, Bangkok.

Medthai. (2020). Mandarin. https://medthai.com. Accessed 25 April 2024.

Ministry of Commerce. (2021). Lime export data. www.moc.go.th/en/page/item/index/id/6. Accessed 25 April 2024.

Ministry of Trade Policy and Strategy Office. (2020). Lime production. https://tpso.go.th/home. Accessed 25 April 2024.

MRG online. (2023). Lime production. https://mgronline.com/local/detail/9650000031901. Accessed 25 April 2024.

Muenmanee, N. (2016). Distribution system of mandarin. In *Postharvest technology of Mandarin*, ed. N. Rattanapanone and D. Boonyakiat, 191–199. O.S. Printing House, Bangkok.

Neyret, M., Robain, H., de Rouw, A., Soulileuth, B., Trisophon, K., Jumpa, K., and Valentin, C. (2018). The transition from arable lands to rubber tree plantations in northern Thailand impacts weed assemblages and soil physical properties. *Soil Use and Management*. doi: 10.1111/SUM.12431

Obsuwan, K., Channon, J., Thepsithar, C., and Thongpakdee, A. (2016). Multiple shoot induction of pomelo [*Citrus maxima* (Burm.) Merr.] 'Thongdee'. *Veridian E-Journal, Science and Technology Silpakorn University* 3(6): 332–339.

Office of Agricultural Economics (2018). Pomelo. www.oae.go.th/view/1/pomelotable

Office of Agricultural Economics. (2023a). Lime. www.oae.go.th/view/1/limetable

Office of Agricultural Economics. (2023b). Mandarin. www.oae.go.th/view/1/mandarintable

Office of Agricultural Economics (2024a). Mandarin. www.oae.go.th/assets/portals/1/files

Office of Agricultural Economics (2024b). Pomelo. https://mis-app.oae.go.th/product/pomelo

Ornwimol, J. (2004). *Citrus orchard*. Kaset Sarn, Nontaburi, Bangkok, 408 (in Thai).

Panyakaew, J., and Naphrom, D. (2010). Effects of plant growth regulators on off-season flowering and fruit quality of mandarin cv. Sai Nam Phung. *Journal of Agriculture* 26: 117–125.

Phunchaisri, K., and Tunsuwan, T. (2001). Effect of nitrogen on quality of lime. *Journal of Agriculture* 17(2): 136–146.

Pimpanuwat, P. (2021). Investigating the potential of Actinobacteria to prevent disease and promoting the growth of *Citrus*. Master Thesis, Naresuan University, Thailand.

Pongsomboon, W. (2003). *Commercial production of lime* [Academic article]. Pichit Horticulture Research Center, 66.

Pongsomboon, W. (2017). *Research and development of pomelo production technology. Full report*. Department of Agriculture, Ministry of Agricultural and Cooperative, 50.

Radanachaless, T., and Thumdee, S. (1999). *Commercial fruit tree varieties in Thailand: Variety selection guide for grower*, 292. Green Wall Publishing. House, Bangkok.

Rasmeethamwong, P. (2005). *Off season lime: A guide of growing off season lime professionally*, 118. Naka Intermedia, Bangkok.

Rodsuwannoi, S. (2007). Influences of Citrus Rootstocks on the Growth of 'Shogun' (*Citrus reticulata* Blanco). Master Thesis, Prince of Songkla University, 80.

Ruamrak, M. (1996). *Fresh lime juice, ready to drink fresh lime juice, frozen lemonade, lime pickled, glass lime, preserved lime*, 2 [Published document]. Institute of Food Research and Development, Bangkok.

Samut Songkhram Provincial Agriculture and Cooperatives Office (2020). Information for pomelo planning. www.opsmoac.go.th/samutsongkhram-dwl-files-421491791138

Sithirangsan, W. (2001). *Marketing strategies of 'Sai Nam Phung' mandarin entrepreneurs in Chiang Mai Province* [Independent study]. Chiang Mai University, Thailand.

Sornsaket, A., Boonplod, N., Dangtungee, R., Spreer, W., Yangkhamman, P., and Naphrom, D. (2024). Effects of silver nanoparticles on fruit quality of oranges in orange trees with greening disease. *Maejo Journal of Agricultural Production* 6(1): 29–35.

Sruamsiri, P., Bhromsiri, A., Srichuwong, S., Tayutivutikul, J., Akrapisarn, A., Naphrom, D., Kunasakdakul, K., Boonchitsirikul, C., Pankasemsuk, T., Shutsrirung, A., Santasup, C., Janbang, Y., and Kumpuen, W. 2009. Sai Num Phung Tangerine: Technology for fruit quality and marketing. 1st Edition. *Wanida Printing. Chiang Mai.* 116.

Supakamnerd, N., Lerdrat, P., and Suwankhatnikom, S. (2005). *Pomelo nutrient requirements evaluation by plant analysis. Final Report.* Agricultural Research Development Agency, Bangkok.

Thawornrat, S. (2002). Influences of Citrus Rootstocks on Development of Neck Orange (*Citrus reticulate* Blanco). Master Thesis, Prince of Songkla University, 69.

Wachirawongsakorn, P., Phiromchom, S., and Junrak, S. (2022). Safety evaluation on cadmium contamination and bioaccumulation in pomelo (*Citrus maxima* (Burm. F.) Merr.) in Phichit province. *Academic Journal Uttaradit Rajabhat University* 18(1): 1–19.

6 Citrus Industry of Bhutan

Nangsel Tshomo, Sangay Gyeltshen, and Phuntsho Wangdi

ABBREVIATIONS

ACMI	Agriculture Commodity Market Initiative
ARDC	Agriculture Research and Development Centre
BAFRA	Bhutan Food and Drug Authority
BCCI	Bhutan Chamber of Commerce & Industry
BEA	Bhutan Electricity Authority.
CDP	Committee for Development Policy
DAMC	Department of Agricultural Marketing and Co-operatives
DAO	*Dzongkhag* Agriculture Officer
DOA	Department of Agriculture
FCB	Food Corporation of Bhutan Limited
MAGIP	Market Access and Growth Intensification Project
NCOA	National Centre for Organic Agriculture
NPHC	National Post Harvest Centre.
NPPC	National Plant Protection Centre
RAMCO	Regional Agricultural Marketing and Cooperatives Office

6.1 INTRODUCTION

Citrus refers to all the plants that are under the *Rutaceae* family and are native to the tropics of Asia.

The citrus industry in Bhutan holds the credit of being the nation's most important fruit crop and production continues to climb year by year. Bhutan citrus fruit production was at a level of 18,540 tonnes in 2022, up from 16,105 tonnes the previous year; this is a change of 15.11% (Citrus Fruit Production in Bhutan, 2023). Currently, the citrus industry is dominated exclusively by mandarin (Citrus reticulata Blanco), which constitutes more than 95% of total citrus production in the country cultivated across 5,048.6 hectares of land (Dorji et al., 2016).

It is grown in as many as 17 districts of the country and is one of the highest-value export crops serving as the main source of income for many households. However, despite its significance, the national yield of citrus in Bhutan is lower than in countries like *Thailand and Taiwan* due to poor technology adoption and traditional management practices.

This lower yield of citrus in Bhutan compared to the above countries poses both questions and an opportunity for improvement of the citrus industry in Bhutan. This chapter of the book will manoeuvre through the current citrus industry in Bhutan, major cultivars of citrus grown in the country, challenges faced, and the way forward.

DOI: 10.1201/9781003584599-6

6.2 MAJOR CULTIVARS OF CITRUS

6.2.1 Mandarin (Table 6.1)

Currently, the citrus industry is dominated exclusively by mandarin (Citrus reticulata Blanco), which constitutes more than 95% of total citrus production in the country cultivated across 5,048.6 hectares of land (Dorji et al., 2016). Some of the Mandarin varieties and cultivars that are present in the country are as follows:

- Okitsu Wase Mandarin is a cultivar with TSS 10.8°B and an acidity of 1.60%. The fruit weight is 120 gm.
- AREP-1 HK Mandarin is a cultivar with TSS 10.8°B and an acidity of 1.1%. The fruit weight is 160 gm.
- Ohtsu-4 Mandarin contains a TSS of 11.2°B and an acidity of 1.4%. The fruit weight is 100 gm.
- Wengkhar Tshalu 1 has a TSS of 11.5°B, 1.15% acidity and it weighs 145 gm.
- Yoshida Ponkan Mandarin is a cultivar containing 13.7°B TSS, and 1.25% acidity and it weighs 155 gm.
- Wengkhar Tshalu 2 is a Mandarin Cultivar having 12.9°B with the acidity of 0.097% and a fruit weight of 150 gm.
- AREP-2 TNK contains TSS and acidity of 11°B and 1.84% respectively. The fruit weighs around 160 gm.
- Encore Mandarin has an acidity concentration of 2.45%, TSS of 15.5°B, and weighs 130 gm.
- Clementine is a cultivar of Mandarin that has a TSS of 11.5°B, acidity of 1% and fruit weight is 140 gm.
- The Tarku cultivar of Mandarin has a fruit weight of 120 gm, 14.3°B TSS, and 1.17% acidity.
- Dorokha Local is a Mandarin cultivar that has 11.6°, 1.24% acidity, and 105 gm of fruit weight.

TABLE 6.1
Mandarin Cultivars and Their Photos

Sr.No.	Name of the cultivar	Photo
1	Okitsu Wase Mandarin	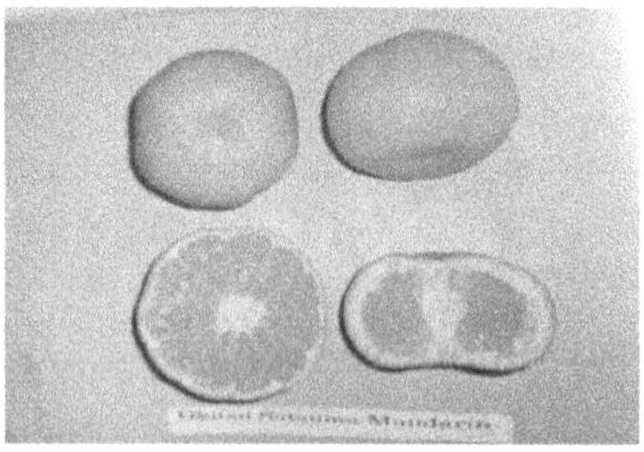
2	AREP-1 HK Mandarin	

(Continued)

TABLE 6.1 (Continued)
Mandarin Cultivars and Their Photos

Sr.No.	Name of the cultivar	Photo
3	Ohtsu-4 Mandarin	
4	Wengkhar Tshalu 1	
5	Yoshida Ponkan Mandarin	
6	Wengkhar Tshalu 2	
7	AREP-2 TNK	
8	Encore	

(*Continued*)

TABLE 6.1 (Continued)
Mandarin Cultivars and Their Photos

Sr.No.	Name of the cultivar	Photo
9	Clementine	

TABLE 6.2
Orange Cultivars and Their Photos

Sr.No.	Name	Photo
1	Navelina	
2	Junar	

6.2.2 Orange (Table 6.2)

- Navelina Oranges have TSS of 13°B, acidity of 1.72% and it weighs 170 gm.
- Junar is a cultivar of Orange that has 1.38% acidity, 10.8°B and 215 gm fruit weight.

6.2.3 Tangors (Table 6.3)

Murcott is the only cultivar of Tangors present in the country. It contains 1.3% acidity and the fruit weighs about 208 gm.

TABLE 6.3
Tangors Cultivars and Their Photos

Sr.No.	Name	Photo
1.	Murcott	

TABLE 6.4
Tangerine Cultivars and Their Photos

Sr.No.	Name	Photo
1.	Thai Tangerine	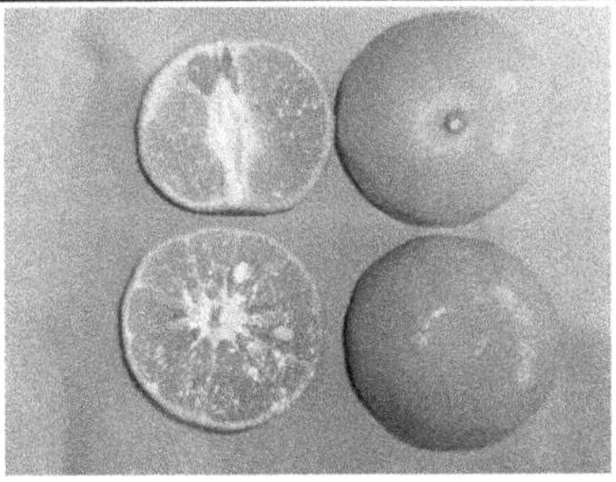

6.2.4 Tangerine (Table 6.4)

Thai Tangerine contains TSS of 14°B, acidity of 1.28% and the fruit weight is 70 gm.

6.2.5 Pummelo (Table 6.5)

Banpeiyu has a TSS of 10.3°B, acidity of 2.16%, and weighs 2,300 gm.

6.2.6 Grapefruit (Table 6.6)

Star Ruby is a cultivar of Grapefruit that has a TSS of 11°B and 3.55% acidity. The fruit weight is 300 gm.

TABLE 6.5
Pummelo Cultivars and Their Photos

Sr.No.	Name	Photo
1.	*Banpeiyu*	

TABLE 6.6
Grapefruit Cultivars and Their Photos

Sr.No.	Name	Photo
1.	Star Ruby	

6.2.7 Lemon (Table 6.7)

- Oroblanco cultivar of Lemon weighs about 140 gm.
- Frost Eureka Lemon has a fruit weight of 160 gm.

6.2.8 Lime (Table 6.8)

Mexican Lime is the most popular Lime cultivar present in the country. The fruit weighs about 60 gm.

6.3 AGRO-CLIMATIC ZONES

Bhutan is characterized into six major agroclimatic zones, which are as follows; alpine, cool temperate, warm temperate, dry sub-tropical, humid sub-tropical, and wet-sub tropical. (Table 6.9) Citrus cultivation is mainly done in dry, sub-tropical, humid sub-tropical, and wet-sub tropical. The key Mandarin growing districts are Dagana, Sarpang, Pemagatshel, Samtse, Tsirang, and Chukha (Awais Ahmad Tipu & A. Fantazy, 2014).

Two major mandarin varieties, i.e., Sikkim mandarin are concentrated in the south-west districts while the Khasi mandarin dominates the south-central and south-east districts. In Bhutan, Citrus

TABLE 6.7
Oroblanco Cultivars and Their Photos

Sr.No.	Name	Photo
1.	Oroblanco	
2.	Frost Eureka	

TABLE 6.8
Lime Cultivars and Their Photos

Sr.No.	Name	Photo
1.	Mexican Lime	

TABLE 6.9
Different Agro-Ecological Zones of Bhutan and Their Altitude Range

Sl. No.	Agro-ecological zones	Altitude range
1	Wet sub-tropical	100–600 amsl
2	Humid sub-tropical	600–1200 amsl
3	Dry sub-tropical	1200–1800 amsl
4	Warm temperate	1800–2600 amsl
5	Cool temperate	2600–3600 amsl
6	Alpine	3600–7500 amsl

cultivation thrives within a temperature range of 20°C to 28°C (68°F–82°F), with light frosts being tolerable.

Most of the zones where citrus cultivation takes in Bhutan have soils that are usually deep, loose, and well aerated with a pH range of 6–7 and soil texture ranging from sandy loam to silt loam. Before planting or establishment a citrus orchard, the soil is checked whether it has a good surface and internal drainage. Digging of pothole 3–4 feet deep and filling it with water is water is recommended to check its internal drainage system. If it takes more than 48 hours to drain, that soil/land should be avoided unless raised planting beds are used.

6.4 SIGNIFICANT TECHNOLOGIES AND PACKAGE OF PRACTICES FOLLOWED IN CITRUS FARMING

6.4.1 Site Selection

The site for the citrus orchard should be selected while keeping in mind the climate, water availability, and soil structure.

6.4.2 Pre-Planting Operations

Removal of weeds and raking of trees is a must when young citrus are to be planted in the orchard. If the orchard establishment is happening in an area where older trees were cultivated before, it should be kept in mind that any old roots and debris should be removed to minimize the spreading of diseases and pest infestations.

6.4.3 Pit Digging

In Bhutan, 3 ft (0.91 metres) is recommended as the pit size for citrus. Pits are usually dug 1–2 months before plantation. While digging, topsoil and subsoil are kept separately. The pits are usually kept exposed to the sunlight for a few days before planting. This serves as a method of sterilization as it kills pests and germs present in the pit.

6.4.4 Pit Filling

A mixture of topsoil, ½ kilogram of bone meal, and ten kilograms of well-rotten farm yard manure is used to fill the pits. Mounding of the soil to the height of 30 cm above the ground level is also done. All of these operations should be done a week or two priors to the soil to set in.

6.4.5 Plantation Time

Under rainfed condition, citrus is mainly planted from June to August month and March under irrigated condition.

6.4.6 Planting Pattern and Spacing

A hexagonal and square system of planting pattern is commonly followed in Bhutan. 6 × 6 m spacing is widely accepted and followed throughout the country and for grafted and high-density plantations, 3 × 3 m is practiced.

6.4.7 Propagation

Citrus can be propagated both vegetatively and through seeds. In Bhutan, exclusively, almost all the citrus planting materials are produced through seedlings.

6.4.8 Nursery

Seeds should be selected from mature, healthy, and disease-free mother plants having consistent fruiting performances. Seeds are rinsed thoroughly and planted immediately in the nursery if not they should be spread evenly across the absorbent paper away from direct sunlight. Seeds are then stored in polyethylene bags at 4–7°C after the surface moisture of the seed is dried.

The seeds are sown in line, approximately 2–3 cm away from each other. It is then covered with sand or soil and a light irrigation is given. They are transplanted to the second nursery once the seedlings give rise to 4–6 new leaves.

Plant-to-plant distance should be around 20 to 30 cm and row-to-row 45 cm in the second nursery. Only true-to-type seedlings having uniform upright, vigorous growth should be kept by uprooting the off-type seedlings. Seedlings being raised as rootstock for budding should have their side branches removed to produce a clear, straight stem. Once the tree turns a year old, it can be used for budding. They are ready for sale between 19 and 24 months old.

6.4.9 Caring of Young Trees

Application of lime or whitewashing the tree trunk is recommended to protect it from sun scalding. For the first two years of the tree, a protective covering of either straw or farm waste materials should be given during the winter since they are susceptible to frost. It should also be kept in mind that the southern side should be kept open to receive sunlight. Pruning is generally not recommended and only dead, damaged, and diseased should be removed. Irrigations should be given frequently.

6.4.10 Water Requirement and Irrigation

The mature citrus tree requires around 1,000 to 1,563 mm of water each year. In Bhutan, *3.5 mm* of water is given to younger plants. Frequent and light irrigation with shorter intervals is recommended rather than *heavy irrigation with longer rainfall.* The irrigation method commonly followed in Bhutan is hand irrigation and drip irrigation.

6.4.11 Manuring and Fertilizer Schedule

FYM and Potassium are given to the tree right before the onset of the monsoon or before the stress period. Nitrogen is applied in split doses, half of it during bed preparation and the other half when the fruit size reaches around 0.5 cm. Recommended doses of nutrient and farm yard manure are given in Tables 6.10 and 6.11 respectively.

It's not advised in Bhutan to dig a trench around the edge of the tree as it harms the small feeder roots, which in turn reduces the tree growth. Rather it is recommended for the fertilizer to be mixed with the soil by light spading or raking.

TABLE 6.10
Nutrient Recommendation for Citrus

Sl no	Age of the tree (years)	Nutrient (g/tree)		
1		N	P2O5	K20
2	Planting pit	50		
3	Year 1–3	50–100	50 from second year	50 from second year
4	Year 4–6	200–250	100–200	100–200
5	Year 7–9	300–350	200–300	200–300
6	Full Bearing	500	300–400	300–400

TABLE 6.11
Farmyard Manure Recommendation

Sl no	Age of the tree (years)	Amount of FYM (kg)
1	2–3	50
2	4–6	50–70
3	7–10	70–100

6.5 INTEGRATED PEST AND DISEASE MANAGEMENT PRACTICES

Citrus cultivation in Bhutan faces challenges from pests and diseases. Some of the major pests and diseases are as follows.

6.5.1 Citrus Leaf Miner (Figure 6.1)

Causal organism: Phyllocnistis citrella
Control measures:

- Cultural method: Pruning of late summer and autumn flush growth and handpicking of infected leaves is carried out.
- Natural method: Several small wasp parasitoids are natural predators such as *Citrostichus phyllocnistoides, Cirrospilus quadristriatus and Ageniaspis citricola.*
- Chemical Control: Dimethoate at the rate of 2 millilitres per 1 litre of water is sprayed in the nursery.

6.5.2 Citrus Greening (Figure 6.2)

Causal Organism: Candidatus Liberibacter asiaticus
Control measures: Removal of infected trees as soon as it is detected.

6.5.3 Citrus Canker (Figure 6.3)

Causal organism: Xanthomonas campestris pv. citri (Hasse) Dye
Control measures:

- Cultural method: Planting of Windbreak and vegetation among the tree rows.
- Chemical method: two to three sprays of copper during the first 90 days after petal fall.

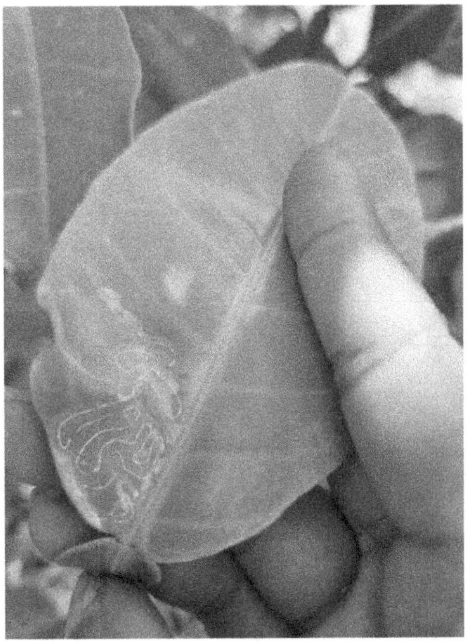

FIGURE 6.1 Leaf miner on pummelo leaf.

FIGURE 6.2 Citrus greening on pummelo leaf.

6.6 METHOD OF HARVEST

In Bhutan, Citrus fruits are harvested manually by pickers. If the tree is tall, ladders are used. When it comes to mandarins, they are cut off using clippers. The pickers are advised to have shorter nails, wear gloves, and use "twist, tilt, and pull" or preferably using clippers.

FIGURE 6.3 Citrus canker on pummelo leaf.

6.7 TRANSPORT

Before transporting, the fruits are drenched in thiabendazole and imazalol to control stem end rot. Chlorine is also added to thiabendazole at 50 ppm and 1,000 ppm respectively. This helps in controlling G. citri-aurantii, Phytophthora species. The optimum range of this solution should be maintained around 6.5–7.5.

Most of the citrus fruits are directly sent to the market after loading them on the truck without storing them in cold storage. Citrus transportation in Bhutan relies heavily on individual farmers (Joshi & Gurung, 2008). As the study indicates, 80% of respondents reported carrying their oranges to collection points (Joshi & Gurung, 2009). If the storehouse or the packing house is first of all fumigated with 1 to 3% formaldehyde if the fruits are going to be stored.

6.8 STATUS OF CITRUS INDUSTRY

6.8.1 Harvesting Window of Major Cultivars Of Citrus

- Sikkim mandarin: November to December
- Khasi mandarin: November to December.

The harvest periods for both mandarin cultivars are similar, with the difference being only a few days apart.

Usually, after being harvested, the fruits reach the market *within a week or a month* at the most. Detailed list of harvest date for Citrus are as given in Table 6.12.

6.8.2 Marketing Network of Citrus

As shown in Figure 6.4, farmers have several options for selling their produce, which are bidders, exporters, processors, middlemen, or directly to the local market. Most of the farmers who are settled in rural areas opt for selling their produce through middlemen since the other options are not readily available to them. Depending upon the market price and demand, the middleman decides

TABLE 6.12
Citrus and Their Harvest Window

Sl no	Name	Harvest period	Type
1	Okitsu Wase	November Last week	Mandarin
2	AREP-1 HK	December second week	Mandarin
3	Ohtsu-4	December First Week	Mandarin
4	Wengkhar Tshalu 1	December Third week	Mandarin
5	Yoshida Ponkhan	December third week	Mandarin
6	Wengkhar Tshalu 2	December last week	Mandarin
7	Encore	March first week	Mandarin
8	Clementine	December third week	Mandarin
9	Navelina	December last week	Orange
10	Dekopon	February last week	Tangelo
11	Tarku	December last week	Mandarin
12	Dorakha Local	December last week to January first week	Mandarin
13	Junar	Mid December to January	Orange
14	Murcott	February third week	Tangors
15	Thai Tangerine	December third week	Tangerine
16	Banpeiyu	February first week	Pummelo
17	Star Ruby	February second week	*Lemon*
18	Oroblanco	January last week	Lemon
19	Frost Eureka	December to February	Lemon
20	Mexican Lime	December to January	Lime

whether to take the produce to the local domestic market or sell it to Exporters or Bidders. The exporters and the bidders are the sole options to get the produce to the international market (especially in India and Bangladesh).

The major promotors present throughout the whole process are as follows:

- BAFRA: Quality assurance
- FCB, BEA: Management of auction yard/Export
- FCB, BCCI, RAMCO, NPHC: Postharvest technologies and training
- DAMC, MAGIP: Market infrastructure
- DOA, ARDC, NCOA, NPPC, DAO: Disease, pest control, and technical advice
- CDP, ARDC, DAO, DOA: Major value chain promoters
- ACMI: Online platform for Local market

6.8.3 Export and Import of Citrus (2014–2023)

As shown in Tables 6.13 and 6.14, both the import and export of citrus in 2020 saw a dip due to the COVID-19 pandemic outbreak. The main citrus markets are Bangladesh and India, and imports are also made from those countries. More than 95% of the citrus imported and exported is mandarin.

In terms of exports, according to the last 10 years, 91.68% of exports were made to Bangladesh and 8.31% to India as shown in Figure 6.5.

In terms of import, according to the last 10 years 99.83% imports were from India and 0.17% for other countries as shown in Figure 6.6.

Due to a lack of cold storage and infrastructure, citrus is not available year-round, and locally produced-citrus follows the seasonal market. This is the sole reason why citrus is imported from other countries.

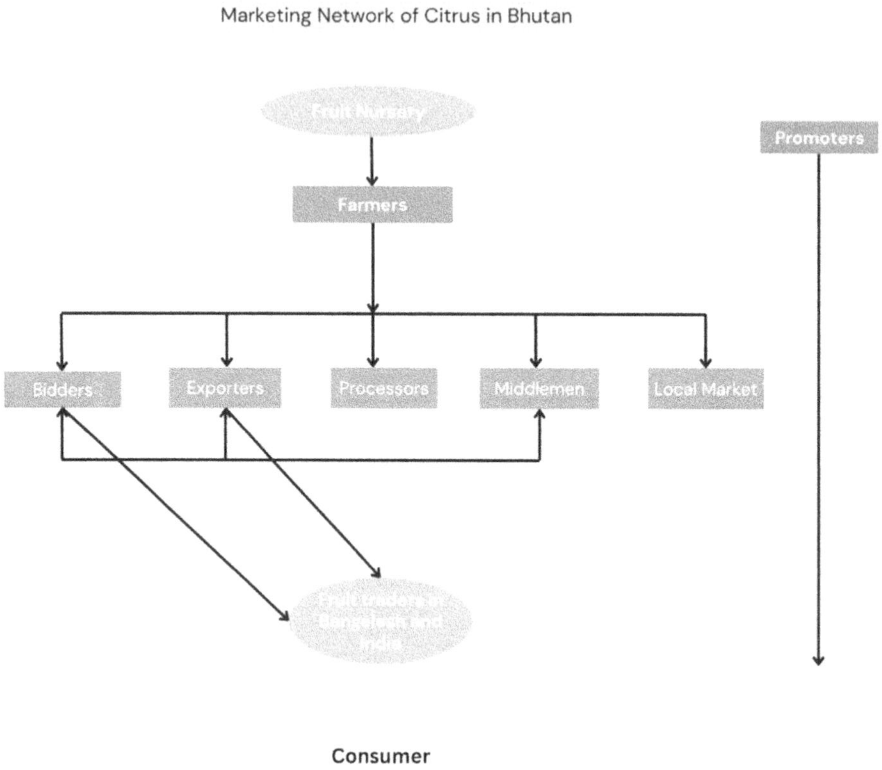

FIGURE 6.4 Marketing network of citrus in Bhutan.

TABLE 6.13
Total Citrus Exported (2014–2023)

Year	Total citrus exported (KGM)	Value (in Nu.)
2023	15,370,956	97,543,996
2022	11,209,177	418,233,622
2021	12,663,591	454,031,030
2020	8,168,528	295,528,239
2019	15,110,573	530,046,763
2018	11,086,857	354,835,369
2017	16,141,589	478,363,991
2016	15,545,690	439,082,416
2015	17,819,759	467,613,977
2014	21,148,159	542,556,550

TABLE 6.14
Total Citrus Imported (2014–2023)

Year	Total citrus imported (KGM)	Value (in Nu.)
2023	719,212	87,857,747
2022	589,629	52,121,641
2021	545,843	42,760,474
2020	232,723	15,202,055
2019	330,216	16,312,368
2018	241,404	13,302,123
2017	263,703	16,139,789
2016	202,172	14,274,231
2015	174,701	8,502,172
2014	76,046	3,212,516

EXPORT DATA

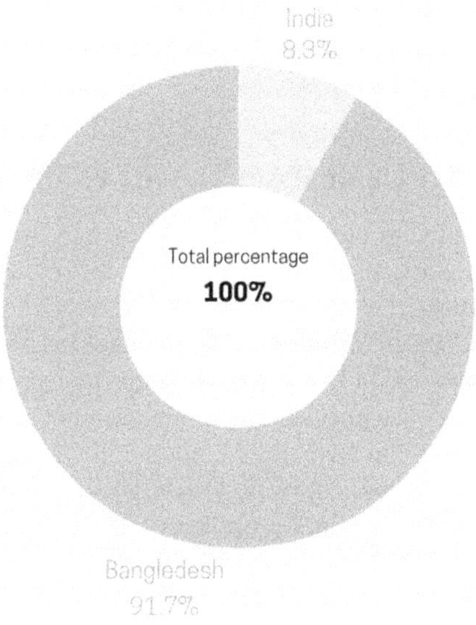

FIGURE 6.5 Export data (in %) (2014–2023).

6.8.4 Status of Citrus Processing

Due to a lack of data, it's tough to come to the conclusion of how much citrus is being processed. According to Joshi and Gurung (2008), the value addition at the producers' level stands at 45.2% of the total value addition as opposed to 16.5% and 38.2% at the contractors' and exporters' levels. Some of the major citrus products produced in Bhutan are as follows:

- Royal Bhutan Orange Juice (RTS)
- Orange squash
- Orange preserve

The agencies involved in citrus's value addition are NPHC, Bhutan Agro Industries Ltd, Bhutan Fruits Products Ltd, and other small cottage industries.

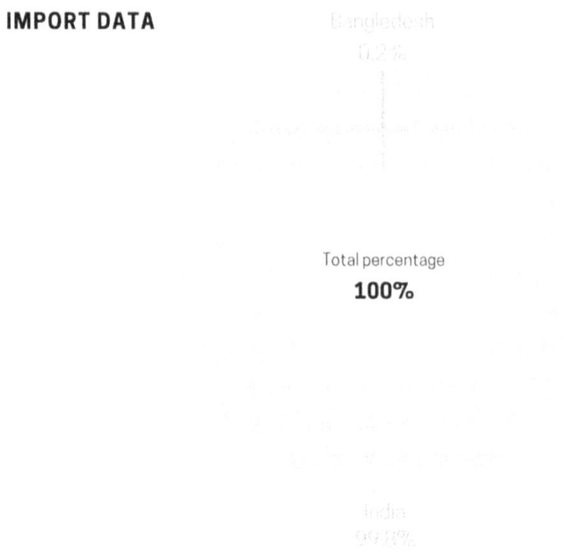

FIGURE 6.6 Import data (in %) (2014–2023).

6.9 STATUS OF CITRUS RESEARCH AND EXTENSION

Citrus being one of the major cash crops in Bhutan, has always been a subject to many researches and studies. These two papers have yielded significant results in learning more about Citrus Status and cultivation in Bhutan.

The paper titled "Adoption of Improved Citrus Orchard Management Practices and Its Impact on Yield and Household Income in Drujegang, Bhutan" by Dorji et al. (2016) gives significant insights into the adoption of improved citrus orchard management practices in Bhutan. The paper is mainly concerned with examining the adaptability, adoptability, and unforeseen potential impact of introducing new technology in citrus cultivation methods. It provided comprehensive analysis and study on pre- and post-intervention technology adoption levels, which emphasized the profound need and role of training and technology transfer. The result of this study showed significant citrus yield, which in turn boosted the overall household income.

Another significant study that was carried out was the Citrus value chain in Bhutan (Joshi & Gurung, 2008) which provided a comprehensive analysis of the industry regarding the value chain of the fruit and its major constraints. strengthening value chain linkages, enhancing market conditions, and providing necessary support services were the key findings of this paper. Suggestions included the establishment of cellar stores and processing units, encouragement of group and cooperative initiatives among growers, improvement of marketing strategies, assurance of quality standards, and advocation and raising of voice for supportive government policies.

Despite all of this, the extension of knowledge and technology transfer is still one of the major issues in the country. The main method or type of extension Bhutan follows to date is still the traditional type in which the researcher/extension personnel dictate the solution to the problem. Almost all the gewogs in the country have an extension officer posted, which serves as a bridge between the farmers and the agriculture centres and researchers. It has been observed that the extension researchers provide is one-sided with researchers coming to the conclusion and providing solutions without any farmers involvement. This results in disconnections and a low adoption rate of new technologies as Dorji et al. (2016) suggest in their paper "Adoption of Improved Citrus Orchard Management Practices and Its Impact on Yield and Household Income in Drujegang, Bhutan"

Currently, the Agriculture Research and Development Centre, Bajo, Wangdue Phodrang, and its subcenter in Tsirang, Bhutan are in the process of researching and conducting evaluation trials on Pumelo varietal release and varietal evaluation on citrus plants in different agro-ecological zone respectively.

A key area that needs researchers' input and study to strengthen the citrus industry in Bhutan is to first of all, release new varieties and species of citrus to diversify the market since Bhutan is being dominated by Mandarin. Secondly, creating a close knitted value chain for the citrus market and finally, finding suitable methods to communicate information on chemical use practices to Bhutanese farmers.

6.10 CHALLENGES AND STRATEGIES

Like any other citrus-growing country, one of the main issues faced by the citrus industry is the recent outbreak of citrus greening/Huanlongbing. Citrus greening disease had a major effect on citrus productivity in Dewathang Geog, with approximately 35% of citrus orchards being impacted by the disease. This resulted in the complete destruction of three orchards and damage to the remaining trees in partially affected orchards (Tshetrim & Chhetri, 2014). The Dewathang Geog citrus productivity report acknowledges the significant knowledge gap observed among citrus growers concerning citrus greening disease. The study revealed that a significant proportion (around 83%) of citrus growers lacked awareness of the disease and failed to address and remove affected trees, potentially allowing the disease to spread to healthy orchards (Batool et al., 2007).

Climate change is also one of the major constraints in citrus production. Extreme weather events like hailstorms or droughts have proven to damage crops and reduce yields and changes in average temperatures have impacted the ideal growing conditions for certain citrus varieties.

Another major challenge faced by growers in the country is their reliance on middlemen to sell their produce. This is because the market is not well-known and difficult to access, especially for small-scale growers.

The way forward for the citrus industry could be to research more on disease-resistant variety, dissemination information related to citrus greening for the farmers, and developing new infrastructures and technologies for a better citrus future in the country. It would also be beneficial to introduce and market new citrus varieties other than mandarin to bring diversification. Promotion of a small-scale cottage industry for value addition of citrus fruits can also be a way forward. Currently, only a few large industries dominate the value-added citrus market.

6.11 CONCLUSION

In summary, despite Citrus being one of the most exported crops in the market, its industry in Bhutan still faces both challenges and opportunities for growth and development. Recent outbreak of Citrus greening, lack of a proper market chain, lack of data, and the significant knowledge gap amongst the growers still prove to be the leading hurdles that trample the growth of the Citrus industry in Bhutan.

To strengthen and secure a brighter future for it, thorough research should be done on identifying and releasing disease-free resistant rootstock variety, communication of correct information to the growers, diversification of citrus varieties, promoting small-scale cottage industries, and improving infrastructures and market chain both within and outside the country.

6.12 ACKNOWLEDGEMENT

I extend my sincere gratitude to Dr Dilip Ghosh, Director of ICAR-CCRI and President of the Indian Society of Citriculture (ISC), for the opportunity to contribute a chapter to the upcoming book "Citrus Industry of Asia".

I am honoured and privileged to contribute my knowledge and insights on the status and way forward for the citrus industry in Bhutan.

Special gratitude to Mr. Phuntsho Wangdi, the Agriculture Research and Development Centre, Bajo, Wangdue phodrang, and its sub centre in Tsirang for providing me with the necessary inputs required to complete this manuscript.

I would also like to take this opportunity to thank ARDC, Wengkhar for giving me the consent and permission to include pictures of the major cultivars of citrus.

Lastly, heartfelt thanks to my family, friends, and my supervisor for their unwavering support and understanding during the writing process.

REFERENCES

Connellan, J., Hardy, S., Harris, A., Wangdi, P., Dorjee, D., Spohr, L., and Sanderson, G. (2008). *Results of citrus farmers survey in Bhutan*. Department of Agriculture, Ministry of Agriculture.

Dorjee, D.D., Chhetri, G., Bockel, L., and Punjabi, M. (2007). *Commodity chain analysis Citrus*. Ministry of Agriculture, Royal Government of Bhutan and FAO.

Dorji, K., Lakey, L., Chophel, S., Dorji, S.D., and Tamang, B. (2016). Adoption of improved citrus orchard management practices: A micro study from Drujegang growers, *Dagana, Bhutan. Agriculture & Food Security*, 5(3). https://doi.org/10.1186/s40066-016-0050-z

Doss, C.R. (2003). *Understanding farm-level technology adoption: Lessons learned from CIMMYT's micro surveys in eastern Africa*. CIMMYT.

Joshi, S.R., and Gurung, B.R. (2008). *Citrus in Bhutan: Value chain analysis*. Ministry of Agriculture and Forests.

Knoema. (2023). *Bhutan Citrus fruit production, 1961–2023*. Retrieved from https://knoema.com/atlas/Bhutan/topics/Agriculture/Crops-Production-Quantity-tonnes/Citrus-fruit-production

Ministry of Agriculture and Forests. (2008). *Citrus production manual*. Horticulture Division.

Ministry of Finance, Royal Government of Bhutan. (2023). *Bhutan Trade Statistics*. Retrieved from www.mof.gov.bt/wp-content/uploads/2023/04/Bhutan-Trade-Statistics-2022-.pdf

Tobgay, S. (2005). *Agriculture diversification in Bhutan*.

Tshetrim, K., and Chhetri, S.B. (2014). Factors contributing to citrus mandarin yield decline. *Bhutan Journal of Natural Resources and Development*, 1(1), 18–23.

7 Citrus Industry of Nepal
Status, Issues, and the Way Forward

Umesh Kumar Acharya and Mohan Bahadur Thapa

7.1 INTRODUCTION

Citrus is one of the important fruit crops grown since time immemorial in Nepal. In the mid-hills, citrus species rule the majority of Nepal's fruit production area. Citrus fruits are also linked to Nepalese customs and rituals, with citron (Bimiro) being an essential item used during Bhaitika, Mahapuja, and Chhat festivals. The mid-hills of Nepal, which range in elevation from 800–1500 m, have favourable agro-climatic conditions for the production of high-quality citrus fruit. Despite the fact that Nepal is home to many different citrus species, only mandarin (*Citrus reticulata* Blanco), sweet orange (*Citrus sinensis* Osbeck), and acid lime (*Citrus aurantifolia* Swingle) are produced on a large scale for commercial purposes. Hill lemon (*Citrus pseudolimon*), sometimes referred to as Nibuwa in Nepali, is a well-liked species used for the production of juice and pickles. Citrus promotes the nation's nutrition, economy, and living standards. From an economic perspective, citrus crops are four to five times more profitable than cereal crops on hill terraces and slopes (Gauchan 2000). Additionally, it offers great opportunities for domestic as well as international trade (Acharya and Shrestha 2021; Adhikari and GC 2021). The Nepali government has been paying attention to research and development of citrus fruit crops for a long time, including in past policy documents such as the Agriculture Perspective Plan (1995) and current documents like the Agriculture Development Strategy (2015).

Citrus fruits, especially mandarins, are typically eaten fresh in Nepal. Fruits are also used to make marmalade, squash, and other flavours. Of the total fruit area in Nepal, citrus accounts for 28.29% (50235 ha) of the area, standing in second position after the mango crop (Table 7.1; MoALD 2022). This shows that citrus is the main fruit grown in Nepal and has an important role in the economic prosperity of Nepali farmers. Citrus is thus a significant agricultural sub-sector in Nepal (Acharya and Adhikari 2022). However, the country's demand for citrus cannot be met by the present level of production (Dahal et al. 2020). In the past 15 years, the productivity of citrus orchards has decreased from 10.86 mt/ha in 2006/07 to 9.67 mt/ha in the 2020/21 harvest (Figure 7.1; MoALD 2022).

7.2 AGRO-CLIMATIC ZONES AND MAJOR CULTIVARS OF CITRUS IN NEPAL

Citrus is cultivated throughout the country in tropical to subtropical regions with a wide range of soils with an optimum pH of 5.5–6.5, receiving an annual rainfall of 120–180 cm, and having mild (frost-free) winters, i.e., temperatures between 5–35°C (Acharya et al. 2019). The mid-hills of Nepal, ranging from 800–1600 masl altitude all across the country, are considered favourable for all types of citrus fruit cultivation. However, pummelo, acid lime, and lemon can be cultivated successfully in the upland conditions of the Terai, inner Terai, foothills, and river basin areas of Nepal.

DOI: 10.1201/9781003584599-7

TABLE 7.1
The Citrus Fruit Production Scenario of Nepal (2015–2021)

Year	Total area (ha)	Productive area (ha)	Production (mt)	Productivity (mt/ha)
2015/16	40554	24854	218447	8.82
2016/17	46328	26759	239773	8.96
2017/18	44424	25946	245176	9.44
2018/19	46411	28406	271908	9.57
2019/20	46715	27339	274140	10.03
2020/21	50235	32188	311188	9.67

(Source: Statistical Information of Nepalese Agriculture 2020/21, MoALD 2022)

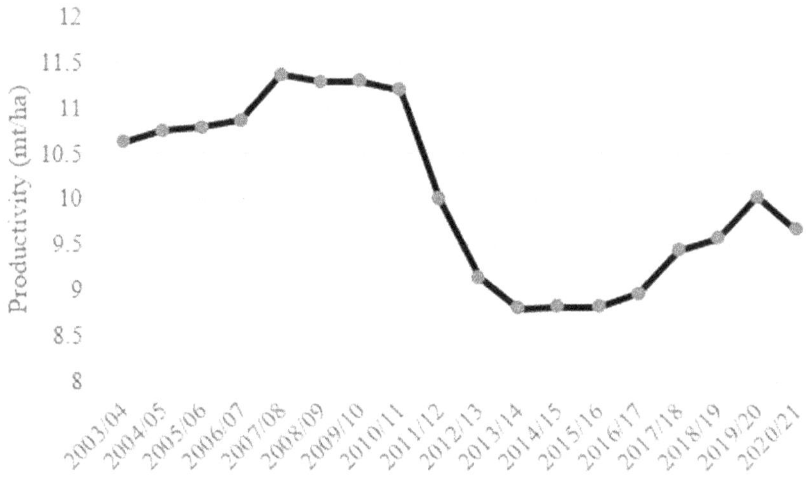

FIGURE 7.1 Change in citrus fruit productivity over 18 years period in Nepal.

Most of the citrus cultivated in Nepal is native to this region (Paudyal et al. 2016). Not only is the productivity of citrus fruits in Nepal low, but the production period is very short due to these native and heirloom varieties. A list of recommended or superior varieties under cultivation in Nepal is shown in Table 7.2.

7.3 PACKAGE OF PRACTICE FOR CITRUS CULTIVATION

The traditional techniques of cultivation of mandarin, sweet orange, lemon, and acid lime, which are adapted to Nepal's climate, have been preserved by farmers for hundreds of years. Most of the farmers are mainly cultivating mandarin fruit using these traditional techniques.

7.3.1 Varietal Development

In the early 1960s, different varieties of mandarin and sweet orange were introduced, and trials were started at the Citrus Research Sub-Centre, Pokhara, and the Citrus Research Centre, Dhankuta (Paudyal et al. 2016). Even though some early and late maturing varieties have been identified through these studies, the registration and commercialization of those varieties have not been done until now, except for the recent registration of Unshiu (Okitsuwase) in the year 2023.

TABLE 7.2
Important Varieties of Citrus Cultivated in Nepal and Their Characteristics

Crop	Variety	Characteristics	Remarks
Mandarin	Khoku local	Fruit size:110 g, bright orange colour, TSS 13%, TA1.15%, 23 t/ha yield, mid hills 1000–1400 m recommendation domain, maturity: December to February	
	Banskhark local	Fruit size:100–120 g, bright orange colour, TSS 13%, TA1.2%, 25 t/ ha yield, mid hills 1000–1400 m recommendation domain, maturity: December to February	
	Paripatle Agaute-1 (Okitsuwase)	Fruit size:100–120 g, greenish orange colour, TSS 9 %, TA0.9%, 12–15 t/ha yield, mid hills 800–1400 m recommendation domain, maturity: September to October	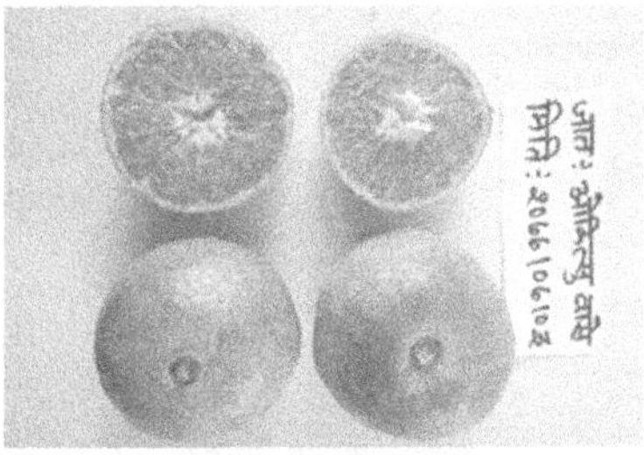
Acid lime	Sunkagati-1	Fruit size:30–40 g, greenish yellowcolour, Juice 49%, TA 7%, 34.5 t/ha yield, 100–800 m recommendation domain, maturity: July to December	
	Sunkagati-2	Fruit size:30–40 g, greenish yellow colour, Juice 49%., TA7%, 26.9 t/ha yield, 100–800 m recommendation domain, maturity: July to December	

(continued)

TABLE 7.2 (Continued)
Important Varieties of Citrus Cultivated in Nepal and Their Characteristics

Crop	Variety	Characteristics	Remarks
	Terhthum local	Fruit size:35–50 g, greenish yellowcolour, Juice 45–50%, TA7-10%, 34.5 t/ha yield, mid hills 1000–1600 m recommendation domain, maturity: October–December	
	NCRP 53	Fruit size:90–120 g, greenish yellow colour, Juice 45%, TA 6.5%, 30.5 t/ha yield, 150–1600 m recommendation domain, maturity: June to December	
Sweet orange	Sindhuli local	Fruit size:100–120 g, bright orange colour, TSS 13%, TA1.2%, 25 t/ha yield, mid hills 1000–1400 m recommendation domain, maturity: December to February	
	Washington Navel	Fruit size:160 g, bright orange colour, TSS 14%, TA1.3%, 12.5 t/ha yield, mid hills 800–1200 m recommendation domain, maturity: November to December	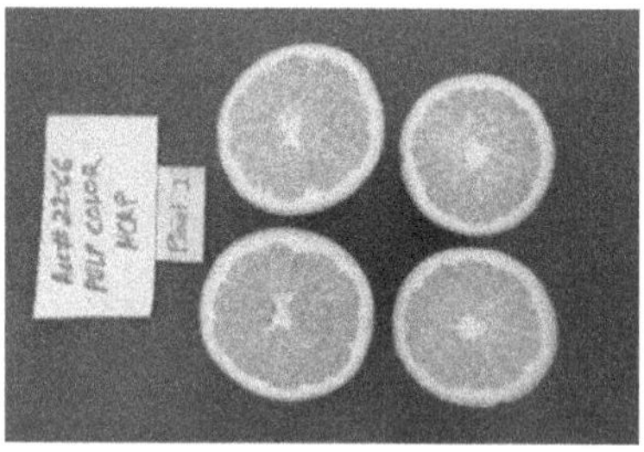
	Valencia late	Fruit size:140 g, bright orange colour, TSS 12–14%, TA1.6%, 28 t/ha yield, mid hills 800–1400 m recommendation domain, maturity: April to March	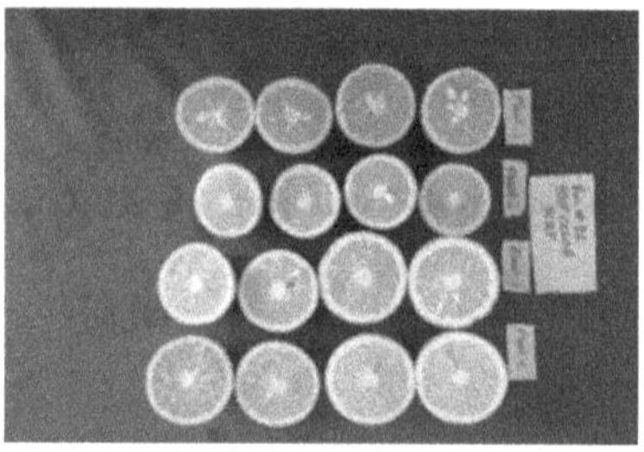

(Source: Acharya et al. 2019a)

TABLE 7.3
Indigenous and Exotic Citrus Germplasms Maintained at National Citrus Research Program (NCRP) Dhankuta and Sub-Tropical Horticulture Centre Kritipur (STHC) in Nepal

Species	NCRP Dhankuta	STHC Kritipur
Mandarin (*C reticulata*)	Bashkharka local, Sikkime, Khoku local, Gorkhalisuntala, Pogan, Kamala, Fortune, Kara, Nova, Pixie, Dancy, Avana, Page, Hernadina, Oraval, Commune, Marisol, Nules, Kinnow, Frutrel early	Clementine, Dekopongan, Frutrel early, Hayaka, Kiyomi, Kinnow, murkott, Ota pongan, Thai tangarin, Yoshida pongan.
Satsuma mandarin (*C unshiu*)	Miyagawawase, Okitsuwase, URSS, Satsumawase, Okitsu	Imamura unshu, Miyagawawase, Okitsuwase, Miyauchi Iyo, Aoshimaunshu, Otsu-4
Tangor (Tangarin X S. orange)	Ellendele, Murcott, Ortanique	
Tangalo (Tangarin X Pumelo)	Minneola, Orlando, Seminole	Orlando
Sweet orange (*C. sinensis*)	Dhankuta local, Blood red, Cara Cara, Delicious seedless, Hamlin, Lane Late, Madame Venous, Malta Blood Red, Meisheu-9, Mosambi, Navalencia, Newhall Navel, Pineapple, Ruby, Salustiana, Samauti, Sevelle Common, Skage Bonanza, Succari, Valencia late, Vanelle, Washington Navel, White Taker, Yoshida Navel, Tomango	Fukuwara-1, Fukuwara-4, Kiyomi, Nepali Junar, Malta Blood Red, New Taracco, New Valencia, Washington Navel, Yishida Navel.
Acid lime (*C aurantifolia*)	20 accessions collected from Nepal. Among them three (Sunkagati-1, Sunkagati-2, Terhthum Local) released, Mexican, Kalamondin	Local, Madrasi
Lemon	ChasmeKagati: Ureka, Panta-1 Nibuwa: Dhanuta collection, Thimura local, BRT Nibuwa, Prembasti local, Lisbon	Lisbon, Ureka
Grapefruit	Shamber, Hendeson, Star Ruby, Reed, Pink ruby	
Kumquat	Oblong type, Round type, Indian, Japanese	Oblong type, Round type
Pomelo	Nam Roi, Phultrac, Phodium	Amanatu, Ohtachibana, Banpeiyu, Kawachibankan, Thai pummelo, Local collections
Rootstocks	Trifoliateorange: Trifoliate (unknown variety), Pomeroy, Flying dragon, Citrange: Carrizo, C-35, Troyer Others: Citrumelo, Rangapur lime, Volkameriana, Rough lemon	Trifoliate orange: Trifoliate (unknown variety), Rubid, USDA Citrange: Carrizo, Troyer

(Source: Modified and adapted from Gotame et al. 2014)

In the year 2004, the National Citrus Research Program, Dhankuta started testing 31 new varieties of mandarin, sweet orange, juntala (hybrid of orange and junar) and grapefruit with an introduction from France. Paudyal and Subedi (2008) evaluated 14 sweet orange varieties of NCRP, Dhankuta for fruit quality and maturity period. Washington Navel and Nevelencia were selected for the early season (November–December), and Valencia Late, Sevelle Common and LeuGim Gong were identified for the late season (March–April) production. But these varieties have not been registered and commercialized yet. Besides these, there are many citrus varieties introduced from India and Australia that are in the testing stage (Table 7.3). Shrestha and Paudyal (2004) evaluated 32 trees (11 from hills and 21 from the Terai) of acid lime grown in farmers' fields. Nine genotypes selected from in situ characterization were further evaluated in a multi-location trial. As a result of this study, two lime varieties, namely Sun Kagati-1 (NCRP-55) and Sun Kagati-2 (NCRP-49), were released from the National Seed Board in 2015 for the Terai and river basin areas.

In 1980, the Kinnow mandarin genotype received from Pakistan was tested in various research centres, but due to the discontinuation of research, the technology related to its commercial production has not been fully developed (Paudyal et al. 2016). Limited quantities of saplings are being distributed from the National Citrus Research Programme, Dhankuta, and small quantities of fresh fruit are in production in some areas of the Terai. In 2015 and 2019, the National Citrus Research Programme, Dhankuta registered three acid lime varieties: Sunkagati 1, Sunkagti 2 (for the Terai and foothills), and Tehrathum local (for hills). Similarly, two mandarin varieties, Khoku local (in 2019) and Banskhark local (in 2021), were registered for the mid-hill region (Gotame et al. 2020). Recently, in June 2023, an early season Unshiu mandarin named Paripatle Agaute-1 (Okitsuwase) was registered for commercial cultivation in the mid-hill region. Early ripening sweet orange varieties: Washington Naval, Valencia, and late ripening varieties: Valencia Late and Lane Late were recommended by the National Citrus Research Program, Dhankuta, but their registration and commercialization have not been finalized yet (NCRP 2022).

7.3.2 Rootstock

Techniques of grafting and budding for the production of grafted plants were standardized in the early 1970s, and currently, the grafting methods are in commercial use. The seeds of trifoliate orange, Troyer citrange, and Carrizo citrange were brought from abroad and tested at NCRP Dhankuta and NCDP Kritipur in early 1980 (Table 7.3). Although these three varieties have been recommended as suitable rootstocks for citrus, only trifoliate orange has been used commercially. A recent study at mid-hill conditions using Citrummelo 4475, old trifoliate, Poncirus-Pomroy, Flying Dragon, Volcameriana, Citrange C-35, Carrizo citrange, and rough lemon as rootstock with Khoku local mandarin and Washington Navel sweet orange found Citrummelo 4475 promising in terms of yield and fruit quality. While rootstock flying dragon was found to be better for acid lime (Terhthum local) for hill farming. In the case of Terai, or plain farming, Rangapur lime was reported to be superior to acid lime varieties Sunkagati 1 and Sunkagati-2 (NCRP 2020). Due to this rootstock and better-performing scion varieties (Sunkagati 1 and 2) acid lime production has been enormously expanded in the terai and inner terai regions of Nepal (Acharya et al. 2021b).

7.3.3 Nursery Management

Technology for quality sapling production with disease-free mother plants, rootstocks, and grafted plants inside screen houses has been standardized (Paudyal et al. 2016). Nursery owners have started using this technology recently. But there are still many aspects that need to be improved. There is also a situation where unnecessary expenses are being spent in the name of high-tech or semi-high-tech nurseries, especially in the climate of Nepal, where simple 40-mesh screen houses do not allow insects to penetrate.

The usual time of citrus grafting in the mid-hills of Nepal is the 3rd week of December to February using Veneer or side grafting techniques with more than 85% success rate (NCRP 2022). Due to this limited window of grafting, the quantity of sapling production has been limited. A three-year study at NCRP Dhankuta for an alternative time for grafting found that Venner and side grafting in May to July also gives a good success rate of 85–90% under containerized nursery conditions in the mid-hills. This technique is especially suitable for acid lime and sweet orange.

7.3.4 Nutrient Management

Even though there are limited studies on soil fertility status of citrus orchard and leaf tissue analysis in the past, some past research experience and a couple of similar studies from neighbouring countries helped to develop a nutrient management Sunkagati-1 guide. A complete package of fertilizer

TABLE 7.4
Fertilizer Recommendation for Traditionally Grown Seedling Mandarin and Okitsuwase Trees under High-Density Planting

Crop	Growth stage	Fertilizer recommendation
Mandarin (landraces)	10–20 years	FYM 100 kg, Urea 500 g, DAP 250, MoP 500 g, Boric acid 20 g, Zinc sulphate 150 g, Copper sulphate 75 g, Magnesium sulphate 75 g/ tree
Mandarin (landraces)	21–40 years tree	FYM 100 kg, Urea 400 g, DAP 200, MoP 400 g, Boric acid 20 g, Zinc sulphate 150 g, Copper sulphate 75 g, Magnesium sulphate 75 g/ tree
Mandarin (landraces)	>40 years tree	FYM 100 kg, Urea 500 g, DAP 250, MoP 500 g, Boric acid 20 g, Zinc sulphate 150 g, Copper sulphate 75 g, Magnesium sulphate 75 g/ tree
Mandarin (Okitsuwase)	High-Density Planting (3 m X 2.5 m)	50 kg FYM + 400 gm Urea + 200 gm DAP + 400 gm MOP + soil application of combined micronutrient (Zn, B, Fe, Mn, and Mo) @125 gm + agri-lime @ 125 gm

Sunkagati-2 recommendation has been developed by NCRP for different citrus crops (Acharya et al. 2021a). However, the use of such recommendation is seldom used by citrus orchardist which is one of the reasons for poor citrus crop productivity in Nepal. Recently, fertilizer trial conducted on senile orchards have made a recommendation for rejuvenation of such orchards (NCRP 2019). The age group and recommendation is as shown in Table 7.4.

7.3.5 Insect Management

Blue beetle, black aphid, brown aphid, red mites, lemon butterfly, stem borer, bark eating caterpillar, green stink bug, Chinese fruit fly, oriental fruit fly, citrus psylla, leaf minors, whiteflies, and scale insects are common citrus insect pests in Nepal (FAO 2011). Research activities initiated after the establishment of NCRP at Dhankuta on insect management include (i) insecticidal trial on leaf minor and (ii) fruit fly control programme district (Paudyal et al. 2016). From a recent survey, Chinese fruit fly, citrus psylla, scale insect, and leaf minors are identified as important pest threats to citrus trees (Adhikari and Acharya 2022). The fruit drop previously thought induced by infestation of oriental fruit fly (*Bactrocera dorsalis*) during September–October was later identified as caused by Chinese Citrus fly (*Bactrocera minax*). The methyl eugenolpheromone previously used to trap male of *B. dorsalis* was ineffective to entrap *B minax* and a protocol to trap female fly using hydrolyze protein has been developed (Acharya and Adhikari 2019). The life cycle of *Bactrocera. minax* in Nepali citrus orchard has been thoroughly studied (Figure 7.2) and the Area wide control programme (AWCP) has been developed and in use in many parts of the country (Adhikari et al. 2022).

The citrus leaf minor (*Phyllocnistis citrella* Stainton) is a potentially serious pest of citrus nursery and it also affects flushes of even grown trees. The previously recommended pesticide Dimethoate (Rogar) was found ineffective and a number of new insecticides were studied to control it under citrus nursery condition. Thiamethoxam (2 gm/lt) and Profenofos (2 ml/lt) were found effective in reducing minor population for 14 days after application (NCRP 2021).

7.3.6 Disease Management

Citrus diseases that are commonly prevalent in Nepal are powdery mildew, wither tip/Anthracnose, damping-off, root rot, foot rot, Gummosis (*Phytophthora* spp.), pink disease, citrus canker, sooty mould, citrus greening, green/blue mould, citrus scab, and tristeza virus (NCDP 2014). These pest management recommendations have been made in the past except citrus greening. Some of the pesticide

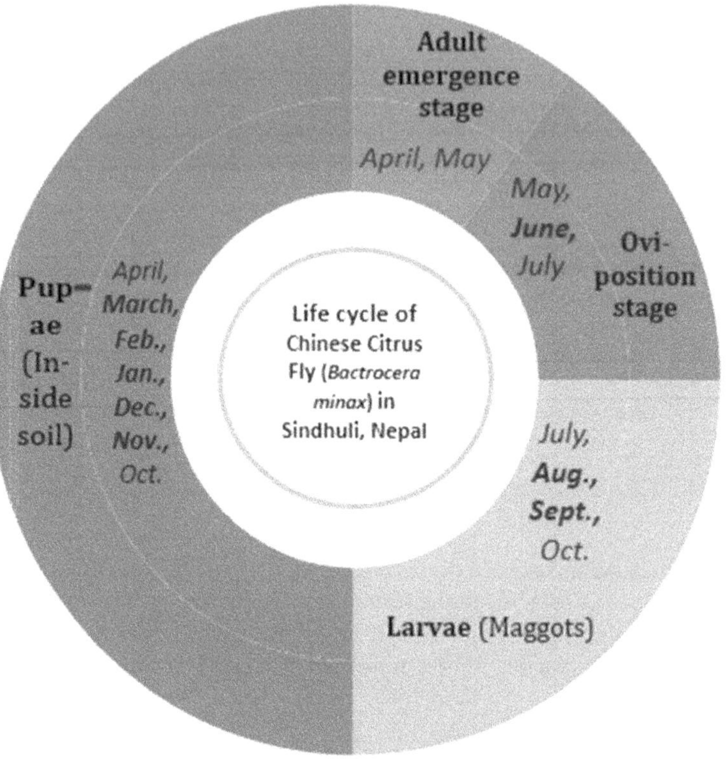

FIGURE 7.2 Life cycle of Chinese fruit fly (*Bactrocera minax*) in Sindhuli district, Nepal. (Adhikari et al. 2018).

recommended before got banned in use and others showed low level of protection and research are focusing on finding good management options (Adhikari and Acharya 2022). A recent study on root and foot rot of acid lime in Terai found that bio-control agent *Pseudomonas fluroscens* @10/ lt as organic measure to control root rot (Gummosis) and 1% Bordeaux mixture or Copper-oxichloride (4 g/lt) as inorganic measure. These solutions should be applied three times a year at 10 litre per tree in January– February, one month after first application and before the rainy season in June (NCRP 2021). For management of HLB disease caused by Asian Citrus Psylla an integrated approach of production of disease-free sapling inside screen house and distribution; identification of diseased plant using PCR method and destroying those positive trees are the approach currently followed in Nepal (Acharya et al. 2019). There was post-harvest decay of up to 25% in mandarin due to green (*Penicillium digitatum*) and blue mould (*Penicilliumitalicum*) in cellar storage and a post-harvest dip of 2–4% calcium chloride for 5 minute has been recommended immediately after harvest of citrus fruit before sending it to cellar and cool rooms with loss up to 8% by the mould (Acharya et al. 2021b).

7.3.7 Weed Management

Weed management is a serious issue in nursery as well as commercial orchards during summer-rainy season. A weed management study under acid lime nursery at NCRP Dhankuta found that Pendimethalin @ 2 ml/lt as pre-emergence and post-emergence was very effective to control weed up to 60 days after spraying (NCRP 2021). Weedicide use is not common in Nepali citrus orchards, rather hand weeding and use of organic mulch is preferred. Recently a few orchardists have been using silver colour reflective plastic mulch. Use of weedicide is not common practice and hand weeding is a preferred method to control weed in citrus orchards (NCRP 2022).

7.3.8 Post-Harvest

In Nepal the post-harvest losses of citrus fruit is around 32% (Khatiwada et al. 2019). A poor pre- and post-harvest handling practice is the main culprit for this loss. Citrus orchardists sell their fresh produce depending upon market demand during Dashin and Tihar festival in September and October months. The eating quality of mandarin and sweet orange fruit are not desirable during this time. Hence, a study has been conducted by NCRP, Dhankuta to find out proper time of harvesting and altitude for quality mandarin and sweet orange production. The study on native mandarin variety under mid hill condition found that the quality of fruit in term of size, juice content, TSS and TA would be much better if grown at the range of 1150–1350 m altitude and it is the best to harvest them after the second week of December (Acharya and Thapa 2021; NCRP 2020). Another similar study on native sweet orange (Junar) in Sindhuli district revealed that 1050–1250 m altitude is the best range in terms of fruit weight, TSS and TA%. The maturity will commence as early as the second week of December with better taste at this range (NCRP 2020). Based on peel colour change Bhusal (2002) recommended that fruits harvested by clipping scissors at 26–50% yellow colour stage to store in underground cellar storage for 3 months. There is a recent recommendation on use of hand-held small agriculture tools to facilitate picking of mandarin and sweet orange fruits. Secateurs, and cut and hold type of manual harvester are found suitable for hilly terrain orchard system which reduces drudgery as well as increase storage shelf life of produce (Khatri et al. 2021, 2022).

Further, there was a lack of a proper grading system for citrus fruit in the past. The size and maturity-based grading system of citrus exists now. Grading of mandarin fruits is carried out based on fruit diameter into extra-large (68–77 mm), large (60–66 mm), medium (55–62 mm) and small sizes (51–56 mm) (Thapa and Shrestha 2000). Similarly, PHLRD (2003) also developed a fruit size based sweet orange grading system in Nepal. Large (76–80 mm; 250 g), medium (71–75 mm; 190 g) and small (66–70 mm; 150 g) are the three grade system in use (Acharya et al. 2021b). Based on various studies on harvesting, packaging, and storage of mandarin, Paudel et al. (2004) recommended modified cellar storage with a precooling chamber to store mandarin fruits up to 120 days treated with garlic extract solution. The fruits harvested at 26–50% yellow stage had minimum weight loss and rotting percentage along with good taste and higher degree of overall acceptability after the storage of 105 days (Bhusal et al. 2007). Bastakoti and Gautam (2007) also conducted storage studies on mandarin in a modified cellar store. They noted that different maturity stages influenced the storability and quality of mandarin fruits. Dipping the fruit in wax (10%) emulsion minimizes weight loss and dipping waxed fruit in Benomyl (0.1%) reduces rotting percentage (Bastakoti and Gautam 2007). However, use of Benomyl has been discontinued in Nepal. A recent study from NCRP (2021) found native Nepali mandarin and sweet orange treated with 10% paraffin wax or commercial wax could last up to three months in a cool bot installed cold chamber. Some of the citrus varieties attain internal maturity while peel colour is green. Guatam et al. (2004) recommended dipping fruits in 100–200 ppm ethylene solution to develop a yellow colour in the peel of mandarin.

The storage of citrus fruit for commercial purpose is a recent practice in Nepal. Couple of ware houses tried it without knowing the post-harvest physiology and incurred more that 50% rotting loss. With more than two years of study on storage of acid lime, sweet orange, and mandarin in various kinds of storage structure at NCRP Dhankuta found that native citrus are sensitive to chilling injuries if storage temperature is below 6^{0}C. The congenial temperature for native citrus storage is 8–12^{0}C with relative humidity of more than 85% (Acharya et al. 2021b; NCRP 2021). The findings from these studies suggested that citrus fruit inside plastic crates could be stored for up to three months in the winter season if treated with wax and/or $CaCl_2$ (4%) stored inside a cold storage facility. The acid lime, however; could be stored for only two months inside such a structure in summer months. There is more than 25% citrus fruit decay due to green and blue mould in cellar and cold storage. A study carried out at NCRP Dhankuta to control decay loss with pre-harvest spray on tree mandarin fruit found the decay loss was as low as 12% when sodium bicarbonate was sprayed at 60, 45 and 30 days before harvesting (NCRP 2018). Later it was found that mandarin and sweet orange fruit dipping immediately after harvest in 4% $CaCl_2$ solution for 5 minute before storing to control both blue and green mould disease (Acharya et al. 2021b). The storage study using modified storage packaging found that 20 micron LDPE plastic bag with 16 holes (5 mm diameter) could be used to store 2-kg acid lime for 90 days under normal storage and 120 days under cold storage facility with 12^0C and 95% RH (NCRP 2021; NHRC 2021).

7.4 STATUS OF CITRUS INDUSTRY

7.4.1 Harvesting Window Of Major Citrus Cultivars

The harvesting window of most of heirloom cultivars is very narrow. However, with varietal development and introduction of exotic cultivars the production season has been expanded.

7.4.1.1 Mandarin

Native varieties of mandarin are mostly cultivated in Nepal. Recently, mandarin cultivars Khoku local and Banskhark local have been registered for commercial cultivation (NCRP 2020). These are also selected from locally cultivated species. The harvesting period of all local varieties cultivated in Nepal is from November to February only (Table 7.5). From the beginning of January, the market shortage of mandarin begins to decrease due to Nepali produce in the market (Paudyal et al. 2016). Kinnow oranges produced in India's Punjab province start entering the Nepali market from January. From January onwards, the import of Indian Kinnow increases rapidly and continues till the month of March. Along with Kinnow, oranges produced in Maharashtra, Indian mandarin known as "Nagpure oranges", are also imported in large quantities. Nagpur mandarin are imported from Nagpur during Dashain and Dipawali festival in Nepal i.e. September–October (AmbeBahar) and after the end of Nepal's production season of mandarin in February–March (MrigBahar). Recently, NCRP has registered a new mandarin cultivar for September– October production named as ParipatleAgaute Suntala-1 (Table 7.2).

7.4.1.2 Sweet Orange (Junar)

Junar is cultivated mainly in Sindhuli and Ramechhap in Nepal. Junar trees planted in the eastern hilly areas have mostly been removed by the farmers due to the outbreak of fruit fly (*Bactrocera minax*) (Paudyal et al. 2016). Junar fruit produced in Nepal is available in the market from November to February (Table 7.6). Since the production is low, even at this time Nepali junar does not come much in big markets including Kathmandu. As a result, Sweet oranges (Mausam) imported from South India's Andhra Pradesh and Maharashtra supplies the Nepali market throughout the year.

Sweet orange is also produced in the Dadeldhura district of Nepal and harvested in March–April, even though the production area is limited. In this area, the sweet orange is called Mausam. The segment of this variety separates easily compared to Sindhuli junar. During the main harvesting

TABLE 7.5
Mandarin Cultivars Produced in Nepal and Their Production Season

Cultivars	April	May	June	July	August	Sept	Oct	Nov	Dec	Jan	Feb	March
Khoku/Banskhark												
Paripatle Agaute-1												
Murcott												

Note: The shaded part in this table shows month of fruit harvest in Nepal.

TABLE 7.6
Sweet Orange Cultivars Produced in Nepal and Their Production Season

Cultivars	April	May	June	July	August	Sept	Oct	Nov	Dec	Jan	Feb	March
Landraces (Sindhuli/ Dadeldhura)								■	■	■	■	
Washington Navel						■	■	■				
Valencia Late											■	■

Note: The shaded part in this table shows month of fruit harvest in Nepal.

TABLE 7.7
Acid Lime Cultivars Produced in Nepal and Their Production Season

Cultivars	April	May	June	July	August	Sept	Oct	Nov	Dec	Jan	Feb	March
NCRP-53		■	■	■	■	■	■	■	■			
Sunkagati-1/2				■	■	■	■	■	■			
Landraces (Terthum Local)							■	■	■			

Note: The shaded part in this table shows month of fruit harvest in Nepal.

season of mandarin, the sweet orange is not sold, hence the farmers keep the orange fruit on the tree until late March and this is called on tree-storage. The junar grown here is not a late ripening variety. It seems that sweet orange is cultivated at a higher altitude (1400–1700 m) in Dadeldhura, which makes it possible to harvest late.

7.4.1.3 Acid Lime/Lemon

In the mid-hill region of Nepal, acid lime and lemon production takes place during the period of October to December (Table 7.7). Until now, Nepali lime was produced only in the mid hills. Recently, Sunkagati 1 and Sunkagati 2 varieties of acid limes and their cultivation techniques suitable for the Terai and foot hills have been developed by NCRP (Paudyal et al. 2016). In the Terai (plains), these cultivars can produce fruit from July to early December. The commercialization of these cultivars is taking place rapidly in the Terai areas. Although it is possible to produce lime from May to December due to the climate of the hills and plains with the use of current varieties, Indian Madrasi acid lime is imported to the Nepali market throughout the year as the current Nepali production is very low. According to the statistics of the Kalimati Fruit and Vegetable Development Committee, about 98 percent of the lime and lemons sold and distributed annually from this market are of Indian origin.

7.4.2 Marketing Network of Citrus

The marketing network existing in the citrus fruit supply chain in Nepal is shown in Figure 7.3. There are four marketing channels reported working in the citrus industry (Pokharel 2019; Baral et al. 2021). Channel I is the major supply chain operating in Nepal, with a 90.7% market share. In place of pre-harvest contractors in Channel I, a new actor, fruit producers' co-operatives, is also working as fruit aggregators and wholesaler in the market.

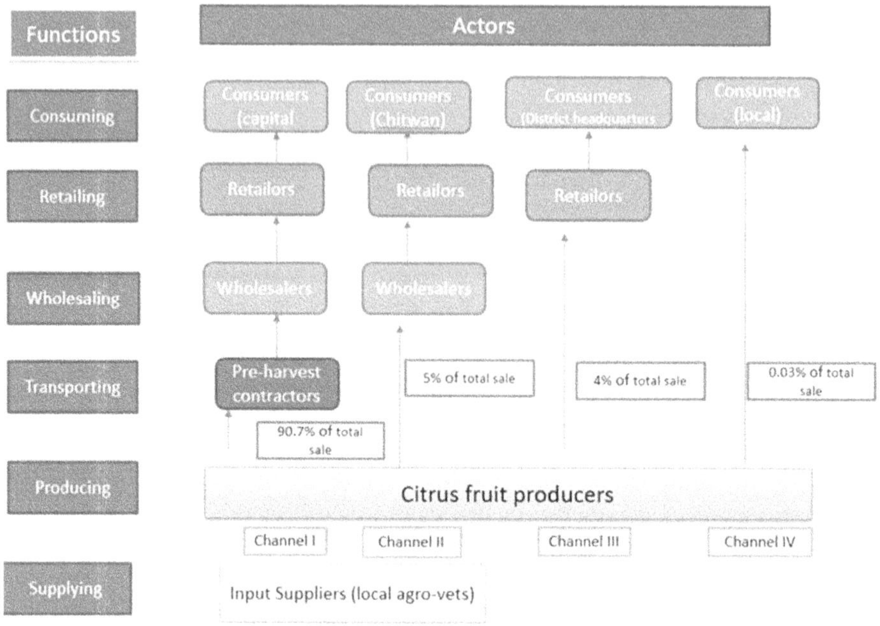

FIGURE 7.3 Market supply chain of citrus fruit in Nepal. (Modified and adapted from Pokharel 2019).

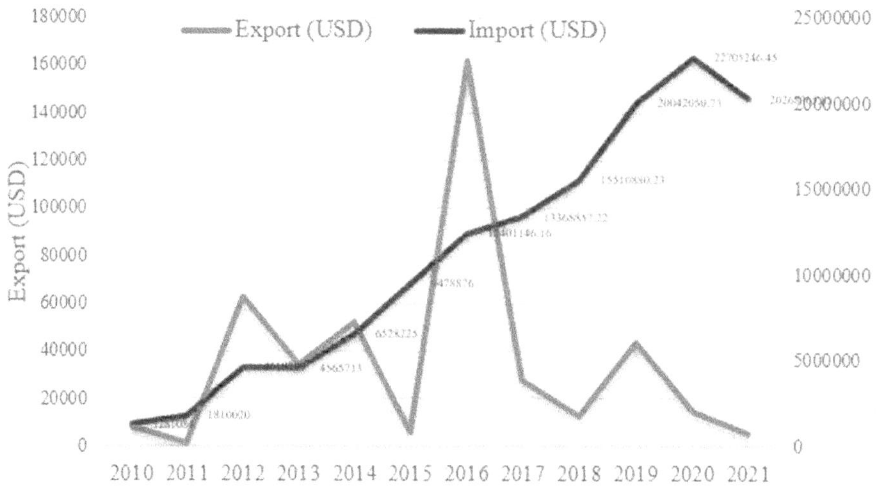

FIGURE 7.4 Import (Y-axis in right; value in USD) and export (Y-axis in left) of citrus fruit (fresh and dried HS code 805) in Nepal. (https://trendeconomy.com/data/h2/Nepal/0805).

7.4.3 Import and Export Status of Citrus

The internal production of citrus fruit does not meet domestic demand. Hence, Nepal imports a large quantity of fruit from abroad (Figure 7.4). The value of imports of citrus fruit totalled 20 million USD in the year 2021. India, Egypt, South Africa, and Uganda are major exporting country to Nepal. Among them India accounts for 97% export, Egypt 2.33%, and rest from South Africa and Uganda in year 2021. Although the production is not enough, Nepal also exports its fresh citrus fruit to India and Bangladesh because of the high quality of the heirloom mandarin cultivar. The top export destination of Nepali citrus fruit is India (99%), followed by Bangladesh. Among the citrus fruit, orange and mandarin account for 56% (11.4 million US$), lemon and lime account for 39% (8.1 million US$), and other citrus fruits, including grapefruit, account for 5% (715 thousand US$) of the total imports to Nepal in the year 2021. There was a decrease of 10.7% imports in the year 2021 as compared to the previous year. Similarly, there was a decrease of 65% in export in the year 2021 as compared to the year 2020 in Nepal.

7.4.4 Citrus Processing Industries

Processed products from citrus fruits range from semi-processed pulp to jam, jelly, squash, nectar, marmalade, and various other products, including industrial as well as homemade kitchen edibles like pickles. Pulp is the intermediate product that is used to produce nectar, squash etc. round the year. It can be preserved for a long time to produce other products, even during the off-season. Wine (cider) is also produced from citrus fruits particularly in hilly regions. The availability of raw materials in both forms i.e. in the fresh fruit form as well as in the semi-processed pulp form cannot fulfil the requirements for processing and manufacturing industries of the country. While the availability of fresh fruits from domestic sources has been in short supply, the production of semi-processed products i.e. pulp to produce final products, is almost non-existent in Nepal (Thapa et al. 2013). However, the current trade statistics from the Department of Custom, Government of Nepal show Nepal exported USD 6.4 million worth of juice, dried products and peel to India, Germany, the USA and Bangladesh in year 2021/22. Almost all the big fruit industries of Nepal are entirely dependent on foreign sources to import pulp and run their production.

The processing industries in Nepal are as below:

- RijalTashi Industries Pvt. Ltd. Itahari, Sunsari
- J B Industries, Pvt. Ltd. Itahari, Sunsari
- Dabur Nepal, Pvt. Ltd. Kathmandu
- Choudhary Group, Sanepa, Lalitpur
- Kabita Fruit Processing Industry, Sallery, Dhankuta
- Dugar Group Industry, Jamal, Kathmandu
- Shiva Baba Juice and Fruit Industry Pvt. Ltd. Birtamod, Jhapa
- Himalayan Food Beverage, Kumaripati, Lalitpur

7.5 RESEARCH AND EXTENSION

Although citrus is a historical fruit in Nepal, systematic research and development of citrus was initiated in 1956, along with the development of periodic plans in Nepal. Citrus R & D was started by establishing a Citrus Research Station in Dhankuta and a Sub-station in Pokhara in 1961 with the support of the Indian Aid Mission (Paudyal et al. 2016). In 1966, the names of Citrus Research Station, Dhankuta, and Sub-station, Pokhara, were changed to Agriculture and Horticulture Research Station, respectively. Both stations were under the Department of Agriculture (DoA).

From a developmental perspective, the National Citrus Development Program (NCDP) was founded in Pokhara in 1972 and relocated to Dhankuta in 1974, giving full authority to the nation's citrus research and development. The Mid-Western Region's Horticulture Research Station, Dailekh, was established in 1977 with a primary focus on citrus in that region. Thus, the major citrus Research and Development centres of Dhankuta, Pokhara, and Dailekh stations were identified. Later, Sindhuli, Palpa, and Dhunibesi Horticulture Farms were also included in the Citrus

Development Program at that time. After the establishment of the Nepal Agricultural Research Council (NARC) in 1991, Dhankuta Agriculture Research Station (Horticulture) was handed over to it and converted into Agriculture Research Station (Horticulture) Dhankuta. The station was later recognized as the National Citrus Research Program (NCRP) as of July 2000. From Dhankuta, the NCDP office was relocated to Kathmandu and brought under the Department of Agriculture in 1993. The Citrus Development Centre, Palpa, was established in 2003, replacing the Horticulture Centre, Palpa.

The Horticulture Development Project (HDP) was started with the support of the Japanese government from 1985 to 1997 with the objectives of increasing the production of citrus (mandarin and sweet orange) and other subtropical and temperate fruits in the mid-hill region of Nepal. At the Kirtipur Horticulture Centre, some exotic citrus species and varieties were imported from Japan and planted in evaluation blocks (Gotame et al. 2014). Studies on rootstock evaluation, cultivation methods, evaluation of local landraces (mandarin, sweet orange, and pummelo), harvesting methods, and storage methods were conducted (Tomiyasu et al. 1997). Other activities conducted by this project during that time were the establishment of demonstration farms, long- and short-term training for JT/JTA and farmers.

The two government organizations solely engaged in citrus research and extension at this time are the National Citrus Research Program (NCRP) under NARC and the National Fruit Development Centre, which was the former National Citrus Development Program (NCDP) under the Department of Agriculture (DOA). At the moment, NCRP's main responsibilities include the collection and conservation of germplasm, variety selection to extend the growing season and improve quality and productivity, nursery management, soil and nutrient management, Chinese citrus fly and other insect control, technology dissemination to farmers' fields, and national coordination of citrus research activities. Six other research stations across the country are helping with collaborative technology development in coordination with the NCRP. The extension of technology is coordinated by the Department of Agriculture and performed under the National Fruit Development Centre (NFDC) at the national level and by the Agriculture Knowledge Centre present in various districts at the provincial level. The primary function of NFDC is to offer regular and special programmes to citrus growers as extension services. Either directly by NFDC (central-level programme) or through Agricultural Knowledge Centre (AKC) offices, regular programmes are carried out. Regular central-level programmes on citrus fruit crops include study and observation, monitoring and supervision, fruit exhibitions, nursery and commercial orchard competitions at the national level, technical support for districts, collection and updating of citrus statistics and information at the national level, and assistance to the government in developing and implementing citrus policies. Regular extension initiatives carried out by AKCs include training, tool distribution, problem-solving campaigns, cellar store establishments, orchard management demonstrations, nursery and commercial gardening establishments, and technical and financial support to farmers for these activities. Further, the agriculture section of municipalities and rural municipalities of local governments are also responsible for technology dissemination. Additionally, there are zones/super-zones in 23 districts under the Prime-minister Agriculture Modernization Project that are dedicated to citrus fruit production and marketing technology extension. There are 900 citrus production pockets across the country that are taken care of by all of the above institutions. The extension programme is mostly concentrated on financial subsidies for the construction of a nursery and screen house, the purchase of seedlings, transportation, crop insurance, the purchase of crates, and the construction of a collection centre, pack house, and cold store. Other production materials (fertilizer, plastic mulch) are also available with financial subsidies to a limited number of orchardists. Additionally, programmes like training and visits to increase the technical capacity of farmers and the establishment of model citrus orchards are also underway.

7.6 CHALLENGES AND STRATEGIES

The government of Nepal has recognized mandarin and sweet oranges as potential export commodities, taking the place of an initiative for exporting sweet orange in Tibet (Adhikari and GC 2021). Nevertheless, the citrus industry is still facing several important problems: traditional practices for crop management, short production season of existing varieties; declining soil fertility and water resources; citrus greening and fruit fly; poor quality and small production scale; poor infrastructures; and legal and institutional mechanism for marketing; and lack of entrepreneurship for this crop.

The majority of farmers are small-scale producers characterized by small land holdings with low investment and risk-bearing capacity (FAO 2011). This is the major reason for poor crop management, which requires a high level of external inputs, high skills, and good crop management knowledge, which are not within the capacity of most farmers. There is a serious shortcoming in crop husbandry practices in most citrus orchards, like manuring, micronutrient management, training and pruning, and disease and pest control, among others. As a result, many orchards are in declining states. Additionally, most farmers have limited access to certified planting materials free of diseases including Phytophthora root rot, citrus greening, Canker, and Tristeza virus (FAO 2011). Similarly, there is a lack of varietal diversity for extending the production season at farmer's fields (Acharya and Pakka 2018). Therefore, the production of existing varieties is limited to a very short period during the normal season. As a result, Nepal imports mandarin, sweet orange, and acid lime worth more than two hundred million annually (Figure 7.4; MoALD 2019). Poor fruit quality due to insect pests and diseases as well as poor orchard management, and physical damage during harvest and transport are some of the important aspects to be considered for the export business in the future (Adhikari and GC 2021).

The citrus industry is still at a traditional level that needs to be transformed into commercial production. Therefore, future strategies should focus on addressing the problems of short production periods for existing varieties, low productivity and production, inferior fruit quality, and citrus decline due to disease and pests, including management factors. Similarly, problems in institutional mechanisms and coordination for marketing and entrepreneurship for this crop should be adequately dealt with by research and development. Moreover, the research focus shall be on citrus-based farming systems utilizing available resources and the socio-economic condition of the farmers.

Therefore, Nepal should prioritize the following research areas for the upcoming years:

- Breeding new varieties for extended harvest period and post-harvest technique and value addition.
- Virus indexing programme should be made compulsory by law with a bud wood certification programme, and it should be followed promptly across the citrus growing areas.
- Quality planting materials (in-vitro technique) free from pathogens and resistant to various insect pests and diseases ought to be made available to the citrus growers.
- The private nurseries should be inspected routinely since the uncertified nursery plants produced from a bud wood of an unknown mother trees decide the future of the orchard.
- Developing disease-resistant rootstock as well as identifying new dwarfing rootstocks for high-density planting.
- Excessive use of fertilizers and chemical pesticides should be checked, and organic citrus farming should be encouraged, especially with the judicious use of bio-fertilizers and bio-control of pests with bio-pesticides.

7.7 CONCLUSIONS

Currently, the Nepal government has acknowledged the significance of the citrus industry and has given it priority status. Mandarin and sweet orange cultivation is widespread in the mid-hills and

Terai regions across the country due to favourable geography and climate. Additionally, acid lime can be grown in the upland areas of the Terai region. The demand for mandarin and acid lime in the domestic market has been steadily increasing in recent years. Consequently, there is immense potential for generating income, employment, and improving nutrition for rural farmers in Nepal. However, the domestic production meets only a small portion of the national demand during the main season, resulting in the annual importation of fresh and processed citrus worth millions of rupees. Therefore, Nepal holds significant potential for commercializing the citrus sector to reduce imports and promote exports. These circumstances create various areas for research and development, encompassing variety development, tree health management, integrated soil management, plant protection, post-harvest practices, processing, and marketing. However, due to limited institutional and technological capabilities, regional collaboration with institutions such as the ICAR/ Central Citrus Research Institute in India and the Asian Food and Agriculture Cooperation Initiatives in Korea could contribute to the development of the citrus sector in this region. Ultimately, such collaborations may facilitate the transformation of the citrus sector into a commercially viable and export-oriented industry.

REFERENCES

Acharya, U. K., and Adhikari, D. R. (2019). Chinese fruit fly (*Bactrocera minax*) management in mid Hills of Nepal. *The Journal of Agriculture and Environment* 20: 47–56.

Acharya, U. K., and Adhikari, D. (2022). *Emerging pest threats of citrus in Nepal*. Paper presented at 2nd International Horticulture Conference, Nepalese Horticulture Society (Unpublished).

Acharya, U. K., Adhikari, S., and Adhikari, D. (2021b). *Post-harvest technology of fruit crops*. National Citrus Research Program, Dhankuta, Nepal (Booklet in Nepali language).

Acharya, U. K., Karki, S., Paudel, K. P., and Pakka, R. (2019b). Huanglongbing (Citrus greening) severity and its management strategies in Nepal. *Proceedings of 10th National Hort Seminar 2019*. Nepal Horticulture Society. 155–160.

Acharya, U. K., and Pakka, R. (2018). Evaluation of mandarin (*Citrus reticulata* Blanco.) cultivars to expand production in eastern hills of Nepal. *Nepalese Horticulture* 13: 34–53.

Acharya, U. K., Pakka, R., Adhikari, D. R., and Joshi, S. L. (2019a). *Citrus fruit production technology*. National Citrus Research Program, Dhankuta, Nepal (Booklet in Nepali language). 1–87.

Acharya, U. K., Paudyal, K. P., and Adhikari, S. (2021a). A journey of acid lime from hills to Terai: The woes and gains. *Proceedings of 12th National Horticulture Seminar, Kathmandu, Nepal*. Nepalese Horticulture Society. 117–122.

Acharya, U. K., and Shrestha, H. K. (2021). Opportunity and Challenges of Sweet Orange (*Citrus sinensis* L., Osbeck) Production in Sindhuli and Ramechhap. *Nepalese Horticulture* 15: 89–96.

Acharya, U. K., and Thapa, B. (2021). Effect of orchard altitude on physical and chemical quality of mandarin fruit at Sindhuli district. In *Proceedings of the Tenth National Horticulture Workshop, February 28–March 1, 2021* (Falgun 16-17, 2077). NARC, NHRC, Khumaltar, Lalitpur, Nepal. 73–77.

Adhikari, D., and GC, Y. D. (2021). Opportunity to export citrus from Nepal to China: Activities accomplished on Plant Quarantine concerned. *International Journal of Agriculture Innovations and Research* 8(5): 438–444.

Adhikari, D., Thapa, R. B., Joshi, S. L., Du, J. J., and Tiwari S. (2022). Biology and Management of Chinese Fruit Fly, *Bactrocera minax* (Enderlein) (Diptera: Tepritidae). *Journal of Agriculture and Forestry University* 5: 1–13.

Adhikari, D., Tiwari, D. B., and Joshi, S. L. (2018). Population dynamics of fruit flies in sweet orange (*Citrus sinensis* L.) orchards in Sindhuli, Nepal. *The Journal of Agriculture and Environment* 19: 9–16.

Baral, S., Marahatta, S., and Shrestha, A. (2021). Economics of production and marketing of mandarins in Parbat and Baglung districts of Nepal. *International Journal of Environment and Food Sciences* 5(3): 323–328.

Bastakoti, B., and Gautam, D. M. (2007). Effect of maturity stage and post- harvest stages on self-life and quality of mandarin orange in modified cellar store. *Journal of Institute of Agriculture and Animal Science* 28: 65–74.

Bhusal, Y. R. (2002). Improvement of post-harvest self-life of mandarin orange (*Citrus reticulata*). M.Sc. Thesis. Department of Horticulture, IAAS Rampur, Chitwan, Nepal.

Bhusal, Y. R., Gautam, D. M., Dhakal, D. D., and Subedi, P. P. (2007). Effect of maturity stages on post-harvest self-life and quality of mandarin orange. *IAAS Research Advances* 1.

Dahal, S., Shrestha, B., Bista, B., and Bhandari, D. (2020). Production and trade Scenario of Citrus Fruits in Nepal. *Food & Agro-business Management (FABM)* 1(1): 47–53.

FAO (Food and Agriculture Organization). (2011). *Training manual for combating citrus decline in Nepal.* www.fao.org/

Gauchan, D. (2000). Economics and sustainability of citrus farming in Nepal: A case study of mid-hills. *Proceedings of the 3rd National Horticulture Research Workshop, 7–8 June, Khumaltar.* Nepal Agricultural Research Council. 7–8.

Gautam, I. P., Paudel, K. B., Upadhyay. K. P., Gautam, D. M., and Khatri, B. K. (2004). De-greening of mandarin orange for early market in the hills of Nepal. *Proceedings of IV National Conference on Science and Technology, March 23–26.* Nepal Academy of Science and Technology. 349–361.

Gotame, T. P., Gautam, I. P., Shrestha, S. L., Shrestha, J., and Joshi, B. K. (2020). Advances in fruit breeding in Nepal. *Journal of Agriculture and Natural Resources* 3: 301–319.

Gotame, T. P., Paudyal, K. P., and Khatiwada, P. P. (2014). *Status of Fruits and Nut Genetic Resources in Nepal, Indigenous and Exotic Varietal Inventory.* Nepal Agricultural Research Council, National Agricultural Research Institute, Horticulture Research Division, Khumaltar, Lalitpur, Nepal. 1–71.

Khatiwada, P., Adhikari, R., and Adhikari, C. (2019). Post-harvest management in supply chain system for profitable citrus fruit production. In P. Khatiwada and K. Paudyal (Eds.) *Technical Policy and directives on Citrus fruit production entrepreneurship development.* Value Chain Development Project, Ministry of Agriculture and Cooperatives, Hariharbhawan, Nepal (in Nepali language).

Khatri, S., Shrestha, S., and Pokharel, K. P. (2021). Evaluation of manual fruit harvesters and storability characteristics of harvested sweet orange stored under ordinary room storage condition. *Sustainability in Food and Agriculture* 2: 84–91.

Khatri, S., Shrestha, S., and Pokharel, K. P. (2022). Investigation of different manual harvesting tools performance for harvesting outputs, mechanical injuries, storability and post-harvest physico-chemical attributes of mandarin fruit in Nepal. *Journal of Horticulture and Post-Harvest Research* 5: 35–52.

MoALD (Ministry of Agriculture and Livestock Development), Government of Nepal. (2021–2022). *Statistical information of Nepalese Agriculture, 2019/20, 2020/21.*

NCDP (National Citrus Development Program), Kirtipur, Nepal. (2014). *Annual Report 2071* (in Nepali).

NCRP (National Citrus Research Program), Dhankuta, Nepal. (2018–2022). *Annual Report 2018/19, 2019/20, 2020/21, 2021/22.*

NHRC (National Horticulture Research Center), Lalitpur, Nepal. (2021). *A simple technology of Acid lime storage (leaflet in Nepali language).*

Paudyal, K. P., and Subedi, H. (2008). Selection of sweet orange (*Citrus sinensis* Osbeck) varieties for production period expansion. *Nepalese Horticulture* 6, Nepal Horticulture Society.

Paudyal, K. P., Shrestha, T. N., and Regmi, C. (2016). Citrus research and development in Nepal. In *Horticulture in last six decades. Silver Jubilee special.* Horticulture Society of Nepal. 113–144.

PHLRD (Postharvest Loss Reduction Division), Government of Nepal. Shreemahal, Pulchowk, Lalitpur. (2003). *Progress Report 2059/060 (2002/2003).*

Pokharel, C. N. (2019). *Analysis of market chain of mandarin in Nepal: A case of Lamjunj District.* Van Hall Lerenstein University of Applied Sciences, Wageningen, the Netherlands. Accessed on 6 June 2023.

Shrestha, R. L., and Paudyal, K. P. (2004). Evaluation of indigenous mandarin (*Citrus reticulate* Blanco) germplasm at different locations of Dhanuka district. *Proceedings of the Fourth National Horticulture Research Workshop, March 2–4, 2004.* Nepal Agricultural Research Council. 37–40.

TE (2022). *Trend economy.* https://trendeconomy.com/data/h2/Nepal/0805 (accessed on June 3, 2022).

Thapa, M. B., Thapa, B. B., and Paudel, S. (2013). *A study report on: Prefeasibility study on processing and marketing of citrus and mango products.* Submitted to Commercial Agricultural Development Project, Department of Agriculture, Hariharbhawan, Lalitpur, Nepal (Unpublished).

Tomiyasu, Y. B., Barma, S. K., and Thapa, D. B. (1997). *Citrus production in Nepal.* National Citrus Development Program, Kathmandu, Nepal (in Nepali language).

8 An Insight Into the Turkish Citrus Industry

Cultivation, Potential, Challenges, and Future Outlooks

Harun Meral, Ali Tuncer, and Ali Ozcan

8.1 INTRODUCTION

Türkiye is located at the northernmost border of the world's citrus production areas. Citrus fruits, which were introduced to Türkiye in the 1930s, constitute one of the sectors of strategic importance for the country and have been at the forefront of the country's fresh fruit and vegetable exports for many years. Due to favorable ecological conditions, citrus cultivation has developed rapidly and there has been a significant increase in the number of citrus trees in recent years. Although the production periods of citrus species and varieties grown in Türkiye vary according to the species, they can start in September and continue until May. The citrus sector has become a sector that provides more than 5 million jobs in Türkiye and has reached an important point in terms of the functionality of the Turkish economy. The Aegean and Mediterranean Regions account for the majority of citrus cultivation in Türkiye, and the cultivation centres are Adana, Mersin, Antalya, Hatay, Muğla, and Osmaniye provinces (Yilmaz & Aydin, 2020).

According to the data from the Turkish Statistical Institute (TÜİK), in 2021, Türkiye produced approximately 5.4 million tonnes of citrus fruits. 84% of orange production, 92% of tangerine production, 93% of lemon production, and 97% of grapefruit production came from the Mediterranean Region. Tangerines constituted 34% of the citrus production, while oranges were 32%, lemons were 29%, and grapefruits made up 5%. There was a production of 1,742,000 tonnes of oranges, 1,819,000 tonnes of tangerines, 1,550,000 tonnes of lemons, 249,000 tonnes of grapefruits, and 2,615 tonnes of bitter oranges. There was a 23% increase in total citrus production compared to the previous year. The most significant increases were seen in orange production with 31% and lemon production with 30%. When considering the citrus production between 2017 and 2021, there was a 12% increase in total citrus production compared to 2017. The most significant increase was seen in lemons with 54%, while the most significant decrease was observed in the Washington cultivar of oranges with 18% as shown in Table 8.1. (TÜİK, 2023).

Türkiye plays a significant role globally in citrus production with its achieved levels. According to the data from the United States Department of Agriculture (USDA) for the 2021/22 season, Türkiye ranks seventh in global orange production with a production volume of 1.8 million tonnes, third in tangerine production with 1.8 million tonnes, fourth in lemon production with 1.4 million tonnes, and fifth in grapefruit production with 250,000 tonnes. When examining global citrus export data, Türkiye ranks fifth in orange exports (6%), first in tangerine exports (28%), second in lemon exports (30%), and third in grapefruit exports (18%) (USDA, 2023).

In Türkiye, the size of citrus orchards reached an area of 1.7 million decares in 2021 (Table 8.2). Compared to five years ago, the total area of these orchards has increased by 19%. Specifically, the total areas of lemon orchards increased by 61% and tangerine orchards by 20%, while there

DOI: 10.1201/9781003584599-8

TABLE 8.1
Citrus Production Statistics in Türkiye (Tonnes)

Product	2017	2018	2019	2020	2021
Orange (Washington)	1.447.971	1.395.550	1.242.000	932.466	1.185.796
Orange (Jaffa)	71.750	72.300	57.000	56.002	60.323
Orange (Others)	430.279	432.150	401.000	345.507	495.881
Sum Orange	**1.950.000**	**1.900.000**	**1.700.000**	**1.333.975**	**1.742.000**
Mandarin (Satsuma)	817.158	853.340	681.000	767.482	801.780
Mandarin (Clementine)	100.025	105.300	90.000	93.117	97.761
Mandarin (King)	5.564	5.810	5.000	7.083	7.152
Mandarin (Others)	627.722	685.550	624.000	717.947	912.307
Sum Mandarin	**1.550.469**	**1.650.000**	**1.400.000**	**1.585.629**	**1.819.000**
Lemon	1.007.133	1.100.000	950.000	1.188.517	1.550.000
Grapefruit	260.000	250.000	249.185	238.012	249.000
Sour Orange	2.124	2.052	2.230	2.609	2.615
TOTAL	**4.769.726**	**4.902.052**	**4.301.415**	**4.348.742**	**5.362.615**

(TÜİK, 2023).

TABLE 8.2
Area Statistics of Citrus Orchards in Türkiye (Thousand Da) and Average Yield Values per Fruit-Bearing Tree (Kg)

	2017		2018		2019		2020		2021	
	Area	Yield	Area	Yield	Area	Yield	Area	Yield	Area	Yield
Sum Orange	**513**	-	**508**	-	**502**	-	**460**	-	**481**	-
Orange (Washington)	366	154	355	149	348	134	310	108	308	140
Orange (Jaffa)	18	130	18	136	18	110	14	114	14	120
Orange (Others)	130	137	135	134	137	125	136	109	159	135
Sum Mandarin	**507**	-	**516**	-	**536**	-	**598**	-	**607**	-
Mandarin (Satsuma)	243	112	243	115	249	91	268	101	255	107
Mandarin (Clementine)	30	107	31	114	30	95	29	104	28	107
Mandarin (King)	2	106	2	113	2	93	2	125	2	123
Mandarin (Others)	232	106	240	114	255	93	299	97	322	107
Lemon	324	121	359	126	402	97	469	107	522	114
Grapefruit	54	200	52	200	52	203	51	201	50	209
Sour Orange	0,1	39	0	39	0	41	0	38	0	38
TOTAL	**1.398**	-	**1.435**	-	**1.492**	-	**1.578**	-	**1.660**	-

(TÜİK, 2023).

TABLE 8.3
Number of Citrus Trees at Fruit-Bearing and Non-Fruit-Bearing Age in Türkiye (Thousands)

	2017		2018		2019		2020		2021	
	Bearing	Non Bearing	Bearing	Non Bearing	Bearing	Non Bearing	Bearing	Non Bearing	Bearing	Non Bearing
Sum Orange	**13.124**	**917**	**13.150**	**681**	**12.985**	**866**	**12.307**	**1.052**	**12.625**	**1.210**
Orange (Washington)	9.422	560	9.394	371	9.247	457	8.645	396	8.447	388
Orange (Jaffa)	553	34	533	29	519	29	490	49	504	45
Orange (Others)	3.148	323	3.223	281	3.220	380	3.171	607	3.674	777
Sum Mandarin	**14.211**	**3.522**	**14.396**	**3.726**	**15.183**	**4.158**	**15.926**	**5.841**	**16.987**	**5.571**
Mandarin (Satsuma)	7.320	1.217	7.410	1.196	7.489	1.733	7.569	2.341	7.491	2.046
Mandarin (Clementine)	934	95	924	100	945	75	899	78	912	68
Mandarin (King)	53	3	51	4	54	5	57	6	58	4
Mandarin (Others)	5.904	2.207	6.011	2.425	6.695	2.345	7.401	3.417	8.526	3.453
Lemon	8.312	2.019	8.733	2.910	9.798	3.238	11.139	4.391	13.540	4.112
Grapefruit	1.298	64	1.249	61	1.229	42	1.184	44	1.189	28
Sour Orange	55	23	53	23	54	23	69	8	69	8
TOTAL	37.000	6.543	37.582	7.401	39.249	8.327	40.625	11.406	44.410	10.92

(TÜİK, 2023).

were decreases in the areas of orange orchards by 6% and grapefruit orchards by 7%. In the same year, tangerine orchards made up 37% of the total citrus orchard area, lemon orchards 31%, orange orchards 29%, and grapefruit orchards 3%. When considering the number of fruit-bearing and non-fruit-bearing citrus trees in Türkiye over the past five years, there was a 20% increase in the number of fruit-bearing citrus trees compared to 2017, and a 67% increase in the number of non-fruit-bearing trees compared to the same year (Table 8.3). The most significant increase in fruit-bearing tree numbers was seen in lemons (63%), while the most significant decrease was observed in the Washington cultivar of oranges (10%). For non-fruit-bearing tree numbers, there was a 141% increase in the number of orange variety trees compared to 2017, and the largest decrease was seen in bitter orange tree numbers with a drop of 65% (TÜİK, 2023).

In the 2020/21 season, Türkiye's citrus consumption was approximately 2 million tonnes. Oranges constituted 41% of the total citrus consumption, tangerines 32%, lemons 24%, and grapefruits 3%. While there hasn't been a significant change in the overall citrus consumption over the past five years, the most substantial increase was seen in lemon consumption at 47%, and the most notable decrease was in orange consumption at 40%. In the same season, the per capita consumption was as follows: total citrus consumption was 24.7 kg, oranges 9.8 kg, tangerines 7.7 kg, lemons 5.8 kg, and grapefruits 0.8 kg. There has been a 7% decrease in per capita citrus consumption over the last five seasons. (TÜİK, 2023).

Citrus exports occupy a significant share of Türkiye's fresh fruit and vegetable exports. According to the fresh fruit and vegetable export data from the Mediterranean Exporter Associations (AKİB) given in March 2022, citrus exports account for 43% of Türkiye's total fresh fruit and vegetable exports, with fresh vegetables making up 37%, and fresh fruits contributing 19% (AKİB, 2023). In 2021, Türkiye exported of 1.9 million tonnes of citrus products. Of this export volume, tangerines made up 45%, lemons 33%, oranges 14%, and grapefruits 8%. Analyzing the data from the last five years reveals a 29% increase in total citrus exports, with an 84% increase in tangerines, 31% in lemons, and 14% in grapefruits. In contrast, orange exports have seen a decrease of 34%. In terms of citrus imports for 2021, the figure stands at 72,000 tonnes, with oranges dominating the imports at a 60% share. Following oranges are tangerines (33%), lemons (5%), and grapefruits (2%). Evaluating data from the last five years, there's an 18% increase in total citrus imports since 2017. The most significant decrease is seen in grapefruit imports at 59%, while the most substantial increase, amounting to a 322% share, is observed in tangerine imports (TÜİK, 2023).

In reviewing the 2021 citrus grower prices in Türkiye, it is observed that oranges were the highest-priced product at 1.90 TL/Kg, while grapefruits were the lowest priced at 1.04 TL/Kg. Compared to 2017, oranges experienced the highest price increase, seeing a surge of 168%. In comparison to the previous year, the grower price for oranges increased by 28%, tangerines by 23%, and grapefruits by 5%, whereas the price for lemons decreased by 27%. When evaluating the trajectory of consumer prices over the past five years, there is a notable increase of 124% in orange consumer prices since 2017, 99% in tangerine prices, and 39% in lemon prices. Comparing with the previous season, tangerine consumer prices increased by 25%, and orange prices by 22%, while lemon consumer prices saw a reduction of 13%. (TÜİK, 2023).

Tangerine, one of the citrus species, can find buyers in the market until the end of the production season due to the presence of cultivars that emerge in early production periods. Its utilization in the processing industry also ensures that consumption extends over longer periods. The importance of lemons, another citrus species, is steadily increasing due to its wider range of fields of usage and year-round consumption. When considering the reasons of that citrus growers prefer citrus production, traditional production habits emerge as the primary choice (Atlı & Sahin, 2021). In Türkiye, the main products people consume as citrus include: fresh fruit consumption, fruit juice and concentrate, dried orange peel powder, orange peel oil, confectionery products, and other by-products arising from consumption behaviours (Karaoğlan et al., 2022).

8.2 MAJOR CITRUS CULTIVARS AND THEIR IMPORTANCE

Citrus varieties and cultivars that have been widely cultivated in Türkiye and have commercial importance are presented below with their characteristics (Barry et al., 2020; Kafa et al., 2010; Ladaniya, 2023).

8.2.1 Oranges

8.2.1.1 Valencia

It is a cultivar native to Spain. It emerged through bud mutation. It forms a spherical shape and dense canopy. The trees are of medium size. The fruits are round, with a slightly rough and medium-thick peel. The fruit pulp is of high quality. Its juice content is about 45%, and it has a balanced sugar to acid ratio. It is a cultivar suitable both for table consumption and industrial use (for juicing). Its yield is consistent, and it isn't selective about its ecology. It can be economically cultivated across a broad spectrum, from cold subtropical climates to desert climates. It's a late-maturing cultivar. It has types that mature at different times, from mid-March to mid-July. Particularly in subtropical conditions, ripe fruit and flowers can be seen simultaneously on the tree. A significant issue with the fruit is "re-greening". The orange colour developed in the winter can disappear from March onwards as the temperature rises and temperature variation decreases, leading to the formation of chlorophyll in the fruit peel. This is a typical characteristic of Valencia. There are various clones of it cultivated worldwide. Significant ones among them are Olinda, Frost, Campbell, Cutter, Midknight, and Delta. Being a late-season cultivar, there's a risk of fruit damage from winter colds in some regions.

8.2.1.2 Washington Navel

This cultivar, which forms the basis of navel oranges, originated as a mutation from the Celesta orange cultivar in the Bahia region of Brazil. It produces medium-sized trees with a rounded canopy. The fruit is round with a slightly rough peel. It's a seedless (parthenocarpic) variety and does not produce pollen. Its maturation time is mid-early (end of December). It is a moderately juicy cultivar. Under appropriate care conditions, it yields consistently. It is sensitive to environmental conditions, especially dry wind and cold. While not as selective as Jaffa, it does show ecological selectivity. One of the best Washington Navel growing regions in the world is the Mersin area. It is sensitive to fruit drops. Once mature, the fruit cannot remain on the tree for an extended period without falling. In Türkiye, it is a primary cultivar sought after for both domestic consumption and export.

8.2.1.3 Navelina

This cultivar, originating from California, emerged as a bud mutation from the Washington Navel. The tree is smaller than the Washington Navel but matures earlier. This precocity is approximately two weeks. In Spain, by mid-October, even though the skin colour is still green, it reaches the minimum acceptable quality standards for harvest. The fruits are slightly smaller and more oval than those of the Washington Navel. Despite being cultivated in various countries, it is primarily grown in Spain. Due to its early maturation, being harvested before winter colds, and its high quality, it is a suitable cultivar for Turkish conditions.

8.2.1.4 Jaffa

The origin of this cultivar traces back to the city of Jaffa in Israel. It emerged as a result of a bud mutation from the Beledi orange in 1844. Given its susceptibility to mutations, it has many types. It forms a large canopy. The leaves are larger and of a lighter colour compared to other oranges. The branches and trunk have a fragile structure. The fruits are medium-sized and oval-shaped, though there might be a flatness towards the stalk end. The rind is slightly rough, with a medium thickness that increases towards the stalk. Commercially, it is considered a seedless cultivar. The flesh of the fruit is crisp, juicy, and very flavorful. It is an ecologically selective cultivar and tends to be periodic. Depending on the ecology and the rootstock used, its quality attributes vary. Its ideal ecological conditions are found in Mersin. It's a mid-season cultivar, with the maturation period occurring from mid-January to early February.

8.2.1.5 Navelate

Originating from Spain, this cultivar has emerged through a natural mutation from the Washington Navel. The tree is slightly larger than the Washington Navel, but it is somewhat thorny. The fruits and the navel are smaller than the Washington Navel, and they are more challenging to peel. The fruits mature in February. Although the internal quality of the fruit reaches an acceptable level, the skin colouring occurs later. The fruits can remain on the tree for four months or even longer without losing their market value. The best-quality conditions for cultivation are in ecologies like those in Spain.

8.2.1.6 Fukumoto

It is one of the significant navel orange cultivars from Japan and California. However, when cultivated outside of Japan, undesirable traits of the tree can manifest. It has a high propensity for mutation. It is highly susceptible to the Foamy bark rot disease. It is an early-maturing cultivar, maturing earlier than the Washington Navel. It ripens concurrently with the Navelina cultivar. Harvesting can be done from October through the end of December. It's a cold-resistant cultivar. Although the external appearance of the fruits resembles the Washington Navel, they are uniformly round.

8.2.2 Tangerines, Mandarins, and Clementines

8.2.2.1 Owari

It gained importance in Japan in the early 20th century and was brought to Spain in 1925. The trees are productive and bear fruit at an early age. The fruits are medium to large and seedless. Under

Japanese conditions, due to the climate being more humid than that of Spain, the fruits are larger and have smoother peels. In Japan, the fruits have a lower acid content and are sweeter in taste. However, in both locations, once mature, the fruits cannot stay on the tree for long without puffiness developing. It tends to show fluctuations in yield proportionally. Generally, it matures by the end of October. The Satsuma cultivar cultivated in our country is mainly "Owari". In Spain, Satsuma production consists of approximately 80% Owari, while in Japan, Owari makes up about 15–20% of the total production.

8.2.2.2 Fremont

This cultivar is a hybrid of the Clementine and Ponkan cultivars. It bears fruit at an early age and has a tendency for upright growth. It's one of the darkest-coloured mandarins. It's a highly productive cultivar, albeit with a slight tendency towards biennial bearing (or periodicity). Despite its high yield, the fruits tend to be small. It's challenging to peel, very juicy, and seeded. After maturation, the fruit can remain on the tree for an extended period, up to three months (until March), without losing its quality. It's primarily cultivated worldwide in the Adana-Mersin regions of Türkiye. It is generally exported to Middle Eastern countries.

8.2.2.3 Nova

This cultivar has been obtained by crossbreeding Clementine and Orlando Tangelo. It is a productive cultivar with minimal tendency towards biennial bearing (or periodicity). The skin is a bright orange colour, slightly rough, and the fruit has a slightly flattened round shape. The number of seeds depends on the pollinator; with weak pollination, it has fewer seeds. However, commercially seedless types have been developed. It is a high-quality fruit cultivar. Mature fruits can remain on the tree without puffing. The maturation period is climate dependent. In subtropical climates like Florida, it matures in November, while in Spain, it matures in mid-December, after the Clementine cultivar. It is prevalent in Spain and Israel and is spreading in the Eastern Mediterranean Region.

8.2.2.4 W. Murcott

Believed to be a hybrid of Murcott and Clementine, this cultivar originates from Morocco. The trees bear fruit at an early age. It is a productive cultivar. The fruits have a flattened shape and are of a dark orange colour. This cultivar is known as W. Murcott in California and Nadorcott in Morocco. The fruit ripening period is in mid to late February.

8.2.2.5 Minneola Tangelo

This cultivar is a hybrid of Dancy mandarin and Duncan grapefruit. The trees are large, with broad leaves. It has a good adaptation capability. Its cold resistance is less than the Orlando Tangelo and it matures slightly later than Orlando. It is a mid-season cultivar, ripening within the month of January. The fruits are large, with a prominent neck. Both the fruit peel and the pulp are dark orange in colour. The fruit is tasty, possessing a unique flavour. The number of seeds depends on the pollination situation and this number can reach up to 20. For optimum yield and fruit size, cross-pollination is needed. It is susceptible to Alternaria disease. While Dancy is especially recommended as a pollinator, the Fremont mandarin can also be used for this purpose.

8.2.2.6 Okitsu Wase

It is the second most cultivated Satsuma cultivar in Japan. It was obtained from Miyagawa via nucellar seedlings in 1940. It began to be particularly cultivated from the 1960s onwards and has recently started to be grown in Spain as well. Young branches have thorns; however, these thorns disappear over time. While the fruits resemble Miyagawa, their shape is flatter, and they are sweeter due to having lower acidity and slightly higher sugar content. It matures about a week earlier than

Miyagawa and about three weeks earlier than Owari. In recent times, it has become widespread in Spain, Argentina, and Uruguay.

8.2.2.7 Miho Wase

This cultivar, a sibling of Okitsu, was obtained from the Miyagawa variety in 1940 and registered in 1963. While it is not commercially significant in Japan, it is more preferred by growers in South Africa. The tree resembles Okitsu but possesses a stronger structure than Miyagawa. It yields as much as Owari; however, the fruit size is smaller than that of Owari.

8.2.2.8 Dobashi Beni

This cultivar originated from the Owari cultivar in K. Dobashi's Garden in 1940. It is a seedless cultivar, and its peel can be easily removed. The fruits are flat in shape. The trees have a medium vigour. Harvest time is during the months of October and November.

8.2.2.9 Mandora (Ortanique)

It is a natural orange-mandarin hybrid and was discovered in Jamaica. It forms large trees. The fruits are medium-sized, with a slightly flattened stylar end and contain a small navel. The peel and fruit texture vary depending on the ecology in which it is cultivated. It matures as late in the season as Valencia. It is suitable for storage. Initially grown in limited quantities in Jamaica, due to its high adaptation capabilities in semi-tropical regions, it has since been cultivated in Central America (Honduras), Australia, Israel, Cyprus, and South Africa.

8.2.2.10 Clementine

There are differing views regarding its origin. According to the first view, it has arisen as a natural hybrid between the Mediterranean mandarin and the Granito orange. Another perspective suggests it originates from the "Canton" mandarin in China, which has similar characteristics. It is believed to have emerged from a bud mutation in Algeria in the 1890s and to have been discovered by a priest named Clement. It is cultivated in the Mediterranean basin, especially in Spain, North Africa, and parts of South Africa. It is the most popular mandarin in Spain and Morocco. In recent times, seedless selections have been developed and can be marketed at high prices. It forms a denser canopy compared to Satsuma. The fruit quality is high, although sometimes the fruits are small. It does not adapt as well to humid subtropical and tropical conditions as it does to the Mediterranean climate. There are selections that mature at various times from September to February. Bekria is a Clementine selection of Moroccan origin. It is very early-ripening, reaching maturity in September while the peel is still green. However, its taste is a bit bland at this stage. Arrufatina originated from a bud mutation of the Nules cultivar. It ripens about two weeks earlier than the Nules cultivar and matures at the beginning of November. Nules came about from a bud mutation of the Fina cultivar. It matures at the end of November. It can remain on the tree without losing its quality until January. It is a cultivar commonly cultivated in Spain. Nour is a selection of Moroccan origin. Although it normally matures in mid-January, it can remain on the tree without losing its quality until March.

8.2.2.11 Robinson

This is an early hybrid mandarin obtained by crossbreeding Clementine and Orlando Tangelo. Compared to Nova, its fruits are more rounded, and the peel is slightly rougher. It is resilient to transport and storage. The difficulty in peeling the fruit's skin is a disadvantage. The ripening period is around November. When ripe, the fruit's skin is a dark orange colour. The branches are thornless but are prone to breaking. The number of seeds varies depending on the pollinator, ranging from 0 to 20.

8.2.3 Pomelos and Grapefruits

8.2.3.1 Star Ruby

It was obtained in Texas in 1959 as a mutation created by irradiating the Hudson cultivar. Among the grapefruits, it has the darkest fruit flesh. This coloration is also reflected in the fruit's skin. It's a seedless mid-season cultivar. Among grapefruits, it is one of the most sensitive cultivars. It grows slowly and is susceptible to root rot and herbicides. Additionally, it is affected by sunburn. It is not suitable for storage. The fruits lack uniformity.

8.2.3.2 Ruby Red

It emerged in Texas in 1926 as an eye mutation from the Thompson cultivar. It is one of the prevalent cultivars. Compared to Marsh Seedless, it has more rounded fruits. The fruit has pink flesh. There's a characteristic of developing a red blush in areas where the fruits touch each other. The fruit quality is high. It matures about two weeks earlier than Marsh Seedless (early January). Due to the growing interest in cultivars with red fruit flesh, its cultivation has been proportionally decreasing.

8.2.3.3 Rio Red

It emerged in Texas as an eye mutation from Red blush. It is one of the most recently developed coloured grapefruits. Its skin colour is like Ray Ruby, and its tree characteristics resemble both red blush and Ray Ruby. The maturation period is January–February.

8.2.3.4 Marsh Seedless

Originating as a chance seedling from the Duncan cultivar in Florida, it was named in 1892. It forms a round and large canopy. It is a consistently productive cultivar. While the fruits are not as large as Duncan's, their seedlessness and the ability to remain on the tree for an extended period have made it a popular cultivar. It's a mid-season cultivar (January–February) and is suitable for storage. The fruit pulp is a light canary yellow and very juicy. It is the most widely cultivated white-fleshed grapefruit in the world. With the introduction of coloured grapefruits, it has started to give way to these newer cultivars.

8.2.4 Lemons and Limes

8.2.4.1 Kütdiken

Although its origin is Italy, it has now become a Turkish cultivar. It belongs to the Femminello subgroup. It is often considered to be the world's highest quality lemon. It's a superior cultivar in terms of internal quality, juice content, and aroma. It does not show periodicity. It is a medium-seeded cultivar (8–12 seeds). It is a mid-season cultivar. Under suitable conditions, it can be stored for up to 9 months. This characteristic has increased its share in cultivation. It holds a significant place in exports. It is a cultivar sensitive to fruit-end rot. Especially in Türkiye, its biggest problem is the fruit-end rot disease.

8.2.4.2 Meyer

It is believed not to be a pure lemon but rather a hybrid of lemon-orange or lemon-mandarin. It was discovered in China in 1908. The fruits are large, round, and have a small nipple. The fruit peel is smooth, yellow orange in colour, and thin. The distinct aromatic oils typically found in lemon peels are absent. The fruit flesh is dark yellow, juicy, and crisp. Its acid content is much lower than that of lemons. When harvested early, it tends to be sour, but this sourness diminishes over time. Due to its smooth peel, the fruit is more susceptible to damage during harvest. Each fruit contains approximately ten small, round seeds. It has greater resistance to both cold and heat compared to regular lemons, which makes its adaptation range broader.

8.2.4.3 İtalyan Memeli (Kara Limon)

Some sources suggest that, considering the fruit shape and the tree's tolerance to dry root rot, it is likely the same selection as Femminello St. Teresa. Globally, it is primarily cultivated only in Türkiye, especially in the vicinity of Mersin. It is one of the main types of lemons grown in Türkiye. A distinctive feature of the fruit is its top-shaped appearance. It is a seeded cultivar. It is a productive, high-quality cultivar that is suitable for transport and storage.

8.2.4.4 Lamas

It is a cultivar especially cultivated in Türkiye. It is productive and of high quality. It has a slight tendency towards periodicity. Due to its selectivity in ecological conditions, it is not widely cultivated. It is suitable for storage. The branches are thorny, and the nipple on the fruit are prominent. It matures starting from November. The cultivar is sensitive to dry root rot.

8.2.4.5 Interdonato

The cultivar is of Sicilian origin and was identified in the 1870s. Rather than being a pure lemon, it is believed to be a hybrid of lemon and wild melon tree. The tree has a round crown and is almost spineless, with apical dominance not as pronounced as in other lemons. The fruits are elongated cylindrical, larger than other lemons, with a very distinct nipple that appears pressed and tilted in one direction. It has a medium-thick rind and is moderately juicy. It's a medium-yielding cultivar. A significant feature is its early ripening. It matures from the middle of October. If left on the tree until the end of December, it tends to drop. It is sensitive to environmental conditions. Compared to other lemons, it is relatively resistant to dry rot. Türkiye is the country where it is most cultivated in the world.

8.2.4.6 Eureka

It is the most commonly cultivated cultivar of lemon. It is also the name of a group of lemons known as the Eureka. It was introduced in California through seeds brought from Sicily in 1858. This cultivar is prevalent in California, Australia, South Africa, Spain, Argentina, and Israel. Its characteristics reflect those of the Eureka group. The tree's crown is almost round and slightly smaller than the Lisbon cultivar. Compared to Lisbon, it is more sensitive to cold. While its yield is less than the Lisbon, its harvest period is longer (from late winter to summer). The leaf colour is lighter than Lisbon, and the nipple formation at the style end is less pronounced. The fruits are smaller than Lisbon's but have a smoother rind. The number of seeds typically ranges from 5 to 10. In the coastal regions of the Mediterranean basin, the fruit quality is very high. Since the fruits tend to cluster, they are more affected by wind and sun compared to Lisbon. There are selections like Cascade, Allen, Cook, and Frost.

8.3 GEOGRAPHICAL DISTRIBUTION AND SCALE

Located at the northernmost boundary of the world's citrus production areas, Türkiye holds a significant position among Mediterranean countries. With approximately 1.8 million tonnes of production, Türkiye ranks seventh in global orange production, third in tangerine production with the same volume, fourth in lemon production with 1.4 million tonnes, and fifth in grapefruit production with 250,000 tonnes (USDA, 2023). The majority of the production originates from the Aegean and Mediterranean Regions. The cities of Adana, Mersin, Antalya, Hatay, Muğla, and Osmaniye are key hubs for citrus cultivation (Figure 8.1). In Türkiye, 84% of the total orange production, 92% of

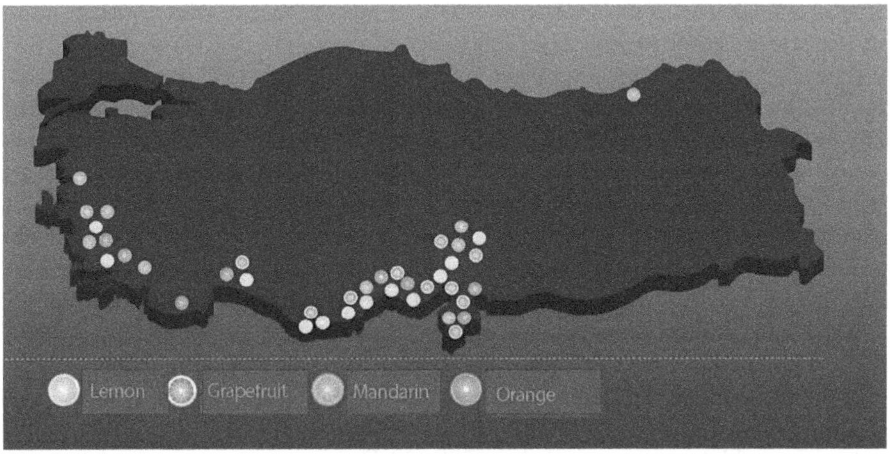

FIGURE 8.1 Citrus cultivation areas in Türkiye. (UTK, 2023.)

tangerine production, 93% of lemon production, and 97% of grapefruit production take place in the Mediterranean Region (TÜİK, 2023).

In 2021, 84% of orange production in Türkiye came from the Mediterranean Region and 16% from the Aegean Region. In that year, out of the total 1.7 million tonnes of oranges produced, Antalya accounted for the highest share with 31%, followed by Hatay (20%) and Adana (18%). The cultivars of oranges consisted of 68% Washington, 28% other cultivars, and 3% Jaffa. Tangerine production totalled 1.8 million tonnes, with 84% sourced from the Mediterranean and 15% from the Aegean Region. In consideration of the cities, Hatay led the tangerine production with a 39% share. Adana (36%) and Mersin (19%) followed Hatay. For that year, the tangerine cultivars grown were 44% Satsuma with a production volume of 802,000 tonnes, 50% other cultivars, and 5% Clementine. Lemon production amounted to 1.6 million tonnes, 93% of which was from the Mediterranean and 7% from the Aegean Region. Considering the cities, Mersin led the lemon production with a 55% share, followed by Adana (26%) and Hatay (9%). In the Eastern Mediterranean Region, lemon cultivars like Kütdiken, Meyer, Italian Nipple, Interdonato, and Lamas were cultivated. The Western Mediterranean Region was notable for growing Interdonato, Black Lemon, and Meyer cultivars. Grapefruit production totalled 249,000 tonnes, with 97% sourced from the Mediterranean and 3% from the Aegean Region. In consideration of the cities, Adana led with an 80% share in grapefruit production, followed by Mersin (9%) and Hatay (5%). Star Ruby, Ruby Red, and Rio Red were the primary grapefruit cultivars cultivated in these regions (Aygoren, 2022; TÜİK, 2023).

Citrus fruits have historically played a vital role in Türkiye's agricultural portfolio, with specific regions of the country emerging as hubs due to their distinct soil and climatic attributes. The soil composition is paramount to the successful cultivation of citrus. For instance, the alluvial soils predominantly found in coastal plains and deltas have become sought after for their impeccable water retention properties, making them a prime choice for many citrus growers. Additionally, loamy and sandy loam soils, recognized for their perfect balance of aeration and drainage, are also prevalent in these regions. Such soils prevent the detrimental root rot disease while ensuring that citrus trees have access to the required nutrients and moisture. On the other hand, areas with dense clay soils are typically sidelined in citrus farming due to their tendency to trap excessive water, which can be harmful to the roots of citrus trees. Citrus fruits require specific soil pH levels, typically ranging from 6.0 to 7.5, for optimal growth. Within Türkiye's citrus cultivation belts, the alluvial soils in coastal plains often exhibit a pH range of 6.5 to 7.2, which is within the ideal range for citrus cultivation. Moreover, these soils demonstrate a field capacity (the water content of the soil after it has been saturated and allowed to drain) of around 20–25%, providing a reliable moisture reservoir for the plants (Kafa et al., 2010; Mattos et al., 2020).

Geographically, Türkiye's Mediterranean and Aegean regions have been recognized as the epicentres of citrus cultivation. In the heart of the Mediterranean belt, cities like Adana, Mersin, and Antalya have carved a niche for themselves. These areas benefit from a climatic rhythm of mild, rainy winters paired with hot, sun-drenched summers, creating a conducive environment for the growth of diverse citrus varieties such as oranges, mandarins, lemons, and grapefruits. Shifting westwards to the Aegean coastline, the region, particularly around the vibrant city of Izmir, showcases a slightly tempered climate. This mildness becomes a boon for certain citrus types, with mandarins, lemons, and specific orange breeds thriving particularly well. The Mediterranean and Aegean regions, pivotal to Türkiye's citrus cultivation, receive sunlight for about 2,800–3,000 hours annually (Kafa et al., 2010). This is crucial because citrus plants typically require between 2,400–3,000 hours of sunshine each year to thrive. The temperatures in these regions range from an average low of 10°C (50°F) in winter to an average high of 34°C (93°F) in summer, which is essential given that citrus trees can be damaged by temperatures below -2°C (28°F) and need temperatures between 25°C to 32°C for optimal photosynthesis (Albrigo et al., 2023; Khalid et al., 2022; Vincent et al., 2020).

Furthermore, the weather dynamics in these regions strike a balance pivotal for citrus cultivation. The trees are somewhat dichotomous in their temperature preferences: while they rely on cool winter spells to trigger a period of dormancy and ensure vigorous bud break in the subsequent season, they are also susceptible to damage from severe frost. The mild winters prevalent in the Mediterranean and Aegean territories thus strike an optimal balance. Precipitation patterns further solidify the regions' suitability for citrus cultivation (Khalid et al., 2022; Vincent et al., 2020). While citrus plants need between 1,000–1,200 mm of rainfall annually, the Mediterranean and Aegean regions receive average annual precipitations of 600 mm to 1,200 mm, respectively. To compensate for any water deficit, especially in the hotter months, irrigation systems are often implemented to maintain soil moisture at around 50–70% of its field capacity. The coastal orientation of these regions provides an added advantage. The consistent sea breezes act as a natural moderator of humidity levels, minimizing the threat of fungal diseases that can plague citrus orchards. To top it all, elevation plays a subtle yet significant role. The majority of Türkiye's citrus orchards are strategically positioned at altitudes below 800 metres, ensuring the trees aren't exposed to the colder temperatures that higher elevations might bring, further accentuating the suitability of the Mediterranean and Aegean coastlines for this vibrant agricultural sector (Kafa et al., 2010). In terms of altitude, research indicates that citrus trees' metabolic processes, like photosynthesis and transpiration, work best below 800 metres. Above this altitude, the cooler temperatures can slow down these processes, impacting fruit quality and yield (Albrigo et al., 2023).

8.4 CULTIVATION PRACTICES AND TECHNOLOGICAL ADVANCEMENTS

Current methods and promising studies conducted in Türkiye, from citrus nursery management to citrus orchard management, citrus harvesting and post-harvest operations are briefly presented below.

8.4.1 Nursery Management

In recent years, citrus sapling cultivation has been carried out using plastic bags or pots. There are two different environments used in this practice. The first is the cultivation medium used in seed sowing benches and crates, and the other is the environment in which the sapling is grown until it is planted in the orchard after transplanting. Mixtures are typically made up of two parts peat to one part pumice or perlite. When peat-based media cannot be used, soils that are free from harmful organisms, high in organic matter, and devoid of weed seeds are prepared, and fine sands are mixed in at specific ratios. Well-composted animal manure can also be added to these mixtures. Citrus seedlings in seed sowing benches and crates remain for a maximum of 4–5 months. In Türkiye, the majority of citrus sapling cultivation predominantly uses rootstocks like Native Sour Orange, Troyer Citrange, Carrizo Citrange, and Native Trifoliate. Trees used for seed sources are always located in breeding plots, and they are specifically checked for virus diseases. Troyer Citrange, Carrizo Citrange, and Native Trifoliate rootstocks can be harvested for seeds from November 15 onwards, while the Native Citrus rootstock can be harvested from the end of November. The fruits harvested from these source trees are split lengthwise, and the seeds in the centre are removed. These seeds can either be sown directly, dried and stored, or sown at a desired time. Before sowing, the seeds must be treated with hot water and then pesticides. In recent years, with the use of seed machines in the nursery industry, citrus seeds can be mechanically sown into viols and then transplanted to bags. A limitation of this method is the inability to select seedlings that develop "gooseneck" deformities since they cannot yield a proper sapling or tree. As the seed sowing and seedling growth periods coincide with winter months, seed sowing greenhouses are heated to accelerate germination and growth. Seedling benches are usually irrigated by misting or sprinkling. Seedlings that reach a 4–5 leaf stage in seed sowing benches and have developed hair roots are uprooted, undergo a slight root trimming, and are then prepared for transplanting. In conventional cultivation, seedlings are planted into commonly used polyethylene bags or pots using planting tools, and the roots are firmly compacted. Seedlings exhibiting a bent root collar or "gooseneck" deformity are not transplanted. Depending on the ambient temperature and nutritive conditions after transplanting, seedlings reach grafting thickness in 4–5 months. The easiest grafting method, requiring the least materials, is grafting by budding. This method is commonly used in citrus sapling cultivation. However, cleft grafts can also be used, especially for changing cultivars on thick rootstocks or for grafting landscape plants. To ensure saplings of citrus cultivars have a straight stem, stakes are used and tied to the sapling as it grows. In modern sapling cultivation, irrigation and fertilization are mostly automatically provided through fertigation. The choice of fertilizer mixtures for citrus saplings and seedlings depends on the composition of the cultivation medium. Slow-release fertilizers, organic fertilizers, and specially produced fertilizers developed in recent years can also be used in citrus sapling cultivation. Saplings are continuously monitored until sale, with suckers constantly removed. Maximum attention is given to combating pests, primarily red spider mites, but also including mealybugs, aphids, whiteflies, and leafworms. These pests not only damage the plant but also serve as vectors for some diseases and pests. Saplings cultivated in greenhouses and that reach sale size are not sold directly to growers. They are kept in mesh greenhouse environments for at least 10–15 days to acclimate to the external environment before being delivered to growers. Inspections of saplings that have reached the necessary size and are ready for sale are carried out by experts from the Provincial Directorate of Agriculture and Forestry and, if necessary, by experts from Research Institutes. Sapling certification is carried out according to the current legislation. If needed, samples are taken from saplings for laboratory testing. Saplings are certified as standard saplings (with a yellow label) or certified saplings (with a blue label) (Özdemir, 2020).

8.4.2 Orchard Management

In Türkiye, citrus saplings can be planted in either the months of October–November or February–April. However, in cold regions with a risk of frost during winter, and especially for frost-sensitive species like lemons, planting is preferred in February–April. Standard planting or ridge planting methods can be employed. In the standard planting method, the appropriate distance between rows and within rows for oranges is 7x7 metres. Yet, by implementing semi-dense planting (7x3.5 metres), double the yield can be achieved. When tree canopies come close to each other, typically around the age of 10–12, it is recommended to uproot the trees and revert to a standard planting system of 7x7 metres, skipping every other row. For mandarins in standard planting, the suggested distances are 5x5 metres or 6x6 metres; for semi-dense planting, it's 5x2.5 metres or 6x3 metres. Lemons and grapefruits, being larger trees, are recommended to be planted at 8x8 metres in standard planting and 8x4 metres in semi-dense planting. In ridge planting, the ridge height can vary between 60–120 cm, with the distance between ridges ranging from 6–6.5 metres. Using potted (grown in bags or pots) saplings for planting is preferred since it increases the chances of successful growth. When planting, the ratio of pollinator cultivars can vary between 1/3 and 1/9, adjusted based on the quality of the pollinator cultivar. Saplings can be planted in two seasons: autumn and spring. In areas where winter cold can be problematic (like the Aegean and Eastern Mediterranean), spring planting might be preferred. In the Mediterranean Region, autumn planting seems more favourable. Autumn planting can be considered more reliable since, after planting, the onset of winter rains minimizes sapling losses and enhances the rate of successful growth. Saplings planted in the autumn tend to exhibit stronger and healthier growth by the following spring (MEGEP, 2016).

Tillage is not recommended and weed control is essential; thus, weed control can be achieved chemically without tilling the soil. However, this method is undesirable in terms of environmental pollution. For this reason, if appropriate conditions can be met, it's recommended not to till the soil in citrus orchards at all. But, if suitable conditions cannot be established, it's advised to till the orchard soil infrequently, superficially, and using light equipment (for instance, a hoeing machine). Giving definitive doses for fertilizing any type of citrus is challenging, and each country provides recommendations based on its own conditions. Leaf and soil analyses are conducted to determine the amounts of fertilizer to be given to the trees. Various factors influence the amount of fertilizer to be applied to trees, such as cultivar, soil structure, previously applied fertilization programmes, the age and size of the tree, and its yield status. As with other fruit types, fertilizers for citrus trees can be applied by surface spreading, burying in the soil, or foliar application. While nitrogenous fertilizers are applied by surface spreading, it is essential that phosphorus and potassium fertilizers be buried in trenches or pits opened within the canopy projection of the trees (MEGEP, 2016).

In citrus orchards of Türkiye, there are four types of pruning: formative pruning, productive pruning, rejuvenation pruning, and skirt pruning. Formative pruning is performed on young seedlings and trees. The shapes that can be given to the trees are lead branch or various lead branch forms. The objective here is to facilitate the tree's natural canopy development and support its growth. The goal is to form a sturdy, balanced, and rejuvenated canopy. In formative pruning, the first canopy is taken at 60 cm. From there, three equally distributed, best-growing branches not emerging from the same point are selected and cut at 25 cm, thus guiding the tree to its second canopy. Shoots emerging from these are cut when they reach 25 cm. The cut is made once the cut location becomes rounded. Strong shoots emerging from the main branch are not allowed. These should be removed while still thin. Strong branches on the second and third branches should be preserved. In citrus, if the interior of a tree's canopy doesn't hold fruit, that tree is not productive. Productive pruning ensures the removal of completed, dry, diseased, unproductive, and parasitic branches within the canopy. The sole aim of this process is to form shoots within the tree canopy and bear fruit. Citrus trees, after reaching a certain age, become unable to produce enough new shoots, leading to decreased yields. This condition is related to the nitrogen/carbon balance. While this ratio is at the desired level in fully productive young trees, it changes adversely with age. Rejuvenation pruning aims to address the age-related decline in productivity. When performing rejuvenation pruning, the primary consideration is not to cut thick branches. Branches thicker than 4–5 centimetres in diameter should not be cut. Practically speaking, branches thicker than the circle formed by touching the thumb and index finger together should not be pruned. Following this type of pruning, the tree's skeleton is exposed and needs protection. After rejuvenation pruning, the pruned area should be painted with white lime paint before spring (before new shoot formation). This protects the tree from the sun. Otherwise, sunburn can cause severe damage, even death, to the tree. Over time, after pruning, the tree canopy fills with new shoots. Skirt pruning involves removing ground-touching main branches and shoots, which should be done as soon after harvesting as possible. The period after fruit set is probably the best time for skirt pruning. In citrus orchards of Türkiye, the pruning time for all citrus except lemons is the winter months. Productivity pruning (removal of diseased, dry, and parasitic branches) is done in a cleaning manner during the winter months when growth halts. For lemons, however, the timing of tip pruning must be chosen carefully. Considering the threat of the tip-drying disease, the right time for tip pruning can be described as mid-October to early November. January is when flower buds start forming in citrus orchards. If tip pruning isn't done in lemons during the autumn and is instead done after winter harvest, the next year's crop is wasted, as the buds turning into fruit are located at the tips; with pruning, these are removed (MEGEP, 2016).

In the Mediterranean and Aegean Regions of Türkiye, where citrus cultivation is widely practiced, the Mediterranean climate prevails, with rains typically ceasing or becoming insufficient after April. This dry period continues until mid-October or November. During this time, it is essential to irrigate the trees. Before planting the saplings, the land is divided into suitable-sized parcels depending on its topography, slope, and size. Each parcel is separately levelled; steep terrains are terraced parallel to levelling curves and prepared for irrigation. An individual water distribution network is planned and established for each parcel according to the irrigation method to be applied. After these preparations, saplings are planted. The annual water requirements for citrus vary between 800–1,200 millimetres, depending on soil, climate, and the physiological state of the tree. However, not all of this amount is provided through irrigation, as some of it is met through rainfall. Research indicates that between May and October, citrus orchards require 650–750 millimetres of water through flood irrigation. These figures reduce to 500–600 millimetres for sprinkler irrigation and 300–400 millimetres for drip irrigation. In citrus-growing areas of the Mediterranean and Aegean regions, after spring fertilization, it is generally not feasible to wait for rain to wet the soil from mid-May or early June, necessitating irrigation. Depending on the early or late cessation of rains in certain years, the start of irrigation can vary by one or several weeks. Regular irrigation continues from the first irrigation, usually until mid-October or early November. The onset of the first effective rains in early or late periods can sometimes alter the last irrigation dates by a week or more. In Türkiye, irrigation methods for citrus can be grouped into four categories: surface (flood) irrigation (basin method, furrow method, and trench method), sprinkler irrigation, drip irrigation, and sub-surface irrigation. The basin method of surface irrigation is preferred in well-levelled, flat orchards with a gradient up to 0.2%. Rows are established with 8–16 trees in each, and irrigation is done row by row. The furrow method is applicable in soils with a slope up to 0.5%. Rows are established with barriers between them, and water is channelled to irrigate each tree. The trench method is suited for orchards with slopes ranging between 0.2–2%. Trench length can range from 70–200 metres. Sprinkler irrigation is preferred in areas where surface irrigation is unfeasible due to insufficient water or difficult topography. In this method, water is supplied through motor pumps or high-water towers and sprinkled over the orchard using sprinklers, which can be tall, medium, or short, or through perforated pipes. Drip irrigation involves slow dripping of water directly to the soil using low-pressure tubes. This system consists of a motor pump or water tower, filters, water and fertilizer tanks, water transport pipes, and drippers. Recent studies in Türkiye indicate that the most suitable and efficient irrigation system for citrus is drip irrigation. Using this system in citrus orchards eliminates labour for irrigation, enables more economical use of water resources, reduces fertilizer labour costs, and hence makes cultivation more profitable (MEGEP, 2016).

Pests and diseases can significantly hinder the quality and yield of citrus crops. Common threats in Türkiye include the Mediterranean fruit fly, aphids, spider mites, and fungal diseases like citrus canker and black spot. Without effective control, these pests and diseases can reduce the market value of the fruit, diminish yields, and threaten the livelihood of local growers. Historically, pesticide application in citrus orchards of Türkiye was predominantly manual, involving labourers using handheld sprayers. This method, while effective to an extent, often led to uneven application and potential overuse or underuse of pesticides. Over-application can result in pesticide residues on fruits, risking consumer health and jeopardizing exports due to stringent international standards. In recent years, there has been a significant shift in the techniques used to apply pesticides in Turkish citrus orchards. Drone technology has been increasingly adopted for precise pesticide application. These unmanned aerial vehicles can cover large areas efficiently, ensure uniform spray, and reduce the risk of over-application. Integrated Pest Management (IPM) has become an essential strategy for safeguarding the citrus orchards of Türkiye, blending traditional knowledge with modern methods. One sterling example is the management of the Mediterranean fruit fly, a notable pest for citrus growers. Instead of broad-spectrum insecticides, many Turkish growers now employ mass trapping using attractants specific to these flies, reducing their populations significantly. Similarly, for controlling the California red scale, a pest that can severely harm citrus trees, the introduction of its natural predator, the parasitic wasp (*Aphytis melinus*), has been practiced. This approach not only keeps the scale's population in check but also reduces pesticide use. Moreover, in regions like Adana and Mersin, regular field monitoring and decision support systems have been integrated to determine the precise timing and necessity of treatments for pests such as spider mites and aphids. Through such targeted and synergistic methods, Turkish citrus growers are ensuring healthier orchards, minimizing chemical residues, and promoting environmental sustainability (TAGEM, 2022).

8.4.3 Harvest and Post-Harvest Management

In Türkiye, nearly all citrus orchards rely on manual harvesting. Teams of workers collect the ripe fruits one by one. Once harvested, the fruits are placed into baskets, which are then emptied into crates or directly into the crates themselves, and subsequently transported to packing houses. While mechanical harvesting is possible for citrus fruits, it isn't widespread in Türkiye, largely because many orchards aren't especially large and labour costs remain reasonably affordable. After harvesting, the fruits are sorted based on their quality, typically into categories such as extra, first class, second class, and sometimes third class, in accordance with the TS34/2007 standards. The process of changing the colour of the green-ripe citrus fruit's skin by applying ethylene gas under appropriate conditions is called "colouring" or "yellowing". This procedure only affects the skin colour in non-climacteric citrus fruits, such as early oranges, satsumas, citrons, and grapefruits, without altering their taste or flavour. Before colouring, fruits are sorted based on their outer skin colour. Each category of fruit is placed in separate chambers for the required duration. The colouring process can be either continuous or intermittent. After classification based on species and cultivar, citrus fruits are packaged for consumers using various materials like cardboard and wooden crates, nets, bags, and pouches. The packaging is designed to optimally display the enclosed product. While citrus fruits can be directly delivered to consumers post-harvest, this method doesn't offer a prolonged shelf life. To achieve long-term storage and cater to distant markets, processing at packing houses becomes essential. In recent years, with the expansion of production areas in Türkiye, citrus exports have captured a significant proportion of the total revenue from fresh fruit and vegetable exports. Consequently, there have been both qualitative and quantitative improvements in citrus packing facilities to meet growing demands. In Türkiye, citrus fruits are preserved and stored in various ways. The most common methods include natural storages, cold storage rooms, and controlled atmosphere chambers. Natural storages, primarily used by growers, can be entirely above ground, semi-underground, or completely underground. Depending on their construction, these storage walls and ceilings are insulated as necessary to prevent excessive cold during harsh winters that might freeze the fruits and excessive heat during autumn that might hasten ripening. In ordinary storage, relative humidity is maintained by frequently wetting the floor and, if needed, by spraying water on the walls. Temperature and humidity in these storages are monitored with thermometers and hygrometers, and ventilation is adjusted based on external temperature readings. Cold storage primarily uses specialized cooling machines to chill the storage's air. To maximize efficiency, the entire storage area is thoroughly insulated. In cold storage, temperature can be maintained at a desired level, controlled by automatic switches. Since these storages lack ventilation windows and chimneys, air washers become crucial and necessary. Additionally, fans might be used to ensure proper air circulation. Controlled atmosphere storages, in addition to cooling, also modify the air composition, primarily by reducing oxygen (O_2) levels and increasing carbon dioxide (CO_2) levels. This controlled atmosphere storage can extend the shelf life of fruits more effectively than just cold storage. Along with temperature and humidity, the CO_2 and O_2 levels are also monitored in these storages. (MEGEP, 2016).

8.4.4 Emerging Technologies in Citriculture

Precision citriculture, a subset of precision agriculture, refers to the tailored management of citrus orchards using advanced technologies to monitor, assess, and address variations in the field. With the evolving demands of the global citrus market and the challenges posed by climate change, the need for more efficient and sustainable cultivation practices has become paramount. Unmanned aerial vehicles (UAVs) equipped with multispectral and hyperspectral cameras can scan vast stretches of orchards, capturing data that reveals plant health, soil moisture levels, and pest infestations. In Türkiye, drones are becoming an indispensable tool in large-scale citrus orchards, providing actionable insights to growers. Furthermore, governmental agencies and agricultural institutions in Türkiye are playing an active role in promoting precision citriculture. Training programmes, subsidies for technology adoption, and research collaborations with universities are accelerating this shift towards data-driven farming. Located in Antalya, BATEM has been a beacon of agricultural innovation in the region. The institute has run multiple projects on precision agriculture, with citrus cultivation being one of the focal points. Through their research, they've developed data-driven models for optimized irrigation, guiding local growers on how to use water resources more efficiently, crucial in a region that's been grappling with water scarcity. Several orchards, especially those in the Adana region, have adopted sensor-driven irrigation systems. These systems, backed by soil moisture sensors and climate data, adjust irrigation levels based on real-time needs. For instance, the fertigation system, which combines fertilization with irrigation, is gaining traction, ensuring that citrus trees get nutrients and water simultaneously, based on their precise needs. Mersin, another citrus cultivation hub, has seen an uptick in the use of drones for orchard monitoring. Local agricultural enterprises employ drones equipped with multispectral cameras to monitor tree health, detect pest infestations, and assess crop maturity. This real-time monitoring helps in the timely application of treatments, ensuring optimum fruit yield and quality. In the coastal regions, particularly around the Aegean area, there have been pilot projects where Geographic Information Systems (GIS) are employed for citrus orchard planning. Using GIS, growers can map out orchard topography, drainage systems, and soil health metrics, facilitating decisions related to planting patterns and infrastructure development. Leading agricultural universities, such as Çukurova University in Adana and Akdeniz University in Antalya, have been actively involved in research and development related to precision citriculture. Through partnerships with local orchards, these institutions test and refine advanced cultivation techniques, ranging from drone imagery analysis to the development of bio-sensors for early pest detection. To ensure the broad adoption of precision citriculture, provincial agricultural directorates, in collaboration with NGOs and international agencies, have organized training sessions for growers. In these sessions, held in places like Adana and Mersin, growers are introduced to new technologies, from soil sensors to data analytics platforms, ensuring grassroots-level adoption (BATEM, 2023; Berberoğlu et al., 2018; Donmez et al., 2021; TAGEM, 2022, 2023; Tercan & Dereli, 2020; Uğuz et al., 2023).

8.5 CURRENT STATUS OF THE INDUSTRY AND SECTOR

The infrastructure, position, vision, and mission of the citrus industry and sector in Türkiye have been examined from its historical trajectory to its present state and projected future. This analysis also encompasses the threats and opportunities faced by the sector, its inputs and outputs, as well as comprehensive data and evaluations under distinct subheadings.

8.5.1 Harvesting Dynamics of Citrus Cultivars

Citrus fruits, predominantly grown in the Mediterranean Region, are now produced in both the Northern and Southern Hemispheres. Citrus fruits produced in the Northern Hemisphere are termed "winter citrus" and their production season starts in October/November and extends until May/June (Figure 8.2). On the other hand, citrus fruits from the Southern Hemisphere are called "summer citrus". Their production season begins in April/May and lasts until November/December (Table 8.4). Türkiye is situated at the northernmost boundary of global citrus production areas. The production periods for the different species and cultivars of citrus cultivated in Türkiye vary; however, they typically start in September and can extend until May (Aygoren, 2022). In citrus cultivation, the development of consumption alternatives is crucial due to the supply surpassing demand after harvest and the rapid perishability of the product. Additionally, when establishing new citrus cultivation areas, market-driven production should be implemented, taking into account the product's market demand, shelf life, and the breadth or scarcity of its uses. To achieve higher economic returns, it is essential to monitor changes related to ecological factors and the product industry both domestically and globally, as well as advancements in production technologies. Research has indicated the necessity of using and developing new cultivars to provide both local and international citrus consumers with fresh and processed products year-round. This has prompted growers to seek new graft scions and rootstocks that would allow them to maintain production throughout the year (Atlı & Söyler, 2018).

8.5.2 The Infrastructure of Citrus Marketing

The citrus supply chain in Türkiye is characterized by a multi-tier system. Growers are the primary producers and often have small to medium-sized holdings. Local aggregators or brokers purchase

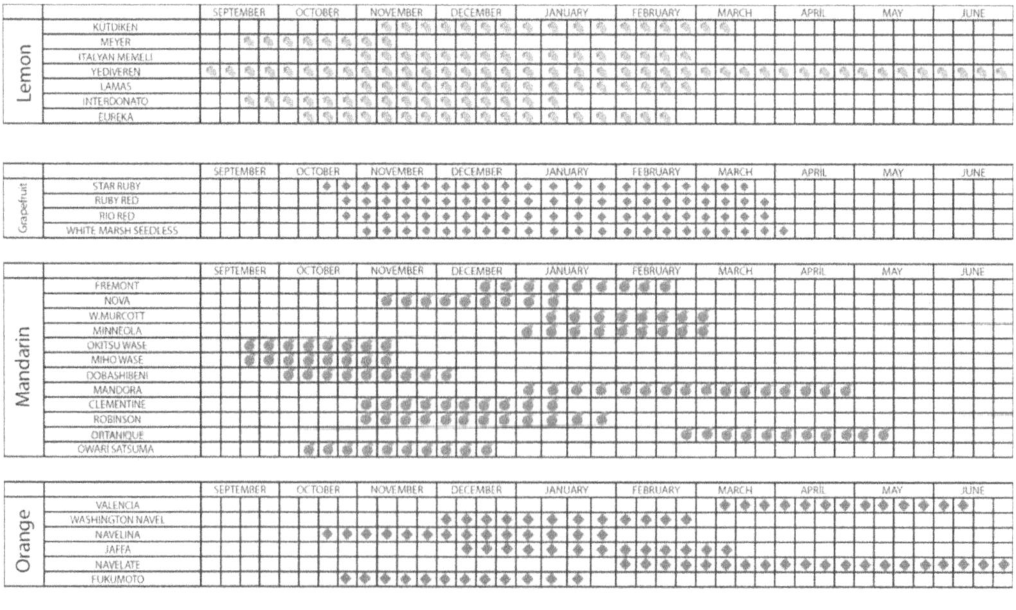

FIGURE 8.2 Cultivation schedule for the citrus species and cultivars prevalently cultivated in Türkiye. (UTK, 2023.)

TABLE 8.4
Harvest and Export Time Data for the Citrus Species and Cultivars Commonly Cultivated in Türkiye for the Year 2022

Cultivar	Harvest	Export
	Lemon	
Meyer	September 8, 2022	September 15, 2022
Interdonato	September 30, 2022	October 6, 2022
Lamas		
	Grapefruit	
Star Ruby	October 14, 2022	October 19,2022
Ruby Red	October 27, 2022	November 3, 2022
M.Seedless	October 27, 2022	November 3, 2022
Rio Red	October 27, 2022	November 3, 2022
	Mandarin	
Okitsu Wase	September 30, 2022	October 6, 2022
Miho Wase	September 30, 2022	October 6, 2022
Primosole	September 30, 2022	October 6, 2022
Early Enn	September 30, 2022	October 6, 2022
Satsuma	October 20, 2022	October 20,2022
Marisol	September 30, 2022	October 6, 2022

(AKİB, 2023).

from multiple growers to accumulate sizable lots for sale. Wholesalers usually operate in larger regional markets and may supply to domestic and international buyers. Processors convert citrus to products like juices, concentrates, and other processed items. Exporters have a significant role, often linked to international buyers or agents. The domestic market's primary channel is through fresh consumption. Local bazaars and supermarkets are the principal sales points. Early maturing, thin-skinned cultivars tend to be more popular for local consumption due to their ease of use and palatability. Türkiye exports a substantial volume of its citrus produce, especially to countries in Europe, Russia, and the Middle East. The thick-skinned cultivars, which have longer shelf lives, are generally preferred for exports. The linkage with the processing industry, especially fruit juice manufacturers, is strong. The demand from this sector can influence farm-level production decisions, particularly when there's a preference for certain citrus cultivars that are juicier or have specific taste profiles. The Turkish government plays a facilitative role through subsidies, grants, and training programmes. Furthermore, trade agreements, phytosanitary standards, and export promotion also fall within the government's domain, thereby influencing market linkages (UTK, 2016). Registered citrus growers in the Turkish Farmer Registration System are entitled to a myriad of subsidies. They receive fuel support of 170 TL ($9.11) per hectare and fertilizer aid of 80 TL ($4.29) per hectare. If they produce top-quality organic lemons, individual growers are granted 1,000 TL ($53.60) per hectare, while groups of growers receive 500 TL ($26.80) per hectare. Conventional lemon growers adhering to the second quality standard are offered a Good Manufacturing Plan support of 400 TL ($21.44) for individual growers and 200 TL ($10.72) for grower groups. Additionally, a soil analysis conducted by an accredited laboratory can be reimbursed at 40 TL ($2.14) per sample. The incentives for planting are set at 1,000 TL ($53.60) per hectare for standard saplings and escalate to 4,000 TL ($214.40) for certified ones. Embracing biological pest control fetches a reward of 700 TL ($37.52) per hectare, and employing pheromones as a biotechnical method to fend off Mediterranean fruit flies yields 1,000 TL ($53.60) per hectare (Sinem Duyum, 2022).

In Türkiye, two of the most significant private bodies actively engaged in the citrus industry and frequently publishing sectoral and technical analyses are the Türkiye Fresh Fruit and Vegetable Exporter Associations and the National Citrus Council (UTK). The Turkey Fresh Fruit and Vegetable Exporter Associations plays a pivotal role in streamlining and augmenting the export endeavours of the fresh produce domain. It amalgamates the industry's key players with the objective of enhancing exports, infiltrating novel markets, and fortifying their stance in established ones. This association not only signifies Türkiye in global expos but also undertakes international market studies, curates educational and informational workshops for exporters, deliberates on industry challenges, and partners with governmental agencies for resolutions. This umbrella organization encapsulates various regional fresh produce exporting sub-groups within Türkiye. While these sub-groups cater to their regional membership, their nationwide actions are harmonized under the aegis of the Türkiye Fresh Fruit and Vegetable Exporter Associations (TİM, 2023). The National Citrus Council (UTK) serves as a central hub for all involved in the citrus industry in Türkiye. Its main responsibilities include identifying challenges faced by the sector, suggesting practical solutions, setting clear strategies for progress, and ensuring that these strategies are effectively put into practice. The council's work is rooted in thorough research, covering areas like the introduction of new citrus cultivars, managing pests and diseases, and improving marketing methods. In addition to research, UTK organizes various educational events, such as seminars and conferences, both nationally and internationally, to share knowledge and promote discussions about the industry's future. With members ranging from growers and exporters to researchers and industry experts, UTK represents a wide spectrum of the citrus community. This broad membership allows UTK to view the industry's challenges from multiple angles and offer well-rounded solutions (UTK, 2023).

8.5.3 Trade Patterns from Import to Export

In citrus industry of Türkiye, thin-skinned early cultivars are generally sold in local markets, while thick-skinned late cultivars are directed towards exports due to the ease they provide in logistics and marketing. Due to strong infrastructure of Türkiye in citrus production, its import operations are primarily based on specific shortfalls related to events during the production year. Recent factors such as the global and national inflation causing an increase in input costs, stagnant producer sale prices, invasions of the Mediterranean fruit fly, regional disease and weed infestations due to climate change, and the direct and indirect impacts of the Covid-19 pandemic have necessitated occasional imports, albeit on a smaller scale. Additionally, citrus imports of Türkiye are predominantly for the fruit juice industry (MATSO, 2020; Sinem Duyum, 2022). In 2021, Türkiye exported a total of 2.9 million tonnes of citrus fruits, with mandarins being the most exported citrus fruit from Türkiye (Table 8.5). Iraq, Russia, and Ukraine are the largest importers of Turkish citrus fruits (Table 8.6). In order to meet the domestic demands, Türkiye imports citrus fruits. In 2021, Türkiye imported a total of 217 thousand tonnes of citrus fruits, with oranges being the most imported citrus fruit (Table 8.7). Turkish Republic of Northern Cyprus is the biggest supplier of citrus to Türkiye (Table 8.8).

8.5.4 Processing Industry and Outputs

Citrus fruits, both due to their various species, cultivar and aroma, as well as their known positive effects on health and nutrition, are fruits that are given importance in terms of consumption in Türkiye. Today, they are used in many areas, primarily in the food industry. Citrus fruits are mainly consumed fresh, but they are also processed to produce a wide range of by-products. The most important of these products include: unpasteurized fresh orange juice, frozen concentrated orange juice, and non-concentrated pasteurized orange juice. Among these, unpasteurized fresh orange juice is particularly preferred by European consumers. In addition to these products, citrus fruits are used in

TABLE 8.5
Data on Citrus Exports from Türkiye (Tonnes)

	Orange	Mandarin	Lemon	Grapefruit	Total
2012	305.532	409.440	395.401	141.282	1.253.447
2013	417.017	535.185	455.352	184.926	1.598.734
2014	399.412	613.648	448.763	152.551	1.689.944
2015	510.897	579.681	445.934	198.107	1.741.021
2016	715.546	716.653	502.584	132.922	2.126.847
2017	1.045.931	770.620	590.162	203.503	2.555.371
2018	731.729	706.361	542.752	149.225	2.271.146
2019	655.901	832.246	417.368	188.189	2.082.168
2020	542.648	908.208	642.125	167.002	2.246.223
2021	874.975	997.084	663.542	157.836	2.817.501

(TÜİK, 2023).

TABLE 8.6
Leading Countries in Citrus Exports from Türkiye by Quantity (Tonnes) and Share (%) for 2021

Orange			Mandarin			Lemon			Grapefruit		
	Quantity	%		Quantity	%		Quantity	%		Quantity	%
Iraq	102.640	40,1	Russia	438.679	52,1	Iraq	174.560	28,1	Russia	47.155	31,7
Russia	91.714	35,8	Ukraine	121.648	14,5	Russia	147.731	23,8	Poland	17.203	11,6
Ukraine	18.955	7,4	Iraq	119.055	14,1	Ukraine	43.525	7	Ukraine	14.359	9,6
Syria	7.663	3	Syria	21.055	2,5	Romania	36.824	5,9	Romania	12.248	8,2
Georgia	7.558	3	Serbia	16.244	1,9	Syria	26.026	4,2	Bulgaria	8.270	5,6
Others	27.373	10,7	Others	125.081	14,9	Others	192.432	31	Others	49.634	33,3
Türkiye	**255.903**	**100**	**Türkiye**	**841.763**	**100**	**Türkiye**	**621.099**	**100**	**Türkiye**	**148.869**	**100**

(TÜİK, 2023).

TABLE 8.7
Data on Citrus Imports in Türkiye (Tonnes)

	Orange	Mandarin	Lemon	Grapefruit	Total
2012	141.120	10.122	9.214	7.239	173.993
2013	111.448	16.523	3.137	6.742	181.340
2014	162.094	18.696	8.017	3.570	230.811
2015	107.655	22.791	3.584	2.984	158.084
2016	160.961	24.958	5.410	3.815	233.658
2017	115.774	25.491	9.874	2.076	177.736
2018	101.002	33.001	1.921	1.938	170.321
2019	131.049	42.475	2.969	1.835	205.254
2020	150.830	39.054	3.824	4.817	238.346
2021	171.035	52.900	5.762	7.932	216.635

(TÜİK, 2023).

TABLE 8.8
Leading Countries in Citrus Imports of Türkiye by Quantity (Tonnes) and Share (%) for 2021

Orange			Mandarin			Lemon			Grapefruit		
	Quantity	%		Quantity	%		Quantity	%		Quantity	%
TRNC	43.377	99,6	TRNC	22.903	94,8	TRNC	2.296	65,4	TRNC	732	63,5
Romania	116	0,3	Russia	355	1,5	Brazil	606	17,3	China	240	20,8
Others	73	0,2	Others	906	3,7	Romania	112	3,2	Bulgaria	46	4
-	-	-	-	-	-	Others	496	14,1	Others	135	11,7
Türkiye	**43.567**	**100**	**Türkiye**	**24.164**	**100**	**Türkiye**	**3.510**	**100**	**Türkiye**	**1.154**	**100**

(TÜİK, 2023).

the production of foods like marmalade, jam, and canned goods. Apart from the food industry, citrus peels are processed and used to flavour various foods, and in the chemical industry, they are used to extract essential oils used in the production of perfumes, soaps, medicines, and similar products. The pulp left over after processing citrus fruits can be used in animal feed. Besides the direct consumption of citrus fruits in Türkiye, there's substantial scope to derive a broader range of high-value products from this fruit, beyond the limited products currently produced. Orange liqueur and pectin stand out as products with significant demand both domestically and internationally. Additionally, orange peels present multiple value-added opportunities (Karaoğlan et al., 2022; Zenginoğlu, 2007).

8.6 STATUS OF CITRICULTURE RESEARCH AND EXTENSION SERVICES

From the perspective of citrus cultivation, Türkiye has several advantages, such as its geographical location, favourable climatic conditions, adequate production volume, and high yields in its production areas. However, there are various challenges in production, marketing, transportation, storage, and packaging. Problems arising from production and marketing prevent higher revenues from both domestic and international markets. To address these challenges, it is targeted to take measures to extend the harvesting and marketing periods due to increased demand, cultivate cultivars that are ecologically suitable to reduce the risk of winter frosts, propagate plantations with dwarf rootstocks considering resistance to biotic and abiotic factors, encourage the application of semi-dense and dense planting techniques in newly established orchards to achieve significant yield increases per unit area, prioritize cultivar development, focus on cultivars suitable for foreign market demands, establish modern processing facilities to increase the consumption of processed citrus fruits, emphasize packaging that meets the product size demands of different markets, and prevent areas primarily suitable for citrus cultivation from being used for non-agricultural purposes like tourism, industrialization, and unplanned urbanization. Citrus products from Türkiye can be exported to many traditional countries that Türkiye traditionally exports agricultural products to. However, many significant citrus-importing countries globally have not yet been accessed. To expand export markets, comprehensive foreign market research and sharing these research findings with all industry stakeholders are primary goals. Most growers own small-scale enterprises and are compelled to sell their product immediately due to capital constraints. This situation often necessitates that growers sell their products on the tree, sometimes even during the flowering stage, highlighting the need for marketing organizations to bolster their competitive strength against buyers. Establishing marketing cooperatives or similar organizations will better facilitate marketing services for citrus products and increase growers' income. Branding and regional image creation are emphasized. Considering the dispersed and small-scale nature of businesses, entrepreneurs are being trained, encouraged, and supported in creating and managing a common brand. The limited knowledge levels of growers negatively influence many cultivation processes, leading to product and quality losses. Due to these reasons, significant research, application, training, and dissemination activities are conducted in citrus cultivation regions by Turkish public and private institutions and non-governmental organizations (Aygoren, 2021).

Since the 1960s, studies conducted in Türkiye regarding the spread of viruses and virus-like diseases have revealed that one of the primary reasons for yield reduction is the prevalence of these diseases in citrus orchards. Many countries have initiated citrus cultivar development programmes encompassing cultivar development, purification, indexing, and introduction quite early, particularly to increase yield and quality and achieve competitive power in the global market. In Türkiye, a programme related to these began in 1988 at the Citrus and Greenhouse Crops Research Institute, supported by the Food and Agriculture Organization (FAO). Within the scope of this project, the first virus-free citrus budwood was produced in 1992, and the first virus-free citrus sapling was produced in 1996. In 1994, the project was renamed the "Türkiye Citrus Cultivar Development Program (TTÇGP)", supported by the General Directorate of Agricultural Research and Policies (TAGEM), and was institutionalized as a continuous programme. Currently, under this programme, between 1994–2020, a total of 287 different citrus cultivars, including 125 oranges, 60 mandarins, 67 lemons, 14 grapefruits, and 21 other citrus species and cultivars, have been incorporated. Production was achieved by obtaining material from purified sources, an initial requirement to mitigate the adverse effects of viruses and virus-like diseases. Techniques such as shoot-tip grafting, biological indexing, molecular analyses (including for Psorosis, Tatterleaf, Tristeza, Cachexia, Exocortis, Stubborn), and heat treatment are utilized. Verified individuals are assessed for accuracy in greenhouses and orchards. Nearly 80 viruses and virus-like diseases affecting citrus have been reported worldwide. These diseases are noted to reduce yields by 10–50%, causing significant economic losses in citrus orchards and even threatening citrus cultivation. In this context, future stages of the project will prioritize the purification of individuals found in Turkish citrus genetic resources, along with cultivars introduced and developed through breeding. Given the widespread use of citrus rootstocks in the country, the threat posed by the tristeza virus, and the need to provide citrus growers with production materials free from viruses and virus-like diseases, these factors emerge as continuous and indispensable topics for the development of the Turkish citrus sector (Kurt, 2020).

Located in the province of Antalya, Batı Akdeniz Agricultural Research Institute (BATEM) was established in 1934 under the name "Warm Climate Plants Research Station." It began its studies on citrus fruits in 1938, focusing on the introduction and adaptation of citrus species and cultivars to Türkiye. Thanks to the adaptation of these cultivars imported from abroad, Türkiye transformed from a citrus-importing country to a citrus-exporting one. Subsequently, through evolutions like the Citrus Research Institute and the Citrus and Greenhouse Crops Research Institute, and now as BATEM, it continues to conduct research and development activities in citrus cultivation and the development of new cultivars, contributing to the innovations needed by the sector. BATEM is the owner of a significant portion of commercially important citrus cultivars in Türkiye and is one of the institutions housing virus-free citrus stock for the purpose of serving citrus nurseries. In this context, it owns 32 citrus cultivars: 7 oranges (Navelina, Washington Navel, Finike Yerli, Alanya Dilimli, Valencia Late, Moro, Jaffa 45-A), 14 mandarins (Okitsu Wase, Miho Wase, Satsuma Clausellina, Satsuma Silverhill, Lee, Nova, Fairchild, Robinson, Fremont, Clementine Fina, Yerli Apireno, Fortune, Kara), 4 lemons (Interdonato, Karalimon, Kıbrıs, İtalyan Memeli), 6 grapefruits (Star Ruby, Rio Red, Redblush, Henderson, Marsh Seedless, Ray Ruby), 1 bergamot (Yerli A-41), and one kumquat (Nagami-oval Kumquat). It also owns three citrus rootstocks: Native Sour Orange, Carrizo Citrange, and Troyer Citrange. Additionally, in 2011, through selection processes, BATEM registered eight more citrus cultivars: 3 mandarins (BATEM Pearl, BATEM Star, BATEM Göral), 3 oranges (BATEM Fatihi, BATEM Şekeri, BATEM Baharı), and two lemons (BATEM Pınarı, BATEM Sarısı), which were then made available to the citrus sector. Following years of rigorous breeding research at the Institute, the "Ertuğrul Bey" cultivar was officially registered in 2020. Representing a significant breakthrough, this cultivar emerged as first seedless mandarin of Türkiye, developed through mutation breeding techniques from the highly-seeded Native (Bodrum) mandarin genotype (Şenay et al., 2023; Turgutoğlu et al., 2023). The Alata Horticultural Research Institute, affiliated with the Ministry of Agriculture and Forestry of the Republic of Türkiye, is located in the city of Mersin. This institution is at the forefront of applied agricultural research, delving into areas such as pomology, viticulture, olericulture, ornamental plant cultivation, soil and water resource management, and apiculture. Within the institute's scope, through mutation breeding techniques, seedless lemon cultivars "Alata", "Gülşen", and "Uzun", as well as the early-maturing "Eylül" lemon cultivar, have been developed and registered. Furthermore, utilizing selective breeding methodologies, four distinct lemon cultivars, namely "Yediveren", "Lamas", "Enter", and "Erdemli33", along with two mandarin cultivars "Toros Kırmızısı" and "Sarıca", were successfully developed and registered (ALATA, 2023). Located in the city of İzmir, the Aegean Agricultural Research Institute, which conducts significant studies on horticultural plants, has developed and registered two mandarin cultivars named "Seferihisar 35" and "Ulubelde 35". Additionally, a trifoliate troyer rootstock by the name "Trixsin35" has been developed and registered within the institute's framework (Ege Tarımsal Araştırma Enstitüsü Müdürlüğü, 2023). In Türkiye, these three agricultural research institutes conduct significant and extensive studies on citrus fruits. Within the plant health departments of these institutes, intensive research and applications are conducted on biotic and abiotic stress factors that pose problems for citrus fruits, especially diseases, pests, and invasive weeds. In the plant health department affiliated with BATEM, 2–2.5 million predatory insects (*Cryptolaemus montrouzieri*) and 1–1.5 million parasitoid insects (*Leptomastix dactylopii*) are produced every year to be used as natural enemies against the citrus mealybug, the most common pest of citrus. With the production of these beneficial insects, contributions are made to the expansion of biological control in citrus cultivation areas located between the provinces of Muğla and Hatay (BATEM, 2023). In addition, the Biological Control Research Institute, located in the city of Adana, conducts significant research on the most critical citrus diseases, pests, and invasive weeds, focusing on biological and integrated control methods. Supported by 15-decare citrus research orchard and state-of-the-art research laboratories within the Institute, the affiliated researchers are dedicated to advancing these studies (Biyolojik Mücadele Araştırma Enstitüsü Müdürlüğü, 2023). Research institutes focused on citrus fruits, affiliated with the Ministry of Agriculture and Forestry of the Republic of Türkiye, operate in coordination with provincial and district agricultural directorates and training centre directorates. Their primary focus is on the implementation of research outcomes and ensuring that these results benefit the growers at the grassroots level. Throughout the cultivation season and beyond, growers are provided with general trainings as well as specialized trainings based on specific requests concerning all aspects of citrus cultivation. Furthermore, services are extended to growers and industry representatives under the scope of the Department of Training and Publication in collaboration with other research and operational units. These services include the provision of farmer handbooks, cultivation guides, brochures, periodic agricultural advisory publications, digital training sessions, and various other educational tools (TAGEM, 2023). Moreover, a new digital era in agriculture has been launched in 2023 with the "TarımCebimde" mobile application under the jurisdiction of the Department of Information Technologies affiliated with the Ministry of Agriculture and Forestry of the Republic of Türkiye. This application, designed to support growers in pre-production, production, and post-production stages, is also tailored for those involved in citrus cultivation. Within this mobile platform, all digital services related to agriculture are offered to the users (Bilgi Teknolojileri Genel Müdürlüğü, 2023).

Numerous agricultural faculties of universities in Türkiye conduct education and research on citrus science and cultivation. Particularly, Akdeniz University, Ege University and Çukurova University stand out in this domain. These universities have various departments and laboratories dedicated to research on citrus. The Subtropical Fruits Research and Application Centre affiliated with Çukurova University is of paramount significance. Drawing inspiration from citrus cultivation within subtropical fruits, the centre is dedicated to monitoring emerging biological and biotechnological agricultural technologies. Its mission is to transfer these innovations to Türkiye, conduct adaptation studies, and shape the results in a format conducive for use by universities and other research institutions. The centre leverages novel technologies in citrus to provide growers with cultivation materials for diverse species and cultivars that are free from viruses and virus-like diseases. Moreover, it stays abreast of cutting-edge methods developed for the diagnosis and research of virus and virus-like diseases, ensuring their effective transfer to the nation (Tiring et al., 2017c, 2017d; Tiring, Satar, & Çimen, 2017b, 2017a). Given the intensity of citrus cultivation in its region, Çukurova University conducts significant research across various facets of the citrus production value chain. The university is also pursuing studies in agricultural digital technologies, such as computer vision and web-based information systems (Berberoğlu et al., 2018; Donmez et al., 2021). In addition, the university conducts significant research in the areas of agricultural biotechnology, plant protection, soil science, and plant nutrition within citrus cultivation (Cetinkaya Yildiz et al., 2022; Hançerli & Uygur, 2021; Uludamar et al., 2021; Yeşiloğlu et al., 2013).

In Türkiye, the most prominent private associations operating extensively in the citrus sector and regularly preparing and presenting sectoral and technical reports are Türkiye Fresh Fruit and Vegetable Exporter Associations and National Citrus Council (UTK). Türkiye Fresh Fruit and Vegetable Exporter Associations is an organization that coordinates and supports the export activities of the fresh fruit and vegetable sector. It brings together all stakeholders of the sector to increase fresh fruit and vegetable exports, penetrate new markets, and operate more effectively in existing markets. The association represents Türkiye at international fairs, conducts overseas market research, organizes training and informational seminars for exporters, seeks solutions to the sector's challenges, and collaborates with the government on these issues. It comprises fresh fruit and vegetable exporter sub-associations operating in different regions across Türkiye. While these sub-associations address the needs of their members regionally, they act in a coordinated manner nationwide under the scope of the Türkiye Fresh Fruit and Vegetable Exporter Associations (TİM, 2023). UTK serves as a platform that brings together all stakeholders of the citrus sector. It is responsible for identifying issues in the citrus sector, developing solution proposals, determining strategies for the sector, and coordinating activities to implement these strategies. The council conducts research related to the citrus industry and carries out projects on subjects like new cultivar development, pest and disease management, and marketing strategies. Additionally, it organizes meetings, seminars, and conferences at both national and international levels concerning the sector. Comprising a spectrum of stakeholders from growers to exporters, researchers to private sector representatives, the council embodies the entirety of the citrus industry. This structure enables it to approach the sector's challenges and needs with a holistic perspective (UTK, 2023).

8.7 CHALLENGES, POTENTIAL STRATEGIES, AND FUTURE OUTLOOK

One of the biggest challenges facing the citrus industry is unpredictable extreme weather conditions. Fluctuations in the temperature causes fruit pre-maturation, drought, and drop in the fruit quality. Secondly, the policymakers are trying to find ways to promote grapefruit production. Even though the domestic consumption for grapefruit is low, there is great potential for exports, especially after the global demand increased following COVID-19 due to the well-known vitamin-C benefits of the fruit. Currently, the most produced grapefruits are bitter and sour in taste, therefore, it is challenging to compete with other types. Thirdly, Russia and Ukraine are two of the biggest importers of Turkish citrus fruits. In recent years, the demand for imports from those countries is not stable. Therefore, the alternative markets are desired. Besides, Turkish fruits imported from Türkiye to EU countries and the UK are subjected to increased strict border inspection for MRLs (maximum residue level) of the pesticides. The EU has a very strict policy on the type and frequency of pesticides used in all agricultural products. Therefore, the growers are facing challenges to maintain the product quality and quantity while minimizing pesticide use. According to the Turkish Agriculture Ministry database, there are 42 different Active Ingredients (AIs) registered to be used for citrus-related disease management (Table 8.9). The EU is targeting pesticide-free agriculture by 2050. Therefore,

TABLE 8.9
The List of Active Ingredients (A.I.s) Approved for Citrus Farming in Türkiye

Name of the A.I.	Approved Use	The Disease	The Target Organism
Spirodiclofen	Acaricide	N.A.	*Phyllocoptruta oleivora and Panonychus citri*
Hydrolized protein	Aerial bait	N.A.	*Ceratitis capitata*
Imazalil	Fungicide	Blue mould rot	*Penicillium italicum*
Azoxystrobin	Fungicide	Green mould rot	*Penicillium digitatum*
Tribasic copper sulfate	Fungicide	Brown rot	*Phytophthora citrophthora*
Pyrimethanil	Fungicide	Green mould rot	*Penicillium digitatum*
2-Phenylphenol	Fungicide	Green mould rot	*Penicillium digitatum*
Fludioxonil	Fungicide	Blue mould rot	*Penicillium italicum*
Thiophanate-methyl	Fungicide	Brown rot	*Phytophthora citrophthora*
Copper oxide	Fungicide	Citrus brown spot	*Alternaria alternata f. sp. citri*
Copper hydroxide + copper oxychloride	Fungicide	Citrus brown spot	*Alternaria alternata f. sp. citri*
Copper hydroxide	Fungicide	Citrus brown spot	*Alternaria alternata f. sp. citri*
Thiabendazole + Imazalil	Fungicide	Green mould rot	*Penicillium digitatum*
Captan	Fungicide	Brown rot	*Phytophthora citrophthora*
Fosetyl-Al	Fungicide	Brown rot	*Phytophthora citrophthora*
Tebuconazole	Fungicide	Citrus brown spot	*Alternaria alternata f. sp. citri*
%80 Sulfur	Fungicide	N.A.	*Panonychus citri*
Metiram	Fungicide	Citrus brown spot	*Alternaria alternata f. sp. citri*
Copper oxychloride	Fungicide	Mal secco	*Phoma tracheiphila*
Trifloxystrobin	Fungicide	Citrus brown spot	*Alternaria alternata f. sp. citri*
Kresoxim-methyl	Fungicide	Citrus brown spot	*Alternaria alternata f. sp. citri*
Glyphosate-diammonium salt	Herbicide	N.A.	N.A.
Glyphosate isopropylamine salt	Herbicide	N.A.	N.A.
Oxyfluorfen	Herbicide	N.A.	N.A.
Carfentrazone- ethyl	Herbicide	N.A.	N.A.
Glufosinate-ammonium	Herbicide	N.A.	N.A.
Glyphosate IPA	Herbicide	N.A.	N.A.
Penoxsulam	Herbicide	N.A.	N.A.
Forasulam	Herbicide	N.A.	N.A.
Diflufenican	Herbicide	N.A.	N.A.
Diquat dibromide	Herbicide	N.A.	N.A.
Spinosad	Insecticide	N.A.	*Ceratitis capitata*
Sulfoxaflor	Insecticide	N.A.	*Planococcus citri*
Cyantraniliprole	Insecticide	N.A.	*Ceratitis capitata*
Bacillus thuringiensis subsp. kurstaki strain SA- 11	Insecticide	N.A.	*Ectomyelois ceratoniae*
Pirimicarb	Insecticide	N.A.	*Toxoptera aurantii*
Spirotetramat	Insecticide	N.A.	*Aonidiella aurantii*
Buprofezin	Insecticide	N.A.	*Dialeurodes citri*
Parahin oil	Insecticide + Acaricide	N.A.	*Planococcus citri*
Mineral oil	Insecticide + Acaricide	N.A.	*Aonidiella aurantii*
Fenamiphos	Nematicide	N.A.	*Meloidogyne* spp.

(General Directorate of Food and Control Department of *Plant Protection Products*, 2023).

immediate actions implementing alternatives to pesticides are needed. Turkish growers commonly use sour orange stock for the grafting of citrus. Even though the practice offers many advantages, the varieties are sensitive to a virus called *Citrus tristeza and* the Mediterranean Citrus Fly. Therefore, there is an immediate demand for citrus cultivars resistant to the disease. Also, new varieties can adapt to increasing temperatures, alkaline soil, and scarcity of water is highly desired (*General Directorate of Food and Control Department of Plant Protection Products*, 2023).

REFERENCES

AKİB. (2023). *Yaş Meyve Sebze Sektörü Türkiye Geneli Değerlendirme Raporu.*

ALATA. (2023). *T.C. Tarım ve Orman Bakanlığı Alata Bahçe Kültürleri Araştırma Enstitüsü Müdürlüğü.* Tescilli Çeşit Listesi.

Albrigo, L. G., Stelinski, L. L., and Timmer, L. W. (2023). Environmental constraints on growth, development and physiology of citrus. In *Crop Production Science in Horticulture*. Crop Production Science in Horticulture. https://doi.org/10.1079/9781845938154.0078

Atlı, H. F., and Sahin, A. (2021). Hatay İli Dörtyol İlçesinde Portakal Üretim ve Pazarlaması. *Türk Tarım ve Doğa Bilimleri Dergisi* 8(3): 834–846.

Atlı, H. F., and Söyler, O. (2018). Dünyada ve Türkiye'de Turunçgil Üretimi ve İhracatının Değerlendirilmesi. *Journal of Academic Social Science* 79*:* 357–366. https://doi.org/10.16992/ASOS.14174

Aygoren, E. (2021). *Turunçgiller Ürün Raporu.* Tarımsal Ekonomi ve Politika Geliştirme Enstitüsü.

Aygoren, E. (2022). *Turunçgiller Ürün Raporu.* Tarımsal Ekonomi ve Politika Geliştirme Enstitüsü.

Barry, G. H., Caruso, M., and Gmitter, F. G. (2020). Chapter 5 – Commercial scion varieties. In M. Talon, M. Caruso, and F. G. Gmitter (Eds.), *The Genus Citrus*, 83–104. Woodhead Publishing. https://doi.org/10.1016/B978-0-12-812163-4.00005-X

BATEM. (2023). *T.C. Tarım ve Orman Bakanlığı Batı Akdeniz Tarımsal Araştırma Enstitüsü Müdürlüğü (BATEM).* Bitki Sağlığı Bölümü.

Berberoğlu, S., Şahingöz, M., Erdoğan, M. A., Dönmez, C., and Çilek, A. (2018). Doğu Akdeniz'de Web-Tabanlı Turunçgil Bilgi Sistemi Oluşturulması. *Geomatik* 3(2): 137–144.

Bilgi Teknolojileri Genel Müdürlüğü. (2023). *T.C. Tarım ve Orman Bakanlığı Bilgi Teknolojileri Genel Müdürlüğü (BTGM).* TarımCebimde.

Biyolojik Mücadele Araştırma Enstitüsü Müdürlüğü. (2023). *T.C. Tarım ve Orman Bakanlığı Biyolojik Mücadele Araştırma Enstitüsü Müdürlüğü.* Hizmetler.

Cetinkaya Yildiz, R., Karatas, A., Horuz, S., and Aysan, Y. (2022). Determination of susceptibility of citrus and stone fruit cultivars against *Pseudomonas syringae* pv. *syringae* in Turkey. *Gesunde Pflanzen* 74(3): 639–646. https://doi.org/10.1007/s10343-022-00638-x

Donmez, C., Villi, O., Berberoglu, S., and Cilek, A. (2021). Computer vision-based citrus tree detection in a cultivated environment using UAV imagery. *Computers and Electronics in Agriculture* 187: 106273.

Ege Tarımsal Araştırma Enstitüsü Müdürlüğü. (2023). *T.C. Tarım ve Orman Bakanlığı Ege Tarımsal Araştırma Enstitüsü Müdürlüğü.* Tescilli Çeşit Listesi.

General Directorate of Food and Control Department of Plant Protection Products. (2023). https://bku.tarimorman.gov.tr/

Hancerli, S., and Uygur, F. N. (2021). Turunçgil Bahçelerinde, Farklı Örtücü Bitki Türlerinin Yabancı Otlanma Üzerindeki Etkisinin Araştırılması. *Turkish Journal of Weed Science* 24(2): 150–165.

Kafa, G., Uzun, A., Turgutoğlu, E., Canan, İ., Öztop, A., and Canıhoş, E. (2010). *Turunçgil Yetiştiriciliği* (G. Kafa & E. Canıhoş, Eds.). T.C. Tarım ve Köyişleri Bakanlığı Yayın Dairesi Başkanlığı.

Karaoğlan, S., Yabacı, G., and Şener, A. (2022). Orange Products and By-products in Turkey. In C. Ozbek and U. Nasibe (Eds.), *First International Traditional Foods and Sustainable Food Systems Symposium*, 213. Toros University, Faculty of Fine Arts, Design and Architecture, Department of Gastronomy and Culinary Arts.

Khalid, M. F., Shafqat, W., Khan, R. I., Abbas, M., Ahmed, T., Ahmad, S., and Hussain, S. (2022). Citrus responses and tolerance against temperature stress. In *Citrus Production*, 149–155. CRC Press.

Kurt, Ş. (2020). *Türkiye Turunçgil Çeşit Geliştirme Programı.* Batı Akdeniz Tarımsal Araştırma Enstitüsü.

Ladaniya, M. (2023). Chapter 2 – Commercial fresh citrus cultivars and producing countries. In *Citrus Fruit* (2nd ed.), 23–91. Academic Press. https://doi.org/10.1016/B978-0-323-99306-7.00012-8

MATSO. (2020). *Narenciye Raporu.* Manavgat Ticaret ve Sanayi Odası.

Mattos, D., Kadyampakeni, D. M., Oliver, A. Q., Boaretto, R. M., Morgan, K. T., and Quaggio, J. A. (2020). Chapter 15 – Soil and nutrition interactions. In M. Talon, M. Caruso, and F. G. Gmitter (Eds.), *The Genus Citrus*, 311–331. Woodhead Publishing. https://doi.org/10.1016/B978-0-12-812163-4.00015-2

MEGEP. (2016). *Turunçgil Meyveleri.* T.C. Milli Eğitim Bakanlığı.

Özdemir, M. (2020). *Turunçgil Fidan Yetiştiriciliği.* T.C. Tarım ve Orman Bakanlığı Tarımsal Araştırmalar ve Politikalar Genel Müdürlüğü (TAGEM) Batı Akdeniz Tarımsal Araştırma Enstitüsü Müdürlüğü (BATEM).

Şenay, K., Turgutoglu, E., and Demir, G. (2023). Dünyada ve Ülkemizde Turunçgil Virüs Arındırma Programları. *Meyve Bilimi* 10(1): 6–11.

Sinem Duyum. (2022). *Citrus Annual.* United States Department of Agriculture Foreign Agricultural Service. https://apps.fas.usda.gov/newgainapi/api/Report/DownloadReportByFileName?fileName=Citrus Annual_Ankara_Turkey_TU2022-0052.pdf

TAGEM. (2022). *Turunçgil Entegre Mücadele Teknik Talimatı* (A. Öztop, Ed.). T.C. Tarım ve Orman Bakanlığı Tarımsal Araştırmalar ve Politikalar Genel Müdürlüğü (TAGEM) Bitki Sağlığı Araştırmaları Daire Başkanlığı.

TAGEM. (2023). *T.C. Tarım ve Orman Bakanlığı Tarımsal Araştırmalar ve Politikalar Genel Müdürlüğü (TAGEM).* Yayınlar/Veriler.

Tercan, E., and Dereli, M. A. (2020). Development of a land suitability model for citrus cultivation using GIS and multi-criteria assessment techniques in Antalya province of Turkey. *Ecological Indicators* 117: 106549. https://doi.org/10.1016/j.ecolind.2020.106549

TİM. (2023). *Türkiye İhracatçılar Meclisi (TİM).* Türkiye Yaş Meyve Sebze İhracatçı Birlikleri.

Tiring, G., Satar, S., and Çimen, B. (2017a). Adana Ekolojik Koşullarında Bazı Satsuma Mandarin (*Citrus unshiu* Marc.) Çeşitlerinin Meyve Kalite Özelliklerinin Saptanması ve Hasat Olumuyla İlişkilendirilmesi. alatarım 2017, 16 (1): 19–25.

Tiring, G., Satar, S., and Çimen, B. (2017b). Bazı Erkenci Göbekli Portakalların Farklı Dönemlerde Meyve Kalite Özelliklerinin Belirlenmesi. *Türk Tarım ve Doğa Bilimleri Dergisi* 4(2): 189–195.

Tiring, G., Satar, S., Yeşiloğlu, T., and Çimen, B. (2017c). Bazı mandarin çeşitlerinin adana ekolojik koşullarında meyve kalite özelliklerinin saptanması. *Turkish Journal of Agriculture-Food Science and Technology* 5(3): 251–255.

Tiring, G., Satar, S., and Yilmaz, B. (2017d). Geç olgunlaşan bazı portakal çeşitlerinin farklı dönemlerde meyve kalite özelliklerinin belirlenmesi. *Türk Tarım ve Doğa Bilimleri Dergisi* 4(3): 331–338.

TÜİK. (2023). *Türkiye İstatistik Kurumu.* https://data.tuik.gov.tr/Bulten/Index?p=Bitkisel-Uretim-Istatistikl eri-2022-45504

Turgutoglu, E., Şenay, K., Demir, G., and Eryilmaz, Z. (2023). Dünyada ve Türkiye'de Turunçgillerin Gelişimi ve BATEM'in Rolü. *Meyve Bilimi* 10(1): 65–70.

Uğuz, S., Şikaroğlu, G., and Yağız, A. (2023). Disease detection and physical disorders classification for citrus fruit images using convolutional neural network. *Journal of Food Measurement and Characterization* 17(3): 2353–2362. https://doi.org/10.1007/s11694-022-01795-3

Uludamar, E. B. K., Cimen, B., and Eleckcioglu, H. (2021). Doğu Akdeniz Bölgesi'nde doğa koşullarında bazı turunçgil anaçlarının Turunçgil nematodu'na [Tylenchulus semipenetrans (Cobb, 1913) (Tylenchida: Tylenchulidae)] karşı dayanıklılığının araştırılması. *Plant Protection Bulletin* 61(1): 41–47.

USDA. (2023). *United States Department of Agriculture.*

UTK. (2016). *Türkiye Turunçgil Sektörü ve İhracat Yapısı.*

UTK. (2023). *Ulusal Turunçgil Konseyi (UTK).* Bilgi Bankası/Raporlar.

Vincent, C., Morillon, R., Arbona, V., and Gómez-Cadenas, A. (2020). Chapter 13 – Citrus in changing environments. In M. Talon, M. Caruso, and F. G. Gmitter (Eds.), *The Genus Citrus*, 271–289. Woodhead Publishing. https://doi.org/10.1016/B978-0-12-812163-4.00013-9

Yesiloglu, T., Cimen, B., Incesu, M., Yilmaz, B., Kacar, Y. A., and ŞİMŞEK, Ö. (2013). Turunçgil sektörünün gereksinim duyduğu yeni çeşitlerin geliştirilmesi. *Tarım Bilimleri Araştırma Dergisi* 2: 127–132.

Yilmaz, H., and Aydin, B. (2020). Comparative input-output energy analysis of citrus production in Turkey: Case of Adana province. *Erwerbs-Obstbau* 62(1): 29–36.

Zenginoğlu, A. (2007). *AB Sürecinde Türkiye Turunçgil İhracatının Yapısı, Ortaya Çıkan Sorunlar ve Çözüm Yolları Üzerine Bir Araştırma*. Ege Üniversitesi.

9 The Citrus Industry of Pakistan

Faisal Shahzad, Muhammad Azam, Muhammad Jafar Jaskani, and Iqrar Ahmad Khan

9.1 STATUS OF CITRUS INDUSTRY

Citrus industry refers to the cultivation/production, marketing, processing, and consumption of a group of fruits known as citrus fruits. Botanically, Citrus is a name for a genus in family Rutaceae. Horticulturally, it refers to all edible and rootstock species of genus Citrus and a few closely related genera of the family Rutaceae. Common edible fruits included in this group are sweet oranges, mandarins, grapefruit, lemon/limes and many minor categories like tangerines, pummelos, tangelos, kumquats etc. There are others used as rootstock for the cultivation of edible types. In addition, there are citrus fruits used as a source of industrial products like perfumes, essence, citric acid etc (Ackon, 2020). Citrus fruits originated in the tropics of Southeast Asia and were spread to the rest of the world through trade routes. Citrus is now cultivated on both sides of the equator, making a citrus belt around 35 degrees North and South, which is extended up to 44-degree latitude in South of France in the Northern Hemisphere and up to 39-degree latitude in Australia in the Southern Hemisphere. Citrus fruits rank on the second position after grapes in the world production of fruits. Global citrus production was about 47,404 thousand Metric tonnes during the production season 2023–2024 (U.S. Department of Agriculture, 2024). Brazil is the leading orange producing country with 15,300 thousand Metric tonnes production on total area of 590 thousand hectares, major production portion ends to juice processing industry (10,932 thousand metric tonnes) and the rest is used for fresh market produce. Citrus stands first in terms of area (≈ 156 thousand hectares) and production (2,229 thousand Metric tonnes) among 30 fruit crops grown in Pakistan (Pakistan Bureau of Statistics, 2024). Citrus fruits share among total value of fruit's exports is one-third (GOP, 2021), and Pakistan generated 77.24 million US$ by exporting citrus fruit in 2023–2024 production season (Ahmad et al., 2018; Trend Economy, 2024). Among different citrus groups (mandarins, grapefruit, oranges, lemons, and limes), mandarins have significant importance in Pakistan. Kinnow mandarin is the most grown citrus variety in Pakistan (>80%) due to its unique characteristics including attractive rind colour, size, easy peeling, high juice content, and excellent taste and aroma (Memon, 2017; Trade Development Authority of Pakistan, 2022).

The characteristic features of Pakistan's citrus industry are: rough lemon as a rootstock, Kinnow as a scion, creating a situation of monoculture; unregulated nursery business is responsible for a total lack of quality control and spread of diseases, no comprehensive disease and pest management strategies, especially, the spread and impact of citrus viruses; management practices in our citriculture are substandard and usually against the recommendations of the experts; as a result, low yields; the income of growers is further reduced due to lack of effective marketing system (most of the groves are sold to the contractors); postharvest losses are more than 20% and at times could go much higher, the processing industry is still in its infancy and needs to grow.

 DOI: 10.1201/9781003584599-9

Taking up on Kinnow again, despite merits, there are serious defects in this variety. For a grower and contractor, it is the longest duration crop. Being late in ripening, it offers a short crushing season to the processors resulting heavy over heads. It is an alternate bearer, i.e., heavy crop in one year and very little or no crop in the second year. Due to very soft peel, fruits are damaged during harvesting and shipping which results in heavy postharvest losses. It is probably the fruit with the highest number of seeds, which is very annoying to the consumer. Other problems easily noticeable with Kinnow are a range of colour and size grades and a trend for deterioration caused by bud mutations and probably due to the combined effect of viruses as well. We are using rough lemon as rootstock. Considering the range of soils and agro ecological zones in which citrus is cultivated, there are rootstocks which could perform better than rough lemon at some of the localities and in certain scion combinations.

9.1.1 Pakistan's Agro-Climatic Zones and Citrus Cultivation

Pakistan is located between 24° and 37° North latitude and 61° to 75° East longitude, spans 1,600 kilometres North to South and 885 kilometres East to West, covering 796,095 square kilometres with a subtropical and semi-arid climate. Numerous agricultural resources, well irrigated plains, seasons, and fertile land are favourable for the citrus sector of Pakistan. In southern plains, rainfall ranges around 125 mm, and 500–900 mm in the sub-mountainous and northern plains, with 70% total coverage in the summer monsoons. Regarding temperature, hot summers can exceed 40°C and mild winters can move just above freezing. Due to the favourable agro-climatic conditions of Pakistan, citrus is grown in all the provinces (Punjab, Sindh, Khyber Pakhtunkhwa, and Baluchistan); however, Punjab stands out as the largest citrus producer that accounts for > 94% of production (Trade Development Authority of Pakistan, 2022). Main citrus producing districts in Punjab are Sargodha, Mandi Bahauddin, Toba Tek Singh, Faisalabad, Layyah, Bahawalpur, Bhakkar, Multan, and D.G. Khan (Figures 9.1 and 9.2). Kyber Pakhtunkhwa is predominant in orange production (Musambi and Blood Red) among all citrus fruits in Pakistan.

FIGURE 9.1 Major citrus producing regions in Pakistan.

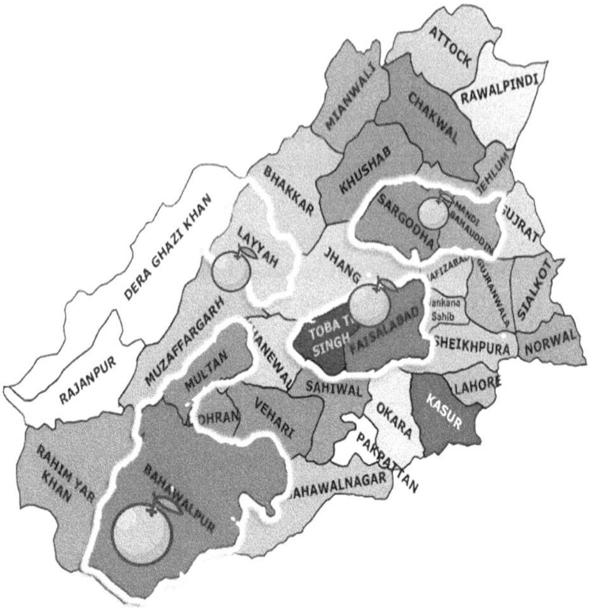

FIGURE 9.2 Major citrus producing districts in Punjab.

Although citrus fruits are known in the Indo-Pak sub-continent for centuries, but citrus culture remained localized, and it is for the last 4–5 decades that efforts to organize its cultivation seemed to be started. It has now become an important item for export. Before independence in United India, citrus acreage comprised of diverse varieties and kinds which were scattered all over the sub-continent. In the twenties, the total area of citrus fruits in the region's now called Pakistan was only a few thousand acres and citrus growing was mainly confined to the canal colonies of the Punjab plains and some area in Peshawar valley. There was very negligible citrus culture in the rest of this region at that time. With the dawn of independence citrus industry got a great boost like almost all other important industries in the country. Chronological sequence of citrus development in Pakistan indicates appreciable growth of citrus industry in this country during all these years (Figures 9.1 and 9.2; Tables 9.1 and 9.2).

Kinnow mandarin was introduced from California to Punjab Agriculture College and Research Institute, Lyallpur (now Faisalabad) in this sub-continent in the year 1943–1944 along with Feutrell's Early from Australia. Both varieties have brought revolution in the citrus industry of this country especially the Kinnow mandarin. Nursery plants of these citrus varieties were given to the growers for commercial production first time in the year 1953. Now citrus fruits dominate both in area (≈ 156 thousand acres) and production (2,229 thousand tonnes) in Pakistan (Pakistan Bureau of Statistics, 2024). The region comprising of Sargodha division is popularly known as California of Pakistan because of citrus production.

9.1.2 Production Trends

Over the last decade, the citrus industry in Pakistan has witnessed notable shifts in production trends, with varying impacts across different provinces. Punjab remains the epicentre of citrus production, contributing the major share to the country's overall output (Akhtar et al., 2023). The province accounts for 94% of Pakistan's total citrus acreage and production (Pakistan Bureau of Statistics, 2024). Over the past ten years, the area under citrus cultivation in Punjab has steadily decreased. This decline can be attributed to the expansion of housing societies, poor grove managements, exotic insects/pest/disease incidence, postharvest storage issues, and substandard marketing etc (Wang, 2019). Citrus

TABLE 9.1
Area under Citrus Cultivation in Pakistan (1000 Hectares)

Year	Punjab	Sindh	Khyber Pakhtunkhwa	Balochistan	Pakistan
2011–12	183.6	4.9	4.1	1.4	194.0
2012–13	183.3	5.1	4.2	1.4	194.0
2013–14	183.2	5.1	4.0	1.4	193.7
2014–15	182.6	4.9	3.8	1.5	192.8
2015–16	181.8	4.9	3.9	1.6	192.2
2016–17	177.2	4.2	3.9	1.6	186.9
2017–18	174.2	4.0	3.9	1.7	183.8
2018–19	170.4	5.5	4.0	1.7	181.6
2019–20	145.2	5.6	3.8	1.7	156.3
2020–21	145.3	5.7	3.7	1.8	156.5
2021–22	145.3	5.8	3.9	1.8	156.8
2022–23	145.1	5.8	3.7	1.6	156.2

Source: Pakistan Bureau of Statistics, 2024

TABLE 9.2
Citrus Production in Pakistan (1000 Metric tonnes)

Year	Punjab	Sindh	Khyber Pakhtunkhwa	Balochistan	Pakistan
2011–12	2076.8	30.9	32.6	7.0	2147.3
2012–13	1930.1	31.4	33.3	6.9	2001.7
2013–14	2097.7	30.9	31.6	7.5	2167.7
2014–15	2328.1	29.7	30.9	6.9	2395.6
2015–16	2276.1	29.5	31.7	6.8	2344.1
2016–17	2116.5	26.3	30.3	7.0	2180.1
2017–18	2289.1	25.0	30.3	6.9	2351.3
2018–19	2397.3	31.6	31.3	8.4	2468.6
2019–20	2288.6	32.9	28.0	8.6	2358.1
2020–21	2548.9	33.5	29.2	10.0	2621.8
2021–22	2296.8	34.1	30.4	10.9	2372.1
2022–23	2160.4	33.7	29.5	5.9	2229.5

Source: Pakistan Bureau of Statistics, 2024

production is gaining popularity among Sindh farmers since the province's warm climate supports the cultivation of various citrus fruits, including sweet oranges, grapefruits, and lemons. However, water scarcity and salinity issues have posed significant challenges. Khyber Pakhtunkhwa (KPK), while not as dominant as Punjab, has also seen growth in its citrus industry (Rana et al., 2022). The province's favourable climatic conditions in certain areas are conducive to citrus cultivation, particularly in the regions of Peshawar, Mardan, and Swat. Balochistan has the smallest share of citrus production among the provinces but holds potential for growth due to its unique agro-climatic conditions. The province primarily grows Kinnow and sweet oranges (Khan, 2019).

Overall, the citrus industry in Pakistan has demonstrated mixed growth trends in citrus production from 2011 to 2023 (Tables 9.1 and 9.2). The national acreage under citrus cultivation has decreased from approximately 194 thousand hectares (2010) to 156 thousand hectares (2023). However, citrus production remains consistent from 1982 thousand Metric tonnes (2010) to 2029 thousand Metric tonnes (2023) (Pakistan Bureau of Statistics, 2024). This growth can be attributed to several factors, including government initiatives aimed at promoting citrus cultivation, the introduction of improved farming techniques, better pest and the disease management, and the expansion of irrigation infrastructure (Raza et al., 2021; Ritenour et al., 2003). Additionally, there has been a growing emphasis on research and development to introduce new citrus varieties that are more resilient to climatic changes and have higher yield potential.

Pakistan has a tremendous potential for the expansion of the citrus industry due to suitability of soil and climatic conditions for citrus production. Any break-through in the production, processing, marketing, and export of citrus will have a direct impact on the rural life. New job opportunities could be created through intensive citriculture for unemployed and under employed rural labour force. Like all major crops produced in this country, the province of Punjab is the leader in fruit production. More than 65% of the total fruit production is concentrated in Punjab and 94% of the citrus is produced here. About 70% of the citrus grown in the province of Punjab is Kinnow mandarin. Most of the citrus orchards belong to the small farmers. The average yield of citrus fruits in Pakistan is a low 14.2 tonnes per hectare. For a simple comparison, the average yields of citrus fruits in Brazil are 26 tonnes per hectare.

9.1.3 Consumption

Pakistan citrus industry is dominant by Kinnow production, used as fresh market produce. Sweet oranges, grapefruits, lemons, and limes are also produced and consumed due to their health benefits, which include being rich in vitamin C, fibre, and pectin, etc (Miller and McDonald, 1999). In addition to direct consumption, citrus fruits are extensively used in Pakistani cuisine to add flavour to dishes, salads, and beverages. Lemons and limes are essential ingredients in traditional recipes and home remedies (Mahajan and Sadana, 2023). The domestic market's demand for citrus is robust, driven by the population's preference for fresh, healthy, and locally produced fruits. This strong local consumption helps to stabilize the industry and provides a reliable market for farmers, complementing the sector's export activities. Fresh fruit market in Pakistan is a vibrant sector characterized by a network of packing houses, processing units, and distribution channels that ensures the delivery of high-quality produce to both domestic and international markets. Packing houses handles the postharvest processing of citrus fruits that are responsible for washing, grading, sorting, waxing, and packing the fruits to meet market standards (Zacarias et al., 2020). A flow chart of the postharvest processing chain is illustrated in Figure 9.3. In recent years, a lot of improvements have been introduced in terms of state-of-the-art machinery and technology to enhance efficiency and output. For local markets, packed fruits are transported to wholesale markets, retail outlets, and supermarkets across the country (Figure 9.4).

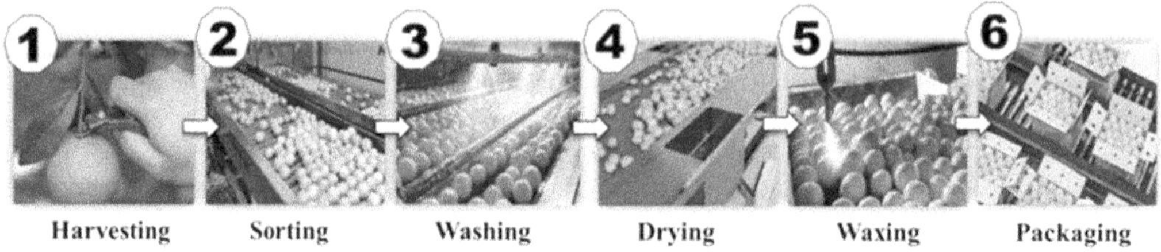

FIGURE 9.3 Packing house processes for fresh market citrus fruit.

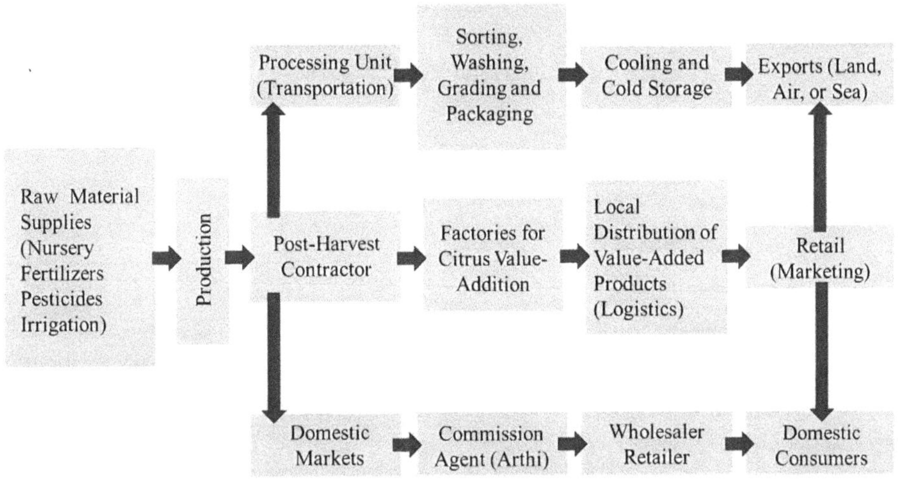

FIGURE 9.4 Citrus value chain in Pakistan.

9.1.3.1 Export Market Citrus Value

According to Trend Economy, in 2023 citrus production season, Pakistan's exports of commodity group 0805, "Citrus fruit, fresh or dried", amounted to $77 million, representing ~ 34% decrease in value compared to production season 2022. This drop was accounted for $40 million compared to 2022, when the total export was $117 million (Table 9.3). In 2022, exports of the commodity group 0805 accounted for 0.38% of Pakistan's total exports, which were valued at 30 billion dollars. This share declined by 0.213% than 2021, when it was 0.594%, with total exports amounting to $28 billion. Citrus is exported to Afghanistan and the United Arab Emirates that accounts major share of 31% and 17%, respectively. Other importing countries are Russia, Phillipines, Saudia Arabisa, Kazakhstan, Oman, Indonesia, Sri Lanka, Qatar, etc. (Figure 9.5).

TABLE 9.3
Pakistan Total Citrus Exports (2012–2023)

Year	Value (US $)	Value growth (%)
2023	77,256515.55	-34.03
2022	117,115,958.40	-31.54
2021	171,075,438.83	-16.72
2020	205,441,946.97	28.21
2019	160,227,144.46	-9.90
2018	177,847,740.52	27.04
2017	139,587,293.17	-18.59
2016	171,471,955.00	-7.02
2015	184,792,505.00	-4.04
2014	192,577,267.00	12.36
2013	171,384,105.00	9.94
2012	155,877,476.00	12.42

Source: Trend Economy, 2024. Available at: https://trendeconomy.com/data/h2/Pakistan/0805

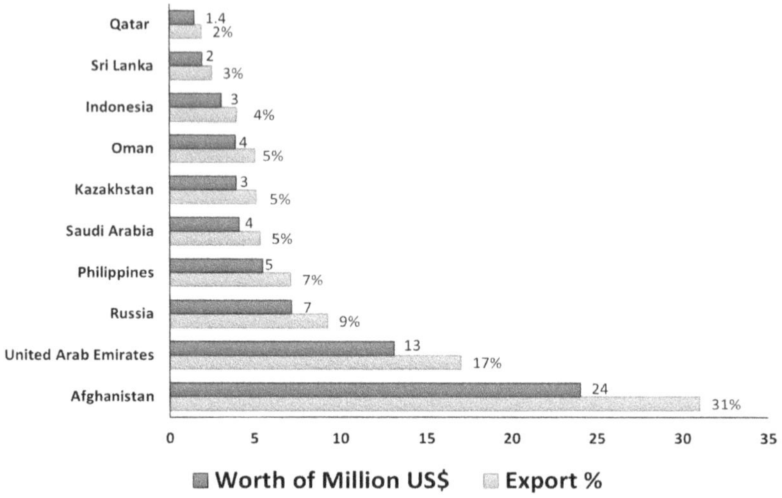

FIGURE 9.5 Countrywide citrus export from Pakistan. (Pakistan Bureau of Statistics, 2024).

9.1.3.2 Processing Industry

Citrus is a non-climacteric fruit, and it is harvested when reaches the desired juice content of 30–40%. There are numerous varieties of citrus fruits in the international market, but commercially, sweet oranges, mandarins, grapefruits, lemons, and limes are preferred due to their unique flavours, ranging from sweet to sour, and consumer preferences. There are 52 factories in Pakistan that process fruit and extract juice, which is only about 6% of total citrus production in the country. Sargodha division in Punjab province is the home to most of these factories. Grade A and B fruits are processed for fresh fruit market, both local and export market while grade C is turned into juice concentrate, which is then distributed to various supermarkets and retail stores (Siddique and Garnevska, 2018). Establishment of the processing industry for Pakistan citrus industry is of dire need. Citrus juice concentrate plants like Citropak at Sargodha have increased the demand for raw material, hence the need for higher production of citrus fruits is now being felt very much by our growers. The area under different fruits in general, no doubt, increased many folds within the last 4–5 decades but per hectare yield unfortunately remained almost static, so the goal of having higher fruit production per hectare level has now gained much importance. It can only be achieved using improved management practices and through new plantation systems. The conventional wide plantation system must be substituted by high-density orcharding with dwarf plants. High-density orchards are claimed to be much more productive, and this technology has proved useful in the case of raising some important citrus fruits.

9.1.4 Citrus Research and Development Infrastructure

Research and development (R&D) plays a critical role in the ongoing transformation of the citrus industry in Pakistan. Agricultural research institutions are working on developing new citrus varieties that are more resistant to pests and diseases, have higher yields with enhanced flavours, and are better suited to the micro-climate regions within the country. Collaborative efforts between research bodies, universities, and international agricultural organizations are fostering innovation and knowledge exchange, driving the citrus industry towards more sustainable and productive practices.

In general, citrus trees are sensitive and prone to biotic and abiotic stresses. Abiotic stresses (drought, salinity, heat) pose serious challenges to citrus growth and productivity due to climate change. Also, in recent years, outbursts of certain insect pests have been observed with higher infestation rate on fruit plantation. Among biotic stresses, pathogens (bacteria, viruses, nematode, fungi) cause citrus diseases such as citrus tristeza virus (CTV), citrus greening/Huanglongbing, phytophthora, citrus scab, melanose, anthracnose, greasy spot, and citrus canker are the most common. In Pakistan, citrus production per unit area is already much lower (14.2 Metric tonnes/ hectare) compared to the developed countries (26 Metric tonnes/hectare) (U.S. Department of Agriculture, 2024), along with reduced tree life span, and compromised fruit quality thus adversely affect export potential. Also, Kinnow mandarin is a late variety with a short harvest window of 2–3 months.

Citrus growers face several issues, including non-availability of disease-free planting material, lack of diversification in citrus germplasm, poor orchard management/plant protection, and improper fruit handling, postharvest management issues, and value addition in the citrus supply chain. To combat with the current challenges relating to citrus productivity enhancement, new germplasm introduction to enhance the citrus harvest window and fruit quality improvement, following institutions are actively involved in collaboration with national/international partners like PARB, ALP, HEC, ACIAR, USDA, CABI etc.

- University of Agriculture, Faisalabad
- PMAS-University of Arid Agriculture, Rawalpindi
- Citrus Research Station, Sahiwal
- Citrus Research Institute (CRI), Sargodha
- Barani Agricultural Research Institute (BARI), Chakwal
- National Agricultural Research Centre (NARC), Islamabad
- Ayub Agricultural Research Institute (AARI), Faisalabad
- Agriculture Research Institute, Tarnab, Peshawar

9.2 CHALLENGES OF CITRUS INDUSTRY

9.2.1 Stagnant Production

Citrus is one of the major fruits in Pakistan especially in Punjab. About 94% of citrus is grown in Punjab in districts Sargodha, Layyah, Toba Tek Singh, Sahiwal, Mandi Bahauddin, Bahawalpur, and Faisalabad. Sargodha is one of the leading cities in Pakistan which produces citrus in larger quantities (~ 45%) than others and exports to different cities. Sargodha is known as the "California of Pakistan" because of its massive citrus production. Citrus fruits include mandarins (Kinnow), sweet oranges, grapefruits, lemons, and limes, with mandarin (Kinnow) being the most important and popular among consumers. Pakistan's annual production of citrus fruit, primarily Kinnow, is approximately 2.0 million metric tonnes. Table 9.1 shows that in 2018–2019 citrus production in Pakistan decreased 181 thousand Metric tonnes (2018–2019) to 156.2 thousand Metric tonnes in 2023–2024 production season (Pakistan Bureau of Statistics, 2024). Table 9.2 shows that citrus production area declined in the last five years (2,448 thousand hectares to 2,229 thousand hectares from the year 2019 to 2024) (Pakistan Bureau of Statistics, 2024). In Pakistan, citrus production per unit area is ~50% lower than major citrus producing countries (U.S. Department of Agriculture, 2024) along with concerns related to reduced tree life span and poor fruit quality at harvest and during postharvest storage.

9.2.2 Monoculture

Pakistan citrus industry is predominant by the monoculture of Kinnow mandarin. Kinnow is admired among consumers due to its attractive peel colour, size, easy peeling, high juice content, and its excellent taste and aroma. However, there are some demerits as Kinnow is a late season variety, have alternate bearing and high seed content. Therefore, the citrus industry of Pakistan needs diversification by adding cultivars for fresh consumption and processing (juice industry). Different methods employed for variety development are hybridization and introduction. Hybridization is a tedious and long-term procedure. For instance, Kinnow, a hybrid between King (*Citrus nobilis*) and Willow leaf (*Citrus deliciosa*) mandarins, was developed in 1915 by Dr. H.B. Frost at Citrus Research Center, University of California, Riverside, USA. After a long evaluation period of 20 years, it was released in 1935 as a new variety for its commercial cultivation. Introducing new citrus germplasm remains the most viable.

9.2.3 Conventional Nursery Propagation System

In Pakistan, the nursery business is based on a traditional practice. Most of the significant diseases of citrus (citrus tristeza virus, citrus greening, phytophthora, etc.) are transmitted through budding/grafting and spread through infected nursery material (Fan et al., 2010; Folimonova, 2020). This ultimately leads to poor-quality fruits and weak, unhealthy orchards (Graham and Feichtenberger, 2015; Hayat et al., 2022).

Commercially, citrus plants are grown with a combination of two parts called rootstock and scion joined through grafting or budding due to various added benefits including fruit quality enhancement (Kamatyanatti and Singh, 2019). A rootstock is the lower part of the plant consisting of the root system which is grown usually from seeds and provides a root system responsible for mechanical support, water, and feed nutrients to the plant helping to grow scion cultivars for adaptation to the environment (Qureshi et al., 2023). Most of the citrus rootstock are not polyembryonic and do not produce nucellar plants, which are not identical to mother plants (Karabiyik and Eti, 2022). Rootstock selection depends upon various factors which may include compatibility with scion variety, ease of availability, degree of polyembryony, tolerance to different types of soils, soil salinity, soil pH, and various abiotic factors. Rootstock used in Pakistan is not well tolerant to biotic stresses like soil-borne pests and diseases (Singh et al., 2018). Rough lemon is commonly used rootstock for Kinnow mandarin in the Punjab province while sour orange in the Khyber Pakhtunkhwa province (Khurshid 2015). Nurseries act as the main source of the spread of diseases in the newly established citrus groves in Pakistan. Hence certified citrus nursery plants are the preliminary step towards the healthy and long productive life of the citrus crop. True to its type and a clean material is very much needed for quality citrus crops.

9.2.4 Poor Agronomic Practices

Pakistan is under serious threat of climate change (Shafqat et al., 2021a; 2021b). Multiple abiotic factors, including extreme temperature change, drought, flooding, poor drainage, salinity of irrigation water and substrates, etc. directly/indirectly influence citrus production. In Pakistan, citrus production is significantly lower compared to major citrus producing countries such as Brazil, China, etc. and achieving an acceptable Kinnow quality (medium size, uniform peel colour, free from blemishes, insects, and pests etc.) remains a major challenge. Square planting system for citrus (18 x 20 ft) remains the accepted geometry in Pakistan in comparison to high-density plantation used in major citrus producing regions worldwide. Like the monoculture of Kinnow, rootstock (rough lemon) is commonly used for propagated citrus plants in Punjab, which has a demerit being vigorous in nature. There exists no major concept about pruning practices, that leads to become tree branches overcrowded, leading to poor air circulation and light penetration, resulting in low fruit yield and quality (Asrey et al., 2013). Citrus nutrition programme needs leaf and soil testing. Soil physical properties related constraints (clay gradient in soil profile, drainage/irrigation/waterlogging) and soil fertility constraints induced by soil pH, salinity (specific ion-and cumulative osmotic pressure effect), calcareousness (pedogenic or non-pedogenic $CaCO_3$) and nutrient depletion are the important pedological factors contributing to citrus decline and low yield. Macro- (N, K, Ca, Mg, and S) and micronutrients (Mn, Zn, B, Fe, and Cu) deficiency or excess can lead to stunted growth, reduced vigour, and poor structural integrity (Ibrahim et al., 2007), making the trees more susceptible to physical damage (Srivastava and Singh, 2009) and vulnerable to pests and diseases due to weak immune responses and poor root development (Ahmad et al., 2024). Imbalance nutrition adversely affects the quality and quantity of fruit, resulting in smaller, misshapen, and less marketable fruits, along with reduced yields (Khalid et al., 2021; Malik et al., 2021). Most commonly irrigation system used in Punjab region is flooding. Flood irrigation systems may seem simple and cost-effective, but can result in waterlogging, suffocating the roots by depriving them of essential oxygen, resulting in root rot and reduced nutrient uptake. Excessive moisture creates a conducive environment for fungal diseases and pests. The soil tillage in the citrus orchards is the most common practice, which in fact disrupts the shallow feeder root system in citrus trees, and is more prone to disease infestation rate; thus floor management should be given prime importance. Like many other fruit crops, the citrus is attacked by various insect pests (citrus psylla, leaf miner, fruit fly, whitefly, mealy bugs, and thrips) which not only hamper the quality and production of Kinnow, but the export value too (Franco et al., 2009; Graham and Timmer, 2003). Major damage is incurred due to attack of citrus psylla, leaf miner, and fruit flies as well as synergistic infestation of both fungal and bacterial diseases (such as phytophthora, CTV, citrus canker, scab, melanose, citrus greening etc.), negatively affect its export quality and quantity. In addition, fruit flies cause severe losses of citrus fruits at the farms (Bar-Joseph et al., 1989). Despite intensive chemical spray, citrus growers are unable to get rid of this menace and are not successful in getting healthy and quality fruits from citrus groves. The quantitative and qualitative fruit losses occur up to 30–40%. *Xanthomonas citri* subsp. citri has caused substantial economic losses, leading to the rejection of citrus shipments by the European Union.

9.2.5 Postharvest Management

Punjab has experienced an increase in scab and melanose cases, especially in Kinnow fruit, leading to higher fruit rejection rates (20–50%) and low export prices. Approximately 50–60% of Kinnow production is consumed domestically, while nearly 30% is lost during postharvest (Rafique et al., 2020). Also, there is a greater number of C grade fruit (low quality with more fruit blemishes) that coincides with production challenges and results in marketable fruit with low prices. Reduced shelf life may also be attributed to poor storage facility, poor harvesting techniques, poor packaging materials, improper training for workers, instability of market prices etc.

9.3 FUTURISTIC APPROACH FOR CITRUS INDUSTRY

9.3.1 Germplasm Enhancement and Diversification

Pakistan's citrus industry is Kinnow based fresh market. Introduction of new citrus germplasm (exotic materials from China, USA, Australia etc.) remains the most viable option, which is easy and short term. New accessions of citrus germplasm can be introduced either as budwood (scion) or seed (rootstock). New germplasm can be evaluated about its adoptability to new agro-climatic conditions. Developing varieties that can withstand extreme weather conditions through genetic modification and selective breeding is crucial. Enhancing the ability of citrus plants to thrive in saline environments, which is becoming increasingly important as freshwater resources dwindle

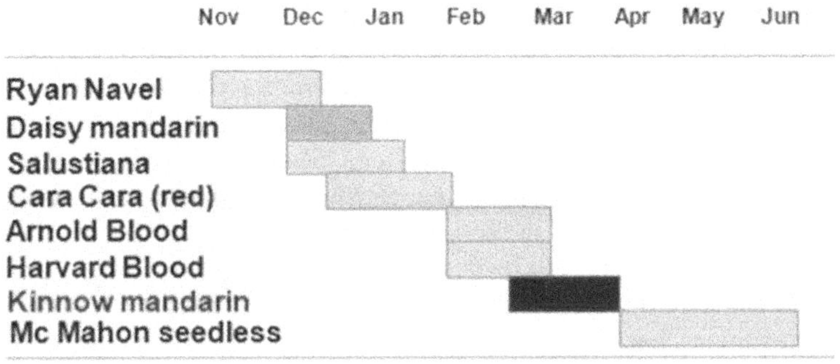

FIGURE 9.6 Expansion of citrus harvest window by introduction and promotion of citrus germplasm.

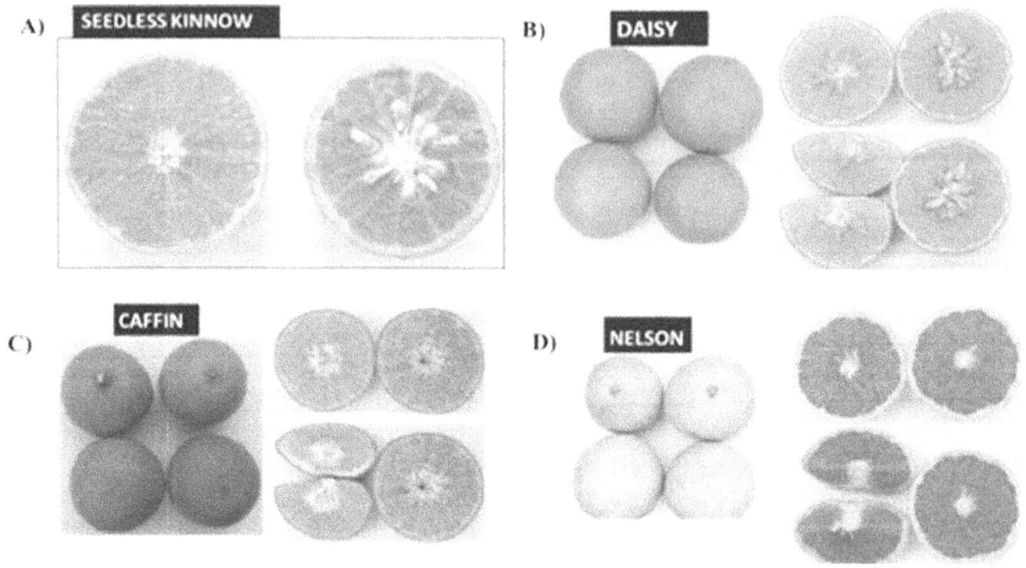

FIGURE 9.7 Citrus cultivars like seedless Kinnow (A), Daisy (B), Caffin (C), and Nelson (D) for new plantations.

(Rauf et al., 2013). Another strategy is top working/variety change by grafting established trees to new scion varieties. To expand the citrus harvest window by introducing new citrus varieties is crucial and already has been introduced from USA and Australia (Figures 9.6 and 9.7) and efforts are now being made to import germplasm from China.

9.3.2 Budwood Certification System

A budwood certification system in citrus is designed to ensure that budwood, which is used for grafting and propagation of citrus plants, is free from diseases and pests (Khan and Jaskani, 1992; Abbas et al., 2008). This system involves several key components and steps. 1. Protected foundation block that includes healthy plants for local and foreign varieties recovered by shoot-tip grafting *in vitro* (Jaskani et al., 1992; Naz et al., 2007). Also, plants are grown in containers inside insect-proof screenhouse. 2. Budwood increase blocks that maintain inside screen- or greenhouses at each nursery, including plants propagated directly from foundation trees to increase the number of buds for propagation of certified trees. 3. Certified nursery trees propagated with budwood from the increase blocks and grown in screen-or greenhouses or in the open field. Similarly, seed for rootstocks originates from a certified seed source tree. Regarding Pakistan citrus industry, conventional citrus nursery production system is in practice; however, it is pivotal to improve future development in citrus sector via budwood certification programme. Hence, universities and institutes engaged in citrus research may establish foundation and multiplication blocks and provide certified nursery plants as a primary source of propagation material. However, during the certification process, consideration is given to 1) selection of mother plants carefully based on their health status and genetic quality. 2) Monitor source trees regularly and test for pathogens such as viruses, bacteria, and fungi using molecular techniques like PCR (Polymerase Chain Reaction). 3) Isolate new source trees or budwood from other plants to ensure disease-freeness before integration into the certification programme. 4) Meet specific guidelines and standards for budwood to be certified by regulatory bodies (Kamatyanatti and Singh, 2019). Once a source tree is confirmed to be disease-free, budwood is harvested. The harvested budwood is handled and stored under conditions that prevent contamination. Certified budwood is distributed to nurseries and growers keeping proper documentation. Figure 9.8 represents the difference between a traditional nursery and a modern budwood certification programme.

FIGURE 9.8 A traditional nursery and modern budwood certification.

9.3.3 Climate-Resilient Rootstocks

Climate-resilient rootstocks in citrus are crucial for maintaining productive and healthy citrus groves amid changing environmental conditions. These rootstocks are selected or developed to withstand various climate-related stresses, such as extreme temperatures, drought, salinity, and diseases/pest infestation, etc (Aslam et al., 2017). Rootstocks that can sustain growth and fruit

TABLE 9.4
Available Rootstock Accessions at University of Agriculture, Faisalabad

S.N.	Rootstock	S.N.	Rootstock	S.N.	Rootstock
1	Standard	17	Madam vinous	33	Gada dehi
2	Kinkoji	18	Flying dragon	34	X 639
3	Troyer	19	Keen sour orange	35	Yuma citrange
4	Nansho dai dai	20	Yuzu	36	Indian
5	Brazillian sour orange	21	Pineapple	37	Benton
6	Frost	22	Var Sanguinea	38	Savage citrange
7	Sour orange	23	Eremocitrus glauce hybrid	39	Corsican
8	Srirampur nucellar	24	Tuningmeng nucellar	40	Glycosmis pentaphylla
9	Bengal	25	Frazer hybrid	41	Improved meyer
10	Perrine	26	Cleopatra mandarin	42	Clausena harmandiana
11	Rich 16-6	27	Cox mandarin hybrid	43	Sun chu shah
12	S-1	28	Gou tou cheng	44	Papuan
13	Philippine	29	Orlando	45	Tangor
14	Schaub	30	Borneo Rangpur	46	Carrizo citrange
15	Valencia	31	Rangpur poona nucellar		
16	Knorr nucellar	32	Swingle		

production with limited water availability are essential in areas prone to drought. They often have deep root systems or other adaptations that allow them to access and use water efficiently. In regions where soil salinity is a problem, salt-tolerant rootstocks can help prevent salt from accumulating in the plant tissues, thus avoiding toxicity and ensuring healthy growth. Certain rootstocks are tolerant to diseases such as citrus greening (Huanglongbing or HLB), phytophthora root rot, and citrus tristeza virus, etc (Bove, 2006; Johnson and Graham, 2015). This tolerance is vital for sustaining citrus production in areas where these diseases are prevalent. Rootstocks that can tolerate extreme temperatures, both high or low, help protect the citrus trees from heat stress and frost damage. Different rootstocks can adapt to various soil types, including heavy clay, sandy, or calcareous soils, enhancing the overall resilience of the citrus orchard. Efficient nutrient uptake, particularly in poor soils, is another critical trait. Rootstocks exhibiting better nutrient uptake efficiency can maintain tree health, productivity, and fruit quality attributes (Qureshi et al., 2023).

Pakistan's citrus industry is dominated by two rootstocks i.e. Rough lemon in Punjab Province and Sour orange in Khyber Pakhtunkhwa province. However, new rootstocks are introduced (Table 9.4) in Pakistan and are under trials. Kinnow mandarin grafted on eight rootstocks (Cox mandarin, Fraser hybrid, Troyer citrange, Cleopatra mandarin, *Poncirus trifoliate*, Benton, C-25, and Carrizo strange) indicated that *Poncirus trifoliata* and Fraser hybrid had the best performance (greater fruit diameter, fruit weight, total soluble solids, vitamin C, and total antioxidants) among treatments, whereas Troyer citrange and C-35 exhibited poor fruit quality and growth traits (Qureshi et al., 2021). In another study, Shafqat et al. (2021a) evaluated ten diverse citrus selections [bitter sweet orange (*Citrus sinensis*), Brazilian sour orange (*Citrus aurantium*), *Citrus obovoideae*, Chakotra (*Citrus maxima*), Sour orange (*C. aurantium*), Keen sour orange (*Citrus aurantium* L.), Gadha dahi (*C. aurantium*), Rough lemon (*Citrus jambhiri*), Savage citrange (*Citrus sinensis* x *Poncirus trifoliata*) and Yuma citrange (*Citroncirus* spp.).] against heat (moderate and high heat were 38°C and 46°C), water deficit (25% and 50% field capacity) and combined (46°C with 25% field capacity and 38°C with 50% field capacity). Overall,

Brazilian sour orange and Keen sour orange performance was excellent that is attributed to reduced desiccation and maintained stomatal conductance, water potential, moisture content thus higher photosynthesis under low soil moisture and high temperature stresses. Furthermore, high expression of *CsTIP2* and *CsTIP1* genes were found in initial 1.5 to 24.0 h under water stress, and *CsTIP1* genes had higher expression pattern under combined effect of water deficit and heat in the initial 1.5 h.

9.3.4 Precise Agronomic Practices

An ideal citrus grove layout should achieve the basic key goals: 1) maximizing the number of trees planted per hectare; 2) ensuring enough space for each tree to grow properly; 3) allowing for easy maintenance and management activities within the grove (Kumar et al., 2021). Different methods (square, triangular, rectangular, high-density plantation, etc.) of citrus plantation are used, all of these have their own advantages and disadvantages. The most prevalent citrus geometry in the world is the high-density plantation system that is efficient for hastening productivity per unit area, effective utilization of land and other resources, better canopy management, farm mechanization, convenient spray of pesticides and harvesting, improved fruit quality and higher net return to farmers (Singh et al., 2022). Among citrus grove management strategies, pruning remains the most vital to improve tree health and productivity, especially in high-density plantation. Proper pruning of citrus can provide many benefits including reduction of fruit blemishes, disease inoculum removal, canopy aeration, fruit quality development, pesticide applications etc (Matias et al., 2023). Proper soil moisture directly influence fruit growth, as it decreases the fruit size if not available in an adequate amount. The water needs of a citrus orchard are influenced by several factors, such as the crop's growth stage and environmental conditions (temperature, rainfall, relative humidity). Maintaining appropriate soil moisture levels is crucial for proper fruit set and achieving optimal fruit growth and development, ensuring optimum and sustainable production. For instance, drip irrigation is the most effective method for watering citrus groves. Drip irrigation delivers water directly to the root zone, slowly through nozzle on rubber pipes. Drip irrigation is ideal for arid regions with salty soil, where canal water can be stored and reused for irrigation. Also, it is beneficial for high-density planted Kinnow groves since water-soluble fertilizers like urea, potassium nitrate, and mono-ammonium phosphate can be effectively applied through this system, meeting the citrus plants' nutritional needs. For existing established orchards, in furrow irrigation, water flows from one end of a raised channel to the other end on both sides of the trees. These channels are connected to a main supply chain. Fruit yield can be increased by 30% using furrow irrigation system when compared to flooding system in Kinnow orchards (Raza et al., 2021). Regarding balanced nutrition programmes, high yield can be maintained using proper fertilizer management. Perennial fruit trees have a high micronutrients requirement, as the quantity of nutrient elements removed by citrus fruits from the soil should be kept in mind. It is estimated that a yield of 40 tonnes/ha of Kinnow mandarin under Punjab conditions removed 96 kg N, 10. 80 kg P and 78.80 kg K per ha from the soil annually (Dhatt et al., 1992). If losses due to leaching and erosion are also taken into consideration, the total annual depletion of tree food is even higher (Shirgure and Srivastava, 2013; Srivastava et al., 2022). General recommendations for N, P, and K are shown in Table 9.5. Balanced nutrition programme may help to boost immunity (systemic acquired resistance), resulting in improving tree health and fruit productivity. Literature suggests that plant growth regulators (auxins, gibberellins, and cytokinins) have potential to improve tree health and fruit quality at harvest in sweet oranges. Regular addition of organic matter helps in improving soil physicochemical conditions and ultimately improves soils health, that leads to improved tree vigour and growth (Khalid et al., 2016). Weed management strategies by mulching

TABLE 9.5
Best Recommended Dose of NPK for Citrus Trees

Age of trees	N (g) per tree	P (g) per tree	K (g) per tree
1 year	100	50	50
2 year	200	100	100
3 year	300	150	150
5 year	500	250	250
10 year	1000	500	1000

(black polyethylene and wood chips), mechanical (rotavator and cultivator) and chemical (glyphosate) have been studied (Sajid et al., 2018). Wood chips compared to mechanical weed control increased tree growth and fruit yield as well as improved soil physical properties by reducing bulk density and soil penetration resistance, improving soil organic matter, hydraulic conductivity, and soil microbial biomass carbon (Krajewski, 2020).

Citrus crops are susceptible to various pests like citrus psyllids, aphids, and citrus leaf miners etc., which can damage trees, reduce fruit quality, and spread diseases (for instance, Huanglongbing/citrus greening disease). Effective pest management strategies involve integrated approaches, including cultural practices, biological controls, and judicious use of pesticides. Timely monitoring, early detection, and intervention are essential to mitigate pest damage and sustain citrus yields, ensuring economic viability for growers while promoting sustainable agricultural practices (Rogers, 2008). Asian citrus psylla (ACP) *Diaphorina citri*, is the most destructive pest of Citrus Groves (under A1 quarantine pest list) as it transmits a phloem-limited bacterium, *Candidatus* Liberibacter asiaticus, which causes the spread of citrus catastrophic disease "Huanglongbing" (HLB) or citrus greening disease and results in decline and death of infected trees. Fruit from HLB-affected trees are small, lopsided, acidic and bitter in taste, which makes them unmarketable. There is no cure for HLB to date and no citrus germplasm is resistant against HLB. Using proper nutrition, irrigation, and integrated pest management/ integrated disease management are mitigation strategies to reduce HLB disease severity symptoms (Qureshi et al., 2009).

9.3.5 Postharvest Practices

Overtime cosmetic quality of fruit has been reduced due to rind blemishes resulting in high rejection rate (reaching 50–60% in some cases) at farm gate, causing very high economic loss to growers. On the other hand, lower volume of A and B grade fruit limits exportable quality fruit. A survey of Kinnow orchard showed that about 47.5% blemishes were related to diseases (mainly melanose + scab and canker), 29.5% linked to insect pest (mainly mites and thrips), and 15% physiological/physical nature (wind scarring, nutrients, branches press, stylar end deformity skin cracking, etc.) (Malik and Khan, 2014). Regular inspection and intensive care are needed to control the spread of diseases and pests before it reaches above the economic threshold level, ultimately improving fruit productivity and profitability. Strategies for managing biotic and abiotic factors to improve citrus quality have been reviewed (Malik et al., 2021). The fruit flies' threat to the citrus industry is considered a major barrier in meeting the national as well as international demand of Pakistani citrus. For sustainable and area-wide management of fruit flies in citrus groves, education and capacity enhancement of citrus growers is needed regarding strategic execution and implementation of fruit flies' integrated pest management (FF-IPM) approach/techniques. According to various reports of Pest Warning & Quality Control of Pesticides Punjab, the infestation of fruit flies is drastically increasing every year imposing great threat to citrus industry. Moreover, most of the farmers are illiterate and uninformed of the strategic Integrated pest management programme (IPM) for the management of fruit flies. By implementing FF-IPM in established citrus groves and postharvest quarantine cold treatment technique, the production and quality of citrus fruit can be improved, ultimately increase exports. To address these challenges, targeted interventions are urgently needed. Enhancing awareness, providing technical support, and promoting sustainable agricultural practices can empower citrus farmers of Pakistan to overcome these constraints, enhance crop resilience, and secure their livelihoods.

9.3.6 Fruit Quality

The quality of fruit in citrus is of paramount importance due to its direct impact on consumer acceptance, marketability, and economic value. High fruit quality ensures desirable characteristics such as taste, juiciness, sweetness, and aroma, all of which contribute to consumer satisfaction and repeat purchases. Beyond sensory attributes, quality also encompasses factors like texture, appearance, and nutritional content, influencing both fresh consumption and processing industries. Proper management practices, including timely thinning, balanced nutrition programmes, plant growth regulators, and effective disease control, are essential to maintain high fruit quality, ensuring desirable appearance, flavour, and overall market value of citrus produce. Improving cosmetic quality (freedom from blemishes on the surface) of "Kinnow" is extremely important for reducing high farmgate rejection, its marketing grade/class and price in domestic as well as international market (Malik and Khan, 2014). Consistency in fruit quality not only builds consumer trust but also supports premium pricing and competitiveness in the market, making it crucial for sustainable citrus production and trade. Alternate bearing is a major challenge for "Kinnow" growers and often fruit thinning is an important practice to maintain the fruit load on trees. Shafqat et al. (2023) evaluated "Kinnow" fruit thinning using different concentrations of chemicals [naphthalene acetic acid (NAA); 3,5,6-pyridyloxyacetic acid (3,5,6-TPA), and ethephon]. The results indicated that all chemical compounds used for fruit thinning increased fruit size and total soluble solids in comparison to the control. Fruit thinning improved fruit quality and fruit crop profitability of "Kinnow" through sustaining the carbohydrate supply and overcoming alternate bearing concerns. Phytohormones application (gibberellic acid (GA_3), 2,4 dichlorophenoxy acetic acid (2,4-D), putrescine, and spermine) before and after colour break stage to improve fruit quality in "Kinnow" mandarins was studied by Khalid et al. (2016). The GA_3 application before colour break significantly decreased reducing sugars, whereas after colour break application increased rind mass and total soluble solids, total and non-reducing sugars in comparison to the control. The 2,4-D improved reducing sugars, and putrescine enhanced mass loss.

9.4 CONCLUSION

The future of citriculture in Pakistan is bright. However, the status of the citrus groves is very shaky and is not expected to live up to expectations, unless investment in research and development is made in an equitable manner. There is a need for the initiation of long-term breeding programmes to develop new varieties and to improve Kinnow itself as well as to diversify the rootstock use, avoiding the risks of monoculture. All major citrus producing countries of the world have citrus budwood certification and clonal protection programmes to regulate the carry business, the point where the deterioration begins. The yields can only be improved by using healthy plants and observing the proper management practices like citrus geometry, irrigation, fertilization, pest and disease control, elimination of intercropping, postharvest management, and value addition etc. The final key to improving the income of the growers is the implementation of an effective marketing system and minimizing the postharvest losses by assigning a greater role to the processing industry and training in postharvest technology according to international standards.

ACKNOWLEDGMENT

The authors extend their sincere appreciation to Higher Education Commission, Pakistan (Project No. CPEC-CRG-2-447) for funds to complete this work.

REFERENCES

Abbas, M., Khan, M. M., Fatima, B., Iftikhar, Y., Mughal, S. M., Jaskani, M. J., Khan I. A., & Abbas, H. (2008). Elimination of Citrus tristeza closterovirus (CTV) and production of certified citrus plants through shoot-tip micrografting. *Pakistan Journal of Botany,* 40(3), 1301–1312.

Ackon, V. (2020). Citrus seeds as a potential source of oil for domestic, industrial and medicinal purposes. M.Phil. Dissertation, Department of Chemistry, School of Physical Sciences, College of Agriculture and Natural Sciences, University of Cape Coast, Ghana.

Ahmad, B., Mehdi, M., Ghafoor, A., & Anwar, H. (2018). Value chain assessment and measuring export determinants of citrus fruits in Pakistan: An analysis of primary data. *Pakistan Journal of Agricultural Sciences*, 55(3), 685–692.

Ahmed, N., Zhang, B., Chachar, Z., Li, J., Xiao, G., Wang, Q., ... & Tu, P. (2024). Micronutrients and their effects on horticultural crop quality, productivity and sustainability. *Scientia Horticulturae*, 323, 112512.

Akhtar, S., Faisal, M., Naseer, M. A. R, Javed, I., Nazir, A., & Raza, M. H. (2023). Revitalizing olive agriculture in Punjab, Pakistan: unleashing potential beyond marketing and production constraints. *International Journal of Agricultural Extension*, 11(3), 253–258.

Aslam, M., Ahmad, K., Akhtar, M. A., & Maqbool, M. A. (2017). Salinity stress in crop plants: effects of stress, tolerance mechanisms and breeding strategies for improvement. *Journal of Agriculture and Basic Science*, 2(1), 2518–4210.

Asrey, R., Patel, V. B., Barman, K., & Pal, R. K. (2013). Pruning affects fruit yield and postharvest quality in mango (*Mangifera indica* L.) cv. Amrapali. *Fruits*, 68(5), 367– 380.

Bar-Joseph, M., Marcus, R., & Lee, R. F. (1989). The continuous challenge of citrus tristeza virus control. *Annual Review of Phytopathology*, 27(1), 291–316.

Bove, J. M. (2006). Huanglongbing: A destructive, newly emerging, century-old disease of citrus. *Journal of Plant Pathology*, 88, 7–37.

Dhatt, A.S., Grewal, G.P.S., Chahil, B.S., & Dhillon, D.S. (1992). Removal of NPK by Kinnow mandarin and Urman ber. *Acta Horticulturae*, 321, 541-544. DOI:10.17660/ActaHortic.1992.321.61

Fan, J., Chen, C., Brlansky, R. H., Gmitter Jr, F. G., & Li, Z. G. (2010). Changes in carbohydrate metabolism in *Citrus sinensis* infected with 'Candidatus Liberibacter asiaticus'. *Plant Pathology*, 59, 1037–1043.

Folimonova, S. Y. (2020). Citrus tristeza virus: A large RNA virus with complex biology turned into a valuable tool for crop protection. *PLoS Pathogens*, 16(4), e1008416.

Franco, J. C., Zada, A., & Mendel, Z. (2009). Novel approaches for the management of mealybug pests. *Biorational Control of Arthropod Pests*, 5, 233–278.

GOP. (2021). Agreement on south Asian free trade area. Ministry of Commerce, Government of Pakistan.

Graham, J., & Feichtenberger, E. (2015). Citrus Phytophthora diseases: Management challenges and successes. *Journal of Citrus Pathology*, 2(1), 27203.

Graham, J. H., & Timmer, L. W. (2003). *Phytophthora diseases of citrus*. University of Florida/IFAS extension Document SL127.

Hayat, A., Haque, E. U., Nasir, M., Afzaal, S., Sabir, S., & Ahmad, S. (2022). Citrus nursery management and modern trends (pp. 95–108). In: Hussain, S., Khalid, M. F., Ali, M. A., Ahmed, N., Hasanuzzaman,

M., & Ahmad, S. (eds), *Citrus Production: Technological Advancements and Adaptation to Changing Climate*. CRC Press, Boca Raton.

Ibrahim, M., Ahmad, N., Anwar, S. A., & Majeed, T. (2007). Effect of micronutrients on citrus fruit yield growing on calcareous soils (pp.179–182). In: Xu, F., Goldbach, H. E., Brown, P. H., Bell, R. W., Fujiwara, T., Hunt, C. D., Goldberg, S., & Shi, L. (eds), *Advances in Plant and Animal Boron Nutrition*. Springer, Dordrecht. https://doi.org/10.1007/978-1-4020-5382-5_18.

Jaskani, M. J., Khan, I. A., & Malik, M.N. (1992). Micrografting of Kinnow mandarin (pp. 373–375). In: Khan I. A. (ed.), *Proceedings of the First International Seminar on Citriculture*. University of Agriculture, Faisalabad, Pakistan. ISBN 969-8237-00-7.

Johnson, E. G., & Graham, J. H. (2015). Root health in the age of HLB. *Citrus Industry*, 98, 14–18.

Kamatyanatti, M., & Singh, S. K. (2019). Recent advances in citrus propagation and planting (pp. 1–12). In: Singh, S., K., & Kaur, S. (eds), *Advances in Horticultural Crop Management and Value Addition*. Laxmi Publications Pvt Ltd, India.

Karabiyik, S., & Eti, S. (2022). Nucellar embryo development in some citrus rootstocks (pp. 33–58). In: Canturk, S., & Gokcen, I.S. (eds), *Current Agricultural Studies in Turkiye Research and Reviews*. Iksad Publishing House, Ankara.

Khalid, M. S., Khalid, S., Farooq, A., Shafique, M., Amjad, A., Razi, M. F. Ud., Sherani, J., & Ali, I. (2021). Characterization of Kinnow mandarin fruit blemishes and tree-fruit-environment (TFE) profile in relation to various blemishes causal agents. *Pure and Applied Biology*, 10(4), 1206–1216.

Khalid, S., Malik, A. U., Khan, A. S., Razzaq, K., & Naseer, M. (2016). Plant growth regulators application time influences fruit quality and storage potential of young Kinnow mandarin trees. *International Journal of Agriculture and Biology*, 18(3), 623–629.

Khan, I. A., & Jaskani, M. J. (1992). A synopsis for citrus budwood certification and clonal protection program (pp. 367–372). In: Khan, I. A. (ed), *Proceedings of the First International Seminar on Citriculture*. University of Agriculture, Faisalabad, Pakistan. ISBN 969-8237-00-7.

Khan, S. (2019). Climate classification of Pakistan. *International Journal of Economic and Environmental Geology*, 10(2), 60–71.

Khurshid, T. (2015). The enhancement of citrus value chains production in Pakistan and Australia through improved orchard management practices. Australian Centre for International Agricultural Research, Australia. https://research.aciar.gov.au/aik-saath/sites/_co-lab.aciar.gov.au.pdf

Krajewski, A. (2020). *Pakistan Citrus Growers' Handbook: Nutrition of Citrus*. International Citrus Technologies Pty Ltd Gingin, Western Australia.

Kumar, A., Ram, S., Bist, L. D., & Singh, C. P. (2021). High- density orcharding in fruit crops: A review. *Annals of the Romanian Society for Cell Biology*, 25(6), 948–962.

Mahajan, M., & Sadana, R. (2023). Citrus-based food products and their shelf life (pp. 367–393). In: Purewal, S. S., Bangar, S. P., & Kaur, P. (eds), *Recent Advances in Citrus Fruits*. Cham: Springer International Publishing.

Malik, A. U., Hasan, M. U., Khalid, S., Mazhar, M. S., Khalid, M. S., Khan, M. N., Saleem, B. A., Khan, A. S., & Anwar, R. (2021). Biotic and abiotic factors causing rind blemishes in citrus and management strategies to improve the cosmetic quality of fruits. *International Journal of Agriculture & Biology*, 25(2), 298–318.

Malik, A. U., & Khan, I. A. (2014). *Kinnow quality: Issues and strategies for improvement, survey report and citrus blemishes resource guide*. University of Agriculture, Faisalabad, Pakistan.

Matias, P., Barrote, I., Azinheira, G., Continella, A., & Duarte, A. (2023). Citrus pruning in the Mediterranean climate: A review. *Plants*, 12(19), 3360.

Memon, N. A. (2017). Citrus fruit (Kino): Punjab produced 98% of production. *Pakistan Food Journal*, November-December 2017 ed., pp. 23–25.

Miller, W. M., & McDonald, R. E. (1999). Chemical composition and physical properties of selected specialty citrus fruits. *Journal of Food Science*, 64(6), 998–1002.

Naz, A. A., Jaskani, M. J., & Abbas, H. (2007). *In vitro* studies on micrografting technique in two cultivars of citrus to produce virus free plants. *Pakistan Journal of Botany*, 39(5), 1773–1778.

Pakistan Bureau of Statistics. (2024). Fruits, vegetables and condiments statistics of Pakistan. Ministry of National Food Security and Research, Islamabad, Pakistan.

Qureshi, J. A., Stansly, P. A., & McCoy, C. W. (2009). Natural enemies of *Diaphorina citri* (Hemiptera: Psyllidae) in Florida and release of *Tamarixia radiata* as a biological control agent. *Florida Entomologist*, 92, 650–658.

Qureshi, M. A., Ashraf, E., Albaayit, S. F. A., Shafqat, W., Shareef, N., Din, S. U., Sadaf, S., Rashid, R., & Tasneem, S. (2023). Rootstock influence on performance of different citrus scion cultivars: A review. *Journal of Global Innovations in Agricultural Sciences*, 11, 273–283.

Qureshi, M. A., Jaskani, M. J., Khan, A. S., Haider, M. S., Shafqat, W., Asif, M., & Mehmood, A. (2021). Influence of different rootstocks on physico- chemical quality attributes of Kinnow mandarin. *Pakistan Journal of Agricultural Sciences,* 58, 929– 935.

Rafique, M., Altaf, F., & Khan, A. (2020). Analysis of post-harvest losses of Kinnow in Sargodha, Pakistan. *Journal of Agricultural Research*, 58(4), 293–300.

Rana, A. W., Raja, S., & Qureshi, T. (2022). Horticulture policy for Khyber Pakhtunkhwa. *PACE Policy Research Paper October 2022. Washington, DC: International Food Policy Research Institute (IFPRI).* https://doi.org/10.2499/p15738coll2.136463

Rauf, S., Iqbal, Z., & Shahzad, M. (2013). Genetic improvement of Citrus for disease resistance. *Archives of Phytopathology and Plant Protection*, *46*(17), 2051–2061.

Raza, A., Warraich, I. A., Nawaz, M. A., Asim, M., Aziz, A., Shireen, F., & Rahman, M. A. (2021). Does furrow irrigation system improve yield and water use efficiency of Kinnow mandarin (*Citrus reticulata* Blanco)? *International Journal of Agricultural Extension*, 8(3), 199–206.

Ritenour, M., Wardowski, W. F., & Tucker, D. P. H. (2003). *Effects of water and nutrients on the postharvest quality and shelf life of citrus.* University of Florida Cooperative Extension Service, Institute of Food and Agricultural Sciences, EDIS.

Rogers, M. E. (2008). General pest management considerations. *Citrus Industry*, 89, 12–17.

Sajid, M., Ahmad, S., Jaskani, M. J., & Yasin, M. (2018). Optimum weed control method increases the yield of Kinnow by improving the physical properties of soil. *Planta Daninha*, 36, e018168903.

Shafqat, W., Fatima, N., Jaskani, M. J., Ikram, S., Amen, R., Chattha, W. S., Haider, M. S., & Ferrarezi, R. S. (2023). Fruit thinning chemical agents improves fruit size and quality in 'Kinnow' mandarin (*Citrus reticulata* Blanco)-Rutaceae. *Pakistan Journal of Botany*, 55(6), 2381–2392.

Shafqat, W., Jaskani, M. J., Maqbool, R., Chattha, W. S., Ali, Z., Naqvi, S. A., Haider, M. S., Khan, I. A., & Vincent, C. I. (2021a). Heat shock protein and aquaporin expression enhance water conserving behavior of citrus under water deficits and high temperature conditions. *Environmental and Experimental Botany,* 181, 104270.

Shafqat, W., Naqvi, S. A., Maqbool, R., Haider, M. S., Jaskani, M. J., & Khan, I. A. (2021b). Climate change and citrus. In: Khan, M. S., & Khan, I. A. (eds), *Citrus: Research, Development and Biotechnology.* IntechOpen. DOI: 10.5772/intechopen.95488

Shirgure, P. S., & Srivastava, A. K. (2013). Nutrient-water interaction in citrus: recent developments. *Agricultural Advances*, 2(8), 224–236.

Siddique, M. I., & Garnevska, E. (2018). Citrus value chain (s): A survey of Pakistan citrus industry. In: Egilmez, G. (ed), *Agricultural Value Chain*. IntechOpen. DOI: 10.5772/intechopen.70161

Singh, N., Sharma, D. P., & Thakur, K.K. (2018). Effect of rootstocks and soil management on growth and physiological parameters in new plantations of apple under replant conditions. *Indian Journal of Horticulture*, 75(3), 392–398.

Singh, P., Kumar, P., Singh, G., & Kaur, L. (2022). Comparison of economic benefits between high density planting and conventional planting of Kinnow. *Indian Journal of Extension Education*, 58(3), 131–134.

Srivastava, A. K., Hota, D., Dahat, S., & Sharma, D. (2022). Citrus nutrition: An Indian perspective. *Annals of Plant Soil Research,* 24(1), 1–15.

Srivastava, A. K. & Singh, S. (2009). Citrus decline: Soil fertility and plant nutrition. *Journal of Plant Nutrition,* 32(2), 197–245.

Trade Development Authority of Pakistan (TDAP). (2022). Pakistan Citrus. Available at https://tdap.gov.pk/wp-content/uploads/2022/05/Citrus-Brochure-compressed.pdf

Trend Economy. (2024). Annual international trade statistics by country. Retrieved on August, 7, 2024 from https://trendeconomy.com/data/h2/Pakistan/0805

U. S. Department of Agriculture. (2024). United States Department of Agriculture Foreign Agricultural Service Citrus: World Markets and Trade. www.fas.usda.gov/data/citrus-world-markets-and-trade

Wang, N. (2019). The citrus huanglongbing crisis and potential solutions. *Molecular Plant*, 12(5), 607–609.

Zacarias, L., Cronje, P. J. R., & Palou, L. (2020). Postharvest technology of citrus fruits (pp. 421–446). In: Talon, M., Caruso, M., & Gmitter, F. G. Jr. (eds), *The Genus Citrus*. Woodhead Publishing.

10 Citrus Industry of Bangladesh

Md. Zahidur Rahman, M.H.M. Borhannuddin Bhuyan, Faruque Ahmed, and Jamilur Rahman

10.1 INTRODUCTION

Citruse fruits are popularly known around the world for their taste, fragrance, refreshing flavours, and health benefits. Like other countries, these fruits are also popular in Bangladesh. Due to the geographic position of this country, several citrus species are available and cultivated in the country. Moreover, Bangladesh lies within the tropical region of South Asia, which makes it highly favourable for growing different species of citrus fruits (Rao et al., 2011). Different kinds of citrus fruits are being cultivated around the country, among which mandarin orange, sweet orange, lemon, lime, pummelo, and citron are very popular in Bangladesh.

In Bangladesh, the cultivation of citrus fruits was started in the 16th century by the Mughals. Though some citruses were being cultivated in the land far before that, the popular species that we see today were brought by the Mughals (Wikipedia, 2024a). On the other hand, the commercial cultivation of these diversified citruses was started in this country just a few decades ago. Recently, a completely new industry has been developed based on fresh citrus and its allied products. Currently, around 1.5 million metric tonnes of citrus fruits are used annually in Bangladesh (BBS, 2023). At the same time, 0.4 million metric tonnes of citrus are being produced annually, and around 50 thousand acres of land are being used for developing citrus orchards which is around 7% of the total land used for the cultivation of fruits (BBS, 2023; DAE, 2024). Additionally, a foreign trade valuing average of $150 million USD is estimated yearly for these fruits and their products (OEC, 2024).

The citrus industry in Bangladesh is mainly concentrated in the hilly and semi-hilly regions of the country, where the climate and soil conditions are very favourable for citrus cultivation (Nazirul et al., 2017). Sylhet, located in the northeast region, is one of the primary citrus-producing areas in Bangladesh. Other important production areas include Chittagong, Bandarban, Rangamati, and Khagrachari districts, and parts of the greater Rajshahi and Rangpur divisions (Hossain et al., 2017). These regions offer suitable agro-climatic conditions characterized by moderate temperatures, well-drained soils, and adequate rainfall, which are conducive for to citrus cultivation (Hossain et al., 2017). Among the cultivated citruses, mandarin, sweet orange, lemon-lime, and pummelo are the main strains. Others like lemon and limes are extensively cultivated for culinary and medicinal use, while pummelos are cultivated for their nutritional value and mandarin and sweet oranges are cultivated for their sweet test and acceptable fragrance (BBS, 2018).

Due to the growing demand for citrus fruits in Bangladesh, the government has taken special projects to increase production and cultivation. As one of the outcomes of the projects, several high-yielding varieties of citrus were released, and till today, around 28 varieties of different citrus species have been developed locally (Azad et al., 2020; Rahim et al., 2011). Besides, the government has also arranged training for the growers and other stakeholders related to the industry in

 DOI: 10.1201/9781003584599-10

the citrus growing zones to develop skilled manpower in the sector. Currently, more than half of a million people are directly involved in this industry (BBS, 2023).

In this chapter, we described and covered different aspects of the citrus industry of Bangladesh, the history of citrus cultivation and its evolution, different citrus fruits and varieties, the current status of the industry, the constraints and achievements, and the future direction of the sector.

10.2 HISTORY OF CITRUS FRUITS IN BANGLADESH

Citrus plants are native to the tropical and subtropical regions of Asia, and as a part of the tropics of south Asia, citrus fruit plants are also commonly found in Bangladesh (Rao et al., 2011). Though some specific members of this genus existed long before in some particular areas of Bangladesh, the organized cultivation of different citrus species was laid on this piece of land by the hands of Mughal nobility during the early 16th to 18th century (Wikipedia, 2024a). It is assumed that different citrus fruits like orange, pummelo, lemon, lime, etc. entered this region through different trade routes as trading products and gifts for royalties (Rao et al., 2011).

The initial citrus varieties that arrived on this land through the Mughals were not well enough to initiate large-scale cultivation, but some of them were reared in royal orchards as well as in the house premises of nobilities, which patronized the adaptation and further spread of these fruit species in the country (Wikipedia, 2024a). Though the citrus fruits were introduced and adapted in this land with the hand of the Mughals the history of this genus did not start from that time; in fact, it started from a long before. According to some recent genetic research, it is assumed that the ancestor of this genus emerged in the north-eastern part of the Indian subcontinent 25 million years ago, just before the incident in which it collided with the continent Asia and formed the Himalayas, and that's why the majority of the wild citrus genotypes are observed in the foothills of Himalayas from where it had spread to different parts of the world (Nicolosi et al., 2000; Scora, 1975; Talon et al., 2020). Wu et al. (2018) performed a genomic, phylogenic, and biogeographical analysis and found some results in support of this theory. From the study, the researchers concluded that the centre of origin of this genus is the southeast foothills of the Himalayas, ranging from Assam, northern Myanmar to western Yunan (province of China), and the whole genus originated and diverged from a common ancestor with *Poncirus trifoliata* (Figure 10.1) (Nicolosi et al., 2000; Scora, 1975).

However, it is reported that a sudden climatic change took place during the late stage of Miocene (11.63 to 5.33 million years ago) which resulted in a precipitate speciation event, and the species like the citrons (*Citrus medica* L.), the pummelos (*Citrus maxima*), the mandarins *(Citrus reticulata)*, kumquats (*Citrus japonica*), change papeda (*Citrus cavalier*) appeared in South Asia, Mainland Southeast Asia and in Southeastern China respectively (Wikipedia, 2024a; Wu et al., 2018). The citrus fruits that are commonly available today are mainly hybrids of four citrus species, namely the citrons (*C. medica*), the pummelos (*C. maxima*), the mandarins (*C. reticulata*) mycratha (Citrus hystrix) (Guerra, 2009; Rao et al., 2011; Wikipedia, 2024a). The most interesting part is that a portion of the Sylhet division of Bangladesh is geographically a fragment of the larger land piece called Assam region which is considered as the centre of origin of one ancestral species i.e., the citrons, and that's why a great range of genetic diversity regarding this species are found in the Sylhet region (Fuller et al., 2017). Notably, a hybrid citrus known as Indian wild orange (*Citrus indica*) which was once considered the most primitive form of citrus is a hybrid of citron and mandarin orange and is exclusively found in Garo hills that lie between Meghalaya (previously Assam) and Sylhet. Thus, this region is marked as the centre of origin of this species (Wu et al., 2018; Wikipedia, 2024a). Additionally, other hybrid species of citron, like lemon are also abundant in the Sylhet region, and the presence of a wide genetic diversity of this hybrid there denotes the allegiance of this region with the evolution and development of this species (Talon et al., 2020). Another hybrid of citron and mandarin called Rangpur (*C. limonia*) which is known as Sylhet lime throughout the

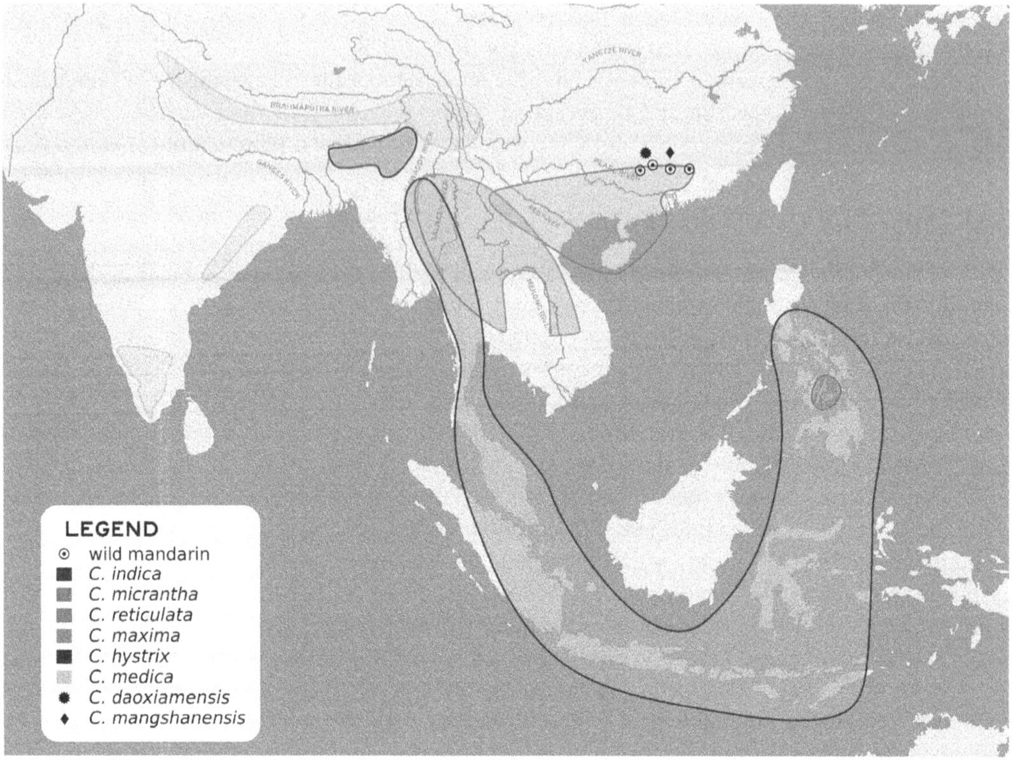

FIGURE 10.1 Center of origin and divergence of different citrus species.

Indian subcontinent is an exclusive species found in the Sylhet region (Wikipedia, 2024c). This evidence as a whole points towards the long connection of this part of land with the citruses.

10.3 MAJOR CITRUS GROWING AREAS IN BANGLADESH

Citrus fruit species are a very common type of fruit in Bangladesh as the climatic condition of this country favours the growth and development of citrus fruits. In every corner of this country, at least one of the citrus species could be found. Though the production of mandarin oranges is confined to specific areas, the lemons and pummelos are scattered all over the country. Since citrus fruits need well-drained soil and ample sunshine to grow well, so the majority of citrus is found to grow well in the hilly areas of Sylhet and Chittagong division (Hossain et al., 2017). However, currently, commercial cultivation of citrus fruits is practiced in almost every region of this country among which Chittagong and Chittagong hill tracts hold the leading position. The major citrus-producing areas of Bangladesh are presented in Figure 10.2.

10.3.1 Sylhet and Comilla Region

The greater Sylhet region is one of the top citruses producing areas in Bangladesh. Due to its geographical presence near to the centre of origin of several citrus species and the hilly land type, a diversified citrus species is found and cultivated in the region (Talon et al., 2020). Almost every type of citrus that grows in Bangladesh is found in this region. Particularly Sylhet and Jaintia Hill is well known for growing citrus fruits in this region (Hossain et al., 2017; Nazirul et al., 2017).

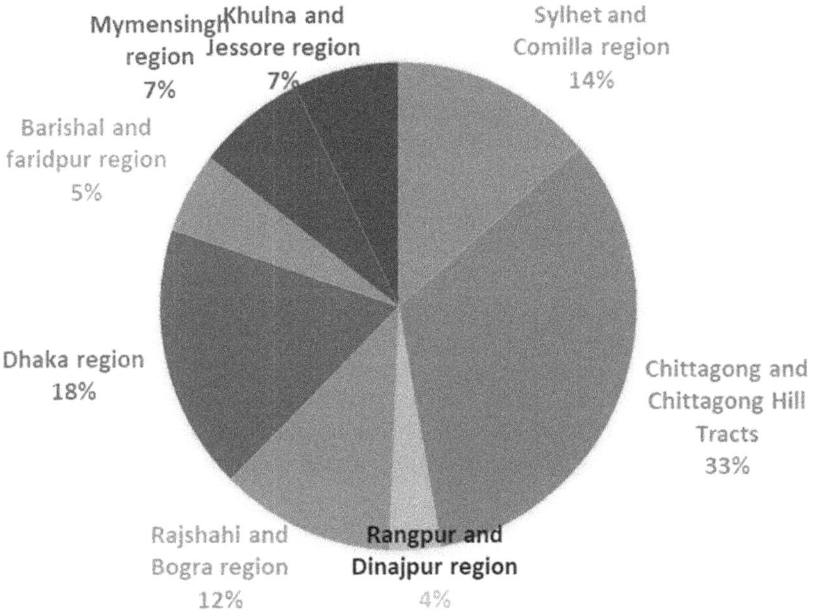

FIGURE 10.2 Main citrus growing areas of Bangladesh.

Like other hilly areas of Bangladesh, the dominant soil type in this area is "Brown Hill Soil" and it varies from brown sandy-loam to clay loam texture, which is well drained as well (FAO, 2000). On the other hand, the climatic pattern is "humid subtropical" which consists of a predominantly hot and humid summer along with a relatively cold winter. Thus, this type of climate and well-drained soil favours the maximum growth of citrus fruits (Huq et al., 2013a).

Within the region, Sylhet, Moulavi Bazar, Sunamganj, and Habiganj are the top-producing areas of all types of citruses (BBS, 2023). Moreover, the production of lemon and pummelo is also found in the Comilla, Brhmanbaria, and Chandpur districts. Besides this, Sankara is an exclusive citrus that is commercially produced and cultivated in Sylhet. In 2022–2023, a cumulative amount of 55,886 M. tonne citrus fruits was produced in this region (DAE, 2024).

10.3.2 Chittagong and Chittagong Hill Tracts

This is another leading region for citrus cultivation in Bangladesh. This region consists of the Chittagong district and three hilly districts viz. Khagrachari, Rangamati hill district, and Bandarban of Chittagong division commonly known as Chittagong hill tracts. This region is the leading producer of citrus throughout the country (Hossain et al., 2017; Uddin et al., 2016).

In this region, two types of land are present, the mainland Chittagong district consists of plain land while the districts of Chittagong hill tracts are made of hilly lands (Alamgir et al., 2015). Currently, in both types of lands, the dominant soil type is brown hilly soil though the soil of hill areas varies from brown sandy-loam to clay loam in terms of soil texture. On the other hand, the soil of the main land Chittagong is sandy-loam, loamy-sand, and silt-loam along with 53–83 percent of sand. In both cases, the soil is highly or well drained (Huq et al., 2013a; FAO, 2000) which favours citrus cultivation. Though the eastern part of the mainland Chittagong is a coastal area and salty in nature but the rest of the mainland and hilly regions have slightly acidic soil (BBS, 2020a). The climatic category of this region is the tropical monsoon with dry and cool weather from November to March and pre-monsoon weather exists during April–May and hot sunny monsoon prevails from June–October (Huq et al., 2013b, 2013a; Banglapedia, 2021) months.

Both of the areas of Chittagong and Chittagong hill tracts are suitable for commercial production of citrus and the weather of this region promotes the growth and development of every type of citrus fruit in Bangladesh. Among the areas, the Chittagong hill tracts are famous for producing all available citrus fruits whereas Chittagong, Lokhipur, Feni, Noakhali, and Cox's Bazar are well known for lemon and pummelo production (Uddin et al., 2016). In 2022–2023 years, a total of 135,659 MT of citrus was produced in this region (DAE, 2024).

10.3.3 Rangpur and Dinajpur Region

This is the northernmost part of Bangladesh and lies close to the Himalayan belts. Like Sylhet and Chittagong regions, the northernmost part does not form with hilly lands, rather it consists of highlands lying 30–179 m above the sea level (Sherozzaman, 2012; Wikipedia, 2024b). This higher alleviation along with the closeness to the Himalayas provided favourable conditions for the growth of several citrus species.

The dominant soil type of this region is sandy clay loam mixed with sand and stones. This soil facilitates proper drainage (Islam et al., 2017). On the other hand, this area experiences a moderate climate having sufficient rainfall found during the monsoon and the winter remains cool and dry. Thus, the soil and weather are highly favourable for growing certain types of citruses (Huq et al., 2013b).

Due to favourable conditions, in recent years an uprising production trend of sweet oranges, lemons, and pummelos has been observed in this area (Ara et al., 2008; Paul & Saha, 2004). In the region, Rangpur, Nilphamari, Kurigram, Lalmonirhat, and Gaibandha are the top-ranked cultivating districts for all types of citrus while Panchagarh, Thakurgaon, and Dinajpur districts also contribute to the production (Paul & Saha, 2004) of citruses. The total production of 2022–2023 in this area was 14,244 MT (DAE, 2024).

10.3.4 Rajshahi and Bogra Region

Rajshahi region, especially Rajshahi district and its neighbouring districts viz. Chapai Nawabganj, Naogaon, and Nator are famous for different tropical fruit cultivation. In previous times citrus fruit cultivation on a commercial scale was scarce in this region due to the superiority of mango and other summer fruits, so citrus fruits were cultivated on a small scale only to meet the demand of local markets. However, with the progress of time and due to the establishment of food processing industries nearby and within the area the production span and produce list increased (Paul & Saha, 2004; Sarker et al., 2019). At present, large-scale cultivation of different citrus is observed in these areas, which fulfils domestic demand as well as gets exported to other countries.

This region is situated on the alluvial plain of the river Padma and is mainly formed with loamy to clay-type soil. The soil is moderately acidic, which is suitable for citrus growth. Besides, the climatic condition of the region is tropical wet and dry or savanna climate which experiences high temperatures, moderate humidity along with moderate rainfall (Huq et al., 2013b, 2013a; Shahid, 2010).

Moderate acidic soil coupled with a savanna climate allows the region to produce different varieties of citrus fruits. Among the citrus-producing areas Rajshahi, Chapai Nawabganj, Naogaon, and Nator districts are notable for the production of all citruses especially sweet orange, lemon, and pummelos. Whereas in the neighbouring districts viz. Pabna, Joypuhat, Sirajgong, and Bogra some citrus fruits are also cultivated (Paul & Saha, 2004; Sarker et al., 2019). In the 2022–2023 season, a cumulative amount of 47,889 MT of citrus fruits was produced in the region (DAE, 2024).

10.3.5 Dhaka Region

In the Dhaka region, Gazipur, Tangail, and Norshingdhi districts are the main hubs for different citrus cultivation. As it is the central part of the country, the transportation and marketing channel is quite favourable (BBS, 2023) for citrus production and cultivation.

This area is comprised of non-calcareous and calcareous brown floodplain soils which are primarily silty loam type. Besides the altitude of these areas is also high. Moreover, the whole region experiences tropical savanna or tropical wet and dry type climate. The climatic condition combined with the soil patronizes the development and growth of some citrus species (Huq et al., 2013a, 2013b).

In spite of having favourable conditions, the citrus production scale of this region was not big enough in the past due to the priority of other field crops. However, the situation is changing now. In previous times only lemons and pummelos were cultivated commercially in Tangail and Gazipur district, but at present sweet orange and some mandarins are also produced in Norshingdhi and other districts. Besides, some citruses are also produced in Dhaka, Manikgonj, Kishoreganj, and Narayanganj (BBS, 2023). In 2022–2023 total production of citrus in this region was 71,730 MT (DAE, 2024).

10.3.6 Barisal and Faridpur Region

Barisal lies in the south-central part of Bangladesh and is crisscrossed by several rivers. Though commercial citrus cultivation has not been started long before yet, now a considerable number of citrus fruits are being cultivated in each year. Today, a considerable amount of sweet orange cultivation is observed in Barisal, Patuakhali, Pirojpur, and Jhalkathi districts (BBS, 2023).

Citrus-producing areas of the Barisal region are situated on a non-calcareous and calcareous brown floodplain soil and experience a climate in between a dry-winter humid subtropical climate and a tropical savanna climate which allows it partially to produce a few types of citrus species (Huq et al., 2013b, 2013a).

At present, Barisal, Patuakhali, Pirojpur, and Jhalkathi districts produce all types of citruses. Along with these above-mentioned districts Faridpur, Madaripur, Gopalganj, Shariatpur, Rajbari, Borguna, and Bhola are also producing lemon, pummelos, and sweet oranges (BBS, 2023). In 2022–2023, a total of 21,693 MT of citrus was produced here (DAE, 2024) in this region.

10.3.7 Mymensingh Region

Mymensingh district is the main hub for different citrus cultivation in this region, as it is situated adjacent to Dhaka and lies beside the Brahmaputra River.

This area is comprised of non-calcareous and calcareous brown floodplain soils which are primarily silty loam type. Besides the altitude of these areas is also high. Moreover, the whole region experiences tropical savanna or tropical wet and dry type climate, which combined with the soil patronizes the development and growth of some citrus species (Huq et al., 2013a, 2013b).

Among the districts, Mymensingh is the major producer of citrus fruits while Netrokona, Sherpur, and Jamalpur also produce a little. In this region, mainly sweet oranges and pummelos are cultivated on a large scale along with a small-scale cultivation of mandarin and lemons. In this area, a total of 29,922 MT of citrus was produced in 2022–2023 (DAE, 2024).

10.3.8 Khulna and Jessore Region

In this region, Jhinaidaha, Chuadanga, Kustia, and Bagerhat are the main citrus-producing districts. This region lies in the southern part of the country including some coastal regions.

TABLE 10.1
Major Citrus Species and Their Growing Areas of Bangladesh

Citrus species	Commercially producing area
Mandarin Orange	Sylhet, Moulavi Bazar
	Chittagong, Chittagong hill tracts, Comilla, Feni
	Chuadanga, Tangail
Sweet orange	Sylhet, Brhmanbaria,
	Chittagong, Chittagong hill tracts, Thakurgaon
	Rajshahi, Chapai Nawabganj, Pirojpur, Patuakhali, Rangpur,
	Gazipur, Mymensingh, Tangail, Chuadanga, Jessore
Lemon	Sylhet, Moulavi Bazar, Habiganj
	Chittagong, Chittagong hill tracts
	Rajshahi, Pabna
	Gazipur, Tangail, Dhaka, Mymensingh
Pomelo	Sylhet, Moulavi Bazar
	Chittagong, Chittagong hill tracts, Cox's Bazar
	Rajshahi, Kustia, Munshigonj, Tangail, Mymensingh, Jamalpur, Bagerhat

This area is comprised of a peat basin and Ganges tidal floodplain. Soils vary from clay to silty loam type. Besides the elevation of these areas is also high. Moreover, the whole region experiences tropical wet and dry type climate which combines with the soil that patronizes the development and growth of some citrus species (Huq et al., 2013a, 2013b). In this region, mainly sweet orange, lemon, and pummelo are grown in Jhinaidaha, Chuadanga, Kustia, and Bagerhat. Additionally, some productions are also noticed in the Khulna, Satkhira, Norail, Meherpur, and Magura districts. In 2022–2023, this region produced 29,108 MT (Table 10.1).

10.4 CURRENT STATUS OF CITRUS INDUSTRY IN BANGLADESH

The history of the citrus industry in Bangladesh is not very old. In spite of being cultivated for a long time in this country, commercial-scale production of citrus fruits has been started a few decades ago. However, within a short period citrus industry has grown extensively. Currently, more or less 0.4 million metric tonnes (MT) of citrus fruits are being produced in Bangladesh per annum and this is increasing by 3–5% each year (BBS, 2023; DAE, 2024). Due to this major leap in citrus production, a distinct industry based on fresh citrus fruits and its products has grown in the country. A market valued at $157.2 million US dollars has grown on the export and import business of citrus and its processed products (OEC, 2024) and the industry will expand with time. In this part of the chapter, we described the commercial aspects of citrus industry in Bangladesh.

10.4.1 Production Trend of Citrus in Bangladesh

The production of citrus fruits in Bangladesh is increasing day by day as the demand for citrus fruit has extended in recent years. It is mentioned earlier that the major cultivated citrus in this country is mandarin orange, sweet orange, lemon, lime, and pummelo. Among this citrus, mandarin orange requires either hilly or high elevated land so the mandarins cannot grow in plain land that's why its production is area-specific as well as low-scaled in comparison to other citrus fruits (Hossain et al., 2017). The rest of the citrus viz. sweet orange, lemon, lime, and pummelo can be cultivated in the general plain land and are also produced in a large amount. The majority share (60%) of the citrus production in Bangladesh is contributed by lemon, which is followed by sweet orange (18%), pummelo (17%), and mandarin orange (8%) (Figure 10.3).

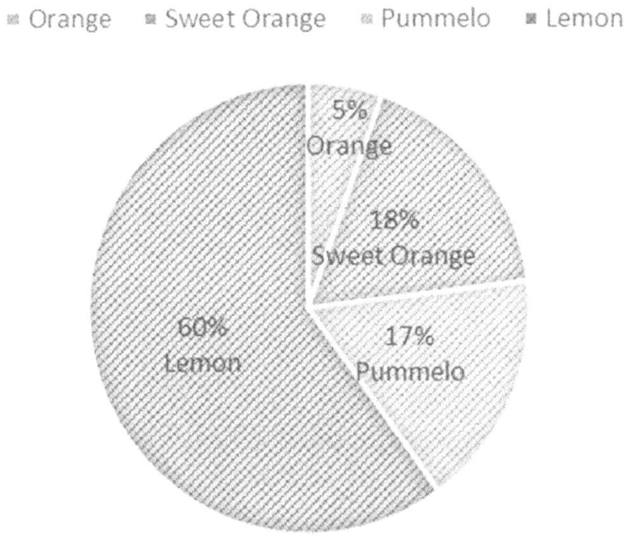

FIGURE 10.3 Individual share of different citrus species in total citrus production of Bangladesh.

In the last few years, the production of lemon and limes has escalated tremendously, which is around 230 thousand MT in the year 2023 in comparison to 50 thousand MT in 2020 (BBS, 2023, 2021). Similarly, in the case of sweet oranges, the annual production increased to around nine thousand MT in comparison to that of 2020, which was only 3.5 thousand MT (BBS, 2023, 2021). These results ultimately point towards the growing production trend of citrus A species-wise production trend of citrus fruits in Bangladesh in the last 7 years is illustrated in Figure 10.4.

The overall production trend of citrus is also expanding at a tremendous rate. The annual production of citrus fruits in Bangladesh was 140 thousand MT, 141 thousand MT, 145 thousand MT, and 146 thousand MT in 2017, 2018, 2019, and 2020 respectively (BBS, 2018, 2019, 2020a, 2021). On the other hand, 170.5 thousand MT, 185.5 thousand MT, 184 thousand MT, and 400 thousand MT annual production was observed in 2021, 2022, and 2023 respectively (BBS, 2018, 2019, 2020b, 2021; DAE, 2024). If both trends are compared then it becomes clear that a significant increase has occurred in the citrus industry of Bangladesh after 2020. A year-wise citrus production trend in Bangladesh is illustrated in Figure 10.5.

10.4.2 Export and Import of Fresh Citrus Fruits and Its Products

In previous years both the demand and production of citrus fruit have increased tremendously. But the demand for the citrus always has surpassed the production. To meet up this gap between production and demand, Bangladesh has to import a huge amount of fresh citrus each year among which the mandarin oranges and the sweet oranges are the main imported citruses. In the previous year, Bangladesh imported fresh citrus fruit valuing more than $150 million from different countries (OEC, 2024). Moreover, the import quantity is increasing each year. As of today, Bangladesh has imported 174.6 thousand MT, 203.9 thousand MT, and 212.9 thousand MT of fresh citrus fruits in 2021, 2022, and 2023 respectively (EPB, 2024). One of the major reasons behind this huge amount of import is the year-round demand for citrus fruits. As a perishable product, these fruits do not last long and cannot be stored beyond a certain period but the demands remain high throughout the year. Thus, to meet the demand huge amount of fruit has to be imported. The major countries from where Bangladesh imports citrus fruits are China, Egypt, Morocco, Pakistan, and Australia (EPB, 2024).

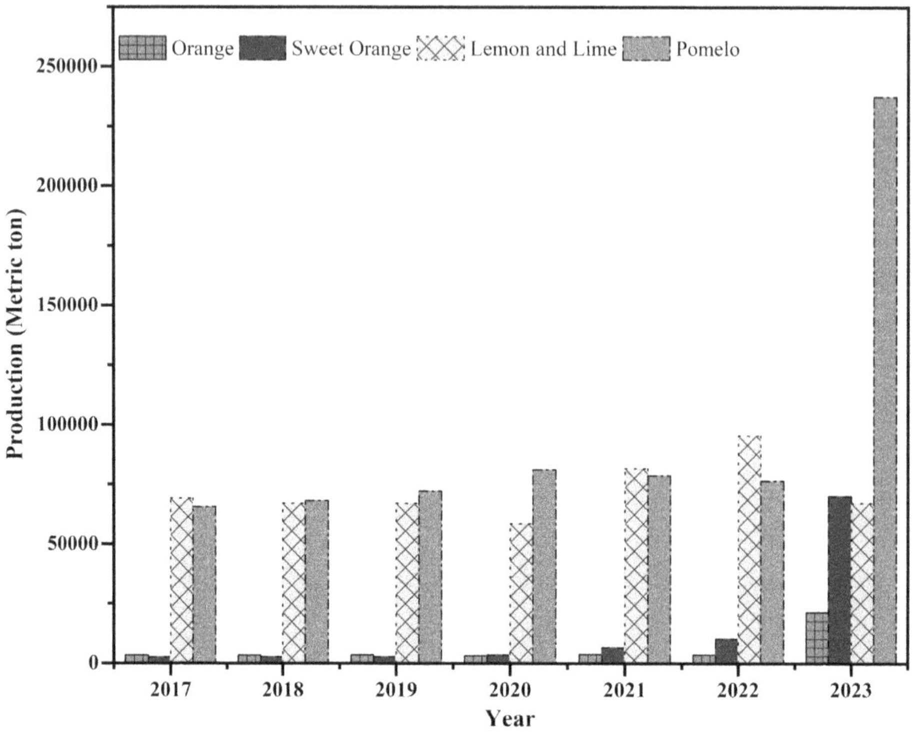

FIGURE 10.4 Species-wise citrus production trend in Bangladesh in last seven years.

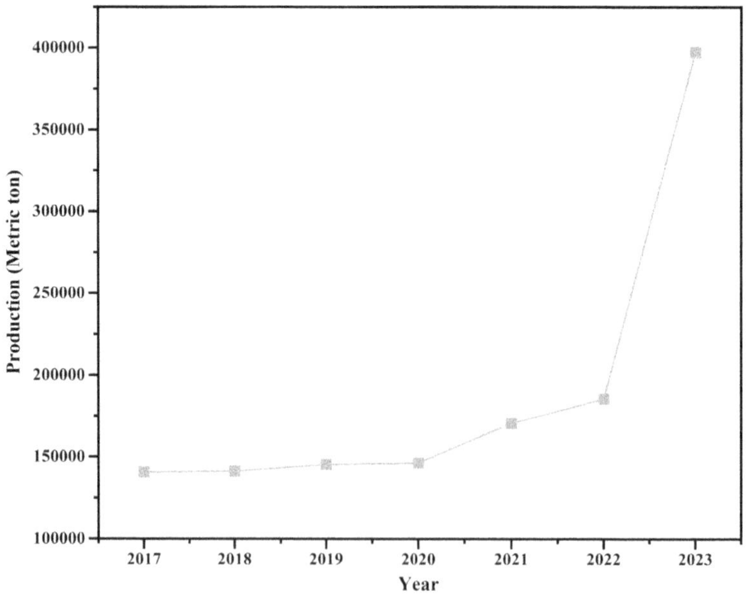

FIGURE 10.5 Year-wise citrus production trend in Bangladesh.

On the other hand, both the export quantity and product list of citrus fruits are limited. Since Bangladesh has to import a lot of fresh fruit to meet the local demand, there is less room left to export fresh fruit outside. However, even after that, a considerable number of citruses are exported to various destinations along with some processed citrus fruit products like dried orange, orange jelly, canned orange juice, orange peel powder, and canned Satkora fruit in the present time. In the previous year, the valuation of this exportation was nearly $2.2 million (OEC, 2024). The main destinations of citrus and its product export from Bangladesh are the United Kingdom (UK), Qatar, Kuwait, Canada, and Singapore (EPB, 2024; OEC, 2024). In the recent time, some products are also exporting to the United States (US) (OEC, 2024). Among the buyers, the United Kingdom is the largest (OEC, 2024). It is estimated that Bangladesh exported citrus products valuing around $807 thousand to the UK in the previous year, on the same time the fastest-growing market for these products is Kuwait (827%) (OEC, 2024).

10.4.3 Harvesting Window and Market Availability of Major Citrus Fruits in Bangladesh

In Bangladesh, there are four major citrus fruits namely, mandarin, sweet orange, or malta, lemon, and pummelo are commonly available in the domestic market and hold the lion's share of the fruit market. The harvesting window highly regulates the availability period of these fruits in the local market. As the harvesting window of these fruits is different, their availability periods in the market are also different.

In the case of mandarin orange, the fruits become mature during the first week of December and the harvesting is continued till the last of the same month (Azad et al., 2020) (Chart 1). As the fruit can be stored for up to 3 months with proper refrigeration the locally produced fruits can be found in the domestic market till March or sometime during the first of April from harvesting (Azad et al., 2020). Here's the interesting fact is that mandarins are year-round available in our local market and that is managed by importing mandarin oranges from foreign countries.

This timeline is slightly different for Malta. The harvesting window of this fruit starts in the last part of September and goes on till the later phase of December (Azad et al., 2020) (Chart 1). Fresh fruit starts to enter the local market and remains available till February. Though long preservation can deter the quality of products sometimes it can be found till mid of March (Azad et al., 2020). Like mandarin orange, sweet orange is also imported to ensure year-round market availability of this fruit.

The lemon and lime are almost year-round fruit. The lemons are harvested year-round, which spans from the first of January to the last of November (Azad et al., 2020) (Chart 1). In the case of lemon, the fruit is not usually harvested during December (Azad et al., 2020). This harvesting period is different for limes which are harvested from the first of January and continue till mid-October (Figure 10.6). Since both of the fruits are produced year-round they are found year-round in the market (Azad et al., 2020).

Sweet lime has a relatively shorter harvesting window, which starts in October and continues till the end of November (Azad et al., 2020; Kundu et al., 2020) (Figure 10.6). The fresh fruit enters the market just after its harvesting in October and remains available in the local market till the first of January (Azad et al., 2020). As sweet lime is a minor citrus in Bangladesh its year-round market availability is maintained on a small scale by import.

Pummelo can be stored on the tree for up to three months hence their harvesting window spans a timeline of three months (Berk, 2016; Kundu et al., 2020). The harvesting window of pummelo starts in early September and continues till late November (Azad et al., 2020) (Figure 10.6). During this period a stable supply of these fruits is noticeable in the local market. Since, this fruit can't be stored for a long period, so their market availability ends within a short time after finishing the harvesting of fruits which can be stretched to early January (Azad et al., 2020).

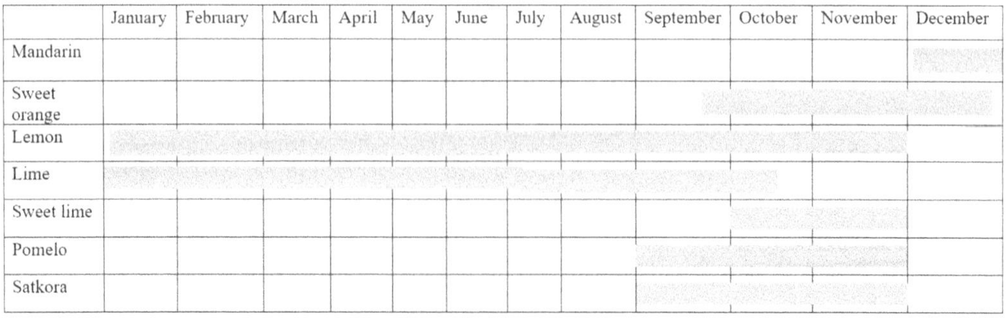

FIGURE 10.6 Harvesting window of major citrus fruits in Bangladesh.

Satkora is harvested on a similar timeline of pummelo. Harvesting starts in early September and goes till late November (Figure 10.6). The fresh fruit is available during this time (Azad et al., 2020). Besides, year-round availability is also found, which is achieved through the canned product.

10.4.4 Marketing Network

The marketing network of citrus fruit in Bangladesh is quite linear since the market is still growing. The market consists of two stages and seven stakeholders with the producer and the consumers standing on each end of the network. The products enter into the channel and reach the consumer after several hand changes.

In the case of the citrus marketing network, the producers grow the citrus fruits, which is the first segment of the network. In the first stage, the producers are approached by the sellers as well as the big industries, and the scattered produce is accumulated in a few spaces. In the second stage the retailers and small-sized industries appear to deal with the sellers on one side and consumers on the other. The retailers supply fresh fruit and processed fruit products prepared by small and big industries. In the case of small industries, they process the fruit products and either sell those products in the local market through retailers or supply the products to big corporations for export. On the other hand, the big industries also process the fruit product and supply it to the local market through their own retail channels or local retailers while they also export the products through their own exporting channels or local retailers while they also export the products through their own exporting channels. In Bangladesh the whole process of marketing is monitored and regulated by the ministry of commerce. A simplified illustration of the marketing network of citrus industry is shown in Figure 10.7.

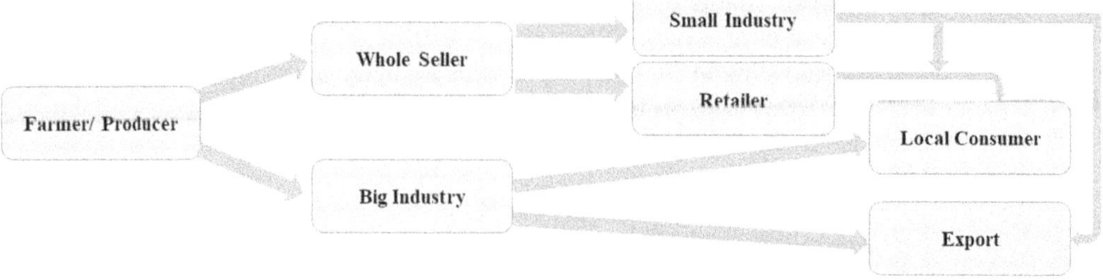

FIGURE 10.7 Marketing network of citrus fruit in Bangladesh.

10.5 STATUS, PRESENT SCENARIO AND FUTURE DIRECTION OF CITRUS RESEARCH, DEVELOPMENT AND EXTENSION IN BANGLADESH

Research activities regarding citrus fruits started in Bangladesh long before. Basically, the land lords during the zamindari era, started collecting different citrus germplasm and planted them in their royal gardens, that initiated the base line of citrus research in Bangladesh. In the year 1947 two independent countries India and Pakistan had emerged from the undivided India. At that time Bangladesh was the part of Pakistan termed as East Pakistan. Immediately after establishment of the country the Government started citrus research and extension activities throughout the country.

In cater to the research and extension needs, a scheme entitled Organization of Horticulture Section under the Directorate of Agriculture was initiated in 1951. Under this scheme four sub-stations were established in different locations of Bangladesh (then East Pakistan) and one of the four sub-stations was established at Chailakhel (Tamabil) which situated about 8 kilometer away from Jaintapur, Sylhet district. The main objective of this sub-station was to collect different varieties of fruit especially citrus fruit plants from various parts of Bangladesh and to evaluate their adaptability for selection and recommendation (Uddin, 1987).

Meanwhile, an orange orchard of 20-ha area was established at Jaintapur in 1960-61 aiming at accelerating the activities in the field of horticultural research, especially on citrus fruits. It was since then commonly known as Citrus Research Station. The main purpose of the station was to collect citrus germplasm from within and outside Bangladesh and to evaluate the germplasm for subsequent selection as varieties and recommendation of the same to the growers. For the extension of 20 ha Orange Orchard, another scheme entitled Horticultural Research Centre was established in 1965, which involved the acquisition of an additional 28 ha acres of land. In 1970 the previous horticultural sub-station at Chailakhel (Tamabil) was shifted to Jaintapur and included in this Citrus Research Station.

After the emergence of Bangladesh as an independent country; from Pakistan, the government emphasized the research activities focusing on citrus research and extension. Bangladesh Agricultural Research Institute (BARI) was established in 1978 and Citrus Research Station was included with BARI at that time. To strengthen the research activities a scheme entitled Citrus and Vegetable Seed Research Centre was started with the funding from Japan International Cooperation Agency (JAICA) in 1978 at this station and additional scientific and other staff was appointed under the project, which continued up to 1981. Under this project, another two sub-stations one at Moulvibazar and another at Ishwardi were established. This is a brief history of the institutional establishment of citrus research in Bangladesh. See Table 10.2.

TABLE 10.2
Milestones of Citrus Research, Development, and Extension in Bangladesh

1951	Organization of Horticulture Section scheme
1954	Functioning of sub-station for citrus research started
1960–1961	Establishment of 20 ha Orange orchard (Citrus Research Station)
1965–1966	Horticulture Research Center scheme
1978	Bangladesh Agricultural Research Institute (BARI) established and Citrus Research Station was included in BARI
1978–1981	Citrus and Vegetable Seed Research Center Project by JAICA
2001–2006	Mandarin Orange and Pineapple development project
2007–2001	Mandarin Orange development project
2013–2008	Citrus Development Project
2019–2024	Extension Management and Production Enhancement of Citrus Fruits Project

From the above discussion, it is very clear that there is a long history of Citrus research, development, and extension in Bangladesh. In the previous section, we discussed about the centre of origin of Citrus fruits. From that discussion, it is obvious that the north-eastern part of Bangladesh is located in the centre of origin of citrus fruits. Therefore, a lot of variation exists especially in the case of sour-tasting citrus fruits. This is the baseline of citrus research in Bangladesh. Some previous reports suggested that there are 78 different types of citrus fruits in Bangladesh. Keeping this view in mind the research activities are mainly based on a collection of germplasm, characterization, evaluation, and finally recommendation of the varieties to the farmers.

In the very beginning, the research and extension activities focused on the varietal development and dissemination of saplings to farmers. In this trend in 1963–1987, a total number of 12,11,718 (Twelve lakhs eleven thousand seven hundred and eighteen) orange seedlings were disseminated to the farmers' level (Uddin, 1987). Meanwhile, there were some other research activities viz. rootstock trial, propagation time and method, the effect of girdling, seed storing, control, and management of disease and insects (BARI, 2022). Physiological studies like a fruit drop, biochemical quality study of fruits, etc. were executed. As a result, 21 high-yielding varieties of eight citrus fruits and more than 30 need-based production technologies were developed and still continuing.

At present the research on citrus fruits extended from BARI to other organizations like Bangladesh Nuclear Agricultural Research Institute (BINA) and other educational institutes viz. Bangladesh Agriculture University (BAU). BINA has already developed three lemon varieties while BAU has developed more than ten citrus fruits varieties. All these varieties and technologies are contributing to the national economy of Bangladesh.

In BARI citrus research is mainly divided into several projects like varietal development through hybridization; selection breeding, propagation technique including rootstock selection, time and method of grafting and budding, cultural management, nutrient management, disease and insect management, abiotic stress study and management, agro-forestry and citrus, urban horticulture and citrus, biotechnological interventions etc. Recently, smart technologies (4iR) are included in citrus research. At present scientists at ten stations and sub-stations are involved in citrus fruit research (BARI, 2024).

In Bangladesh, the Department of Agricultural Extension (DAE) is mainly responsible for disseminating new and problem-based technologies to farmers. Through different projects, DAE extends the technologies to farmers. DAE is the largest organization in Bangladesh having offices at every upazila (sub-district) of Bangladesh. Farmers came to their offices to take advice on different issues like high-yielding varieties and technologies. They also visit farmers' fields in order to advise them on modern technologies. Besides, DAE establishes demonstration orchards at farmers' fields to extend the production area as well as to aware the farmers about the newly invented technologies. Moreover, DAE arranges training sessions and field days to teach the farmers about various technologies. With their efforts in the fiscal year 2022–2023, 23.41% area under citrus production increased along with a 12% increase in citrus fruit production compared to the fiscal year 2018–2019. From the financial support from Extension Management and Production Enhancement of Citrus Fruits Project DAE trained 58,130 farmers, who are now involved in citrus fruit production. The farmers are also provided with fertilizers and other agricultural equipment (secateurs, budding knives, and sprayers) for their orchards. Besides, DAE in their 20 Horticulture Centres produces citrus planting materials and disseminates them to the farmers. Furthermore, DAE forms some common interest groups (CIG) of farmers in their area. These groups are often taken to other orchards of successful farmers and research organizations for motivational visits. In these ways, DAE is boosting the citrus production in Bangladesh.

Research organizations also do some extension work although this is not their mandate. The research organizations mainly BARI and BINA produce citrus planting materials of their recommended varieties. They also establish some mother orchards in farmers' fields so that other farmers can see the new technologies. This approach motivates neighbouring farmers and they establish their own orchard. Moreover, when a new technology is generated, research organizations arrange seminars, workshops, training sessions, and field days with stakeholders like farmers, extension workers, nursery owners, GO and NGO personnel, exporters, and importers to disseminate the technologies.

10.6 AVAILABLE CULTIVARS OF CITRUS FRUITS IN BANGLADESH

Bangladesh lies in the tropical region of South Asia, which makes the country highly favourable for growing different species of citrus fruits (Zhong & Nicolosi, 2020). In addition, the close presence of this country to the centre of origin of different ancestral citrus species also favoured this country to hold a diverse range of genetic materials of this genus (Talon et al., 2020). But, for commercial purposes, a handful of varieties of mandarin orange, sweet orange, lemon, pomelo, lime, and sweet lime are cultivated in the country. Besides these, some other species viz. citrons, Sankara, and ada lebu are also cultivated here but on a small scale hence they become minor citrus in terms of production (BBS, 2023). The following section describes the features of different citrus cultivars currently cultivated in Bangladesh and summarizes the varieties in Table 10.3.

TABLE 10.3
Citrus Species and Their Varieties Available in Bangladesh along with Their Special Features

Mandarin orange	BARI Komola-1	High yielding
	BARI Komola-2	Resistance to pest and diseases
	BARI Komola-3	High edible portion
Sweet orange	BARI Malta-1	Suitable for cultivation all over the country
	BARI Malta-2	High edible portion
	BAU Malta-1	Less number of seeds
Lemon	BARI Lebu-1	Elachi Lebu
	BARI Lebu-2	Year-round fruiting
	BARI Lebu-3	Suitable for cultivation all over the country
	BARI Lebu-4	Tolerant to the water stagnancy as well as pest and canker disease
	BARI Lebu-5	High edible portion
	BARI Lebu-6	Suitable for cultivation all over the country
	BAU Lebu-2	High yielding
	BAU Lebu-3	High yielding
Pomelo	BARI Batabilebu-1	Suitable for cultivation all over the country
	BARI Batabilebu-2	Suitable for cultivation all over the country
	BARI Batabilebu-3	Suitable for cultivation all over the country
	BARI Batabilebu-4	Sweet pulp
	BARI Batabilebu-5	High yielding
	BARI Batabilebu-6	Sweet pulp
	BAU Jambura -1	Year-round fruiting
	BAU Jambura -2	Sweet pulp
Lime	BARI Kagzilebu-1	Rich in vitamin C
	BAU Kagozi lebu-1	Early bearer
Sweet lime	BARI Mistilebu-1	High yielding
Citron	BARI Jaralebu-1	Year-round fruiting
Satkara	BARI Satkara-1	Highly scented
Ada lebu		Possesses ginger-like flavour and fragrance

10.6.1 Mandarin Orange

Mandarin orange (*Citrus reticulata*) which is also known as “Komola” in Bangladesh is the most popular citrus fruit in the country. It is very desired among people due to its sweet taste, good fragrance, and its higher possession of vitamin C. Around the country, three varieties of mandarins are usually cultivated in approximately 427 ha of land. These varieties are developed by the Bangladesh Agricultural Research Institute (BARI) and named as BARI Komola-1, BARI Komola-2 and BARI Komola-3. All these varieties possess sweet pulp (Azad et al., 2020; Rahim et al., 2011).

10.6.1.1 BARI Komola-1

This is the first of the mandarin orange variety released by BARI. It was released in 1996 through the selection of locally available germplasm. This is a medium-height plant with regular and early fruit-bearing habit. A single plant bears almost 300–400 fruits per plant in a season. The fruit size is big and each fruit possesses around 180–200 g weight. The rind is loose and the pulp is highly sweet, which contains a TS value of around 10.2%. It is a high-yielding variety that produces around 20–25 M. tonnes of fruit per hectare (ha) (Azad et al., 2020; Mia, 2016a). See Figure 10.8.

10.6.1.2 BARI Komola-2

This variety is also introduced by BARI and is one of the major varieties of mandarin orange cultivated throughout the country. Like the BARI Komola-1, this variety is also regular fruit-bearing and possesses medium height. The fruit is roundish and small sized with an average 38 g weight. The rind is loose and the pulp is highly sweet (TS- 12%). The pulp of this variety becomes dark yellow at ripening. Its small-sized seeds are one of the major perks of this variety. Moreover, this variety is also popular due to its higher resistance to pests and diseases (Azad et al., 2020; Mia, 2016a) (Figure 10.9).

10.6.1.3 BARI Komola-3

Like the other two varieties of Mandarin released by BARI, this variety is also a regular fruit-bearing type. But it differs from the other two in leaf size, which is long whereas the plant is bushy type.

FIGURE 10.8 BARI Komola-1.

FIGURE 10.9 BARI Komola-2.

FIGURE 10.10 BARI Komola-3.

The fruit is round and big with an average weight of 170 gm. The rind is loose and bright yellow coloured which possesses a dark orange colour upon ripening. The pulp is juicy and highly sweet (TS- 11.5%). Additionally, the edible portion of this variety is 79.8% (Azad et al., 2020; Mia, 2016a) (Figure 10.10).

10.6.2 Sweet Orange

Sweet orange (*Citrus sinensis*) is called as "Malta" in Bangladesh. This is another popular citrus fruit which has a high demand in the local market and the demand is increasing day by day. It is highly sweet and differs from mandarin in its rind, which is tight, and its pulp, which is not segmented under the rind. Due to its higher adaptability, it can be cultivated both in plain land with acidic soil and hilly areas. Due to its high adaptability nature, the cultivation of this citrus fruit has increased rapidly in Bangladesh (Curk et al., 2022). Currently, three varieties of this orange species are used extensively for commercial production around the country. Among the three varieties, 2 varieties namely BARI Malta-1 and BARI Malta-2 were developed by BARI whereas the other variety BAU Malta-1 was developed by Bangladesh Agricultural University (BAU) (Azad et al., 2020; Mia, 2016b).

10.6.2.1 BARI Malta-1

BARI Malta-1 is an introduced variety that originated from Pakistan (Mia, 2016b). This is the first locally released orange variety in Bangladesh. This is a regular fruit-bearing variety. Its plant is short and wide in nature. The variety is a high-yielding type which can produce around 300–400 fruits per plant during the season. The fruit is medium-sized with a round shape and weighs 150 gm per fruit. The rind is medium thick and the fruit possesses a green colour at ripening. The pulp is juicy and very sweet (TS- 7.8%). The variety yields around 20 million tonnes per ha and it is suitable for cultivation in all the districts of the country (Azad et al., 2020; Mia, 2016b). See Figure 10.11.

10.6.2.2 BARI Malta-2

BARI Malta-2 is another variety released by BARI in Bangladesh in 2017. This variety was developed through selection from the local germplasms collected from different areas of the country. It is a regular fruit-bearing variety. The plant grows rapidly and is a bushy type. The fruit is elliptical and bigger in size with an average weight of 179 g. The fruits appear in clusters. The rind of the fruit is smooth and green and the pulp is medium sweet (TS- 7.5%). Its edible portion is 72.95% (Azad et al., 2020; Mia, 2016b) (Figure 10.12).

10.6.2.3 BAU Malta-1

This variety was released in 2009 by Bangladesh Agricultural University. This is an exotic-selected variety. The plant is a dwarf in size and bushy. The fruit is medium-sized with an average weight of

FIGURE 10.11 BARI Malta-1.

FIGURE 10.12 BARI Malta-2.

140 gm per fruit. The rind is greenish whereas the pulp is very juicy and sweet (TS- 17–21%). The fruit possesses a very low number of seeds, which makes it highly demanding among consumers (Mia, 2016b).

10.6.3 Lemon

Lemon (*Citrus limon*) is known as "Lebu" in Bangladesh. It is the most popular and common citrus species around the country and grows almost everywhere. In rural areas of Bangladesh, a lemon plant can be found in almost every house premises. In spite of being a fruit, it is used mainly for culinary and beverage-making purposes. It is famous among consumers for its juice that is used for preparing lemonade known as "Sharbat" locally. It is highly rich in vitamin C. Currently, there are nine released varieties of lemon found in Bangladesh. Among the varieties six are released by BARI, two are released by BAU and one is released by Bangladesh Institute of Nuclear Agriculture (BINA) (Azad et al., 2020; Mia, 2016c).

10.6.3.1 BARI Lebu-1

It is the first ever released lemon variety in Bangladesh by BARI. The variety was developed by selection from 30 local landraces of Bangladesh. This variety is also called "Elachi Lebu" due to its fragrance similar to cardamom. The leaves and the fruit of this variety are large. A plant bears around 100–150 fruits per season and can yield twice per annum. The fruit shape is elliptical and each weighs around 250–270 gm. The rind is thick and uneven. The pulp is less juicy and holds 25% juice. The fruit yield per ha of this variety is 15 M. tonnes. It can be grown almost everywhere in the country but it shows better growth in hilly and rainy areas (Azad et al., 2020; Mia, 2016c) (Figure 10.13).

FIGURE 10.13 BARI Lebu-1.

10.6.3.2 BARI Lebu-2

Like BARI Lebu-1, BARI Lebu-2 is also developed through selection from local germplasm. Its unique character is that it gives fruit year-round, and also has regular fruit-bearing habits. The fruit is round shaped weighing 75–85 gm. The rind is thin and smooth. The pulp is very juicy (32–34%) and the taste is medium sour. This variety is high-yielding and each plant holds 180–190 fruits in a season. The variety yields 12 M. tonne per ha and is suitable for cultivation everywhere around the country (Azad et al., 2020; Mia, 2016c) (Figure 10.14).

10.6.3.3 BARI Lebu-3

BARI Lebu-3 is also developed by selection from the locally available germplasm. The variety has medium plant height and smaller deep green leaves. The fruit-bearing habit of this variety is regular. The fruit shape is round and the rind is thin and smooth. Each fruit weighs around 50–60 g. The pulp is very juicy, nearly 37.7%. A seven-year-old plant can bear 200–220 fruits per season and yields around 10 M. tonne per ha. It grows well in the rainy season and is suitable for all the districts of the country (Azad et al., 2020; Mia, 2016c) (Figure 10.15).

10.6.3.4 BARI Lebu-4

This variety is also developed by selection from locally available germplasm. It is a high-yielding variety and gives fruit year-round. The plant is bushy in nature and bears elliptical deep green medium-sized leaves. The fruit is round and weighs around 82 g per fruit. The fruit is greenish and appears in clusters or singly. Its edible portion is 70%. It contains fewer seeds and is tolerant to water stagnancy as well as pest and canker disease (Azad et al., 2020; Mia, 2016c) (Figure 10.16).

FIGURE 10.14 BARI Lebu-2.

FIGURE 10.15 BARI Lebu-3.

10.6.3.5 BARI Lebu-5

BARI Lebu-5 locally known as "Colombo Lebu" is a high-yielding regular fruit-bearing variety of Colombo lemon. It was developed from local germplasms through selection and evaluation. The plant grows rapidly, has a medium height, and is highly branched with a bushy architecture. The leaves are long, elliptical, and dark green coloured. The fruit is elliptical and long with a bright green colour. Fruit size is big with a weight of 268 gm per fruit. In addition, the fruit has a unique fragrance and appears in clusters. Its rind doesn't taste bitter hence the rind is also edible. The total edible portion of this fruit is 82% (Azad et al., 2020; Mia, 2016c) (Figure 10.17).

FIGURE 10.16 BARI Lebu-4.

FIGURE 10.17 BARI Lebu-5.

10.6.3.6 BARI Lebu-6

This variety is a regular fruit-bearing type variety. It gives fruit year-round. The plant has a medium height with long leaves. The fruit is round and large sized. Each fruit weighs around 163.8 gm. It is highly scented and 67.18% portion of the fruit is edible. The variety yields around 300 fruit per plant during the season. This variety is suitable for cultivate anywhere in the country (Azad et al., 2020; Mia, 2016c) (Figure 10.18).

FIGURE 10.18 BARI Lebu-6.

10.6.3.7 BAU Lebu-2

This variety of lemons is released by BAU. This variety is also elachi-type as it bears scents like cardamom. The variety was developed through local germplasm screening. The plant is bushy type and the fruit is oval in shape and is highly scented. Fruit is small and weighs around 70–150 gm. The thickness of the rind is very low and the edible portion is also low around 30%. It possesses a high number of seeds inside the fruit. This variety yields around 30–35 M. tonne per ha (Mia, 2016c).

10.6.3.8 BAU Lebu-3

It is also developed from locally collected germplasms by BAU. It is a seedless variety that bears fruit on a regular basis. The plant is medium-sized. Fruit is elongated and weighs around 50–130 gm. The rind is medium thick. Its edible portion is above 50%. The average number of seeds per plant is six. It can yield around 20–30 M. tonnes of fruit per ha. in the season (Mia, 2016c).

10.6.3.9 BINA Lebu-4

It is also developed from locally collected germplasm by BINA. It is a seedless variety and bears fruit on a regular basis. The plant is medium-sized and the fruit is round shaped. The rind is thin. The plant starts to give fruit by 11 months from planting. A mature plant of this variety gives around 150–180 fruits per season (Mia, 2016c).

10.6.4 Pummelo

Pummelo (*Citrus maxima*) is another popular citrus fruit in Bangladesh. Locally it is called "Batabi Lebu" or "Jambura". This fruit is rich in vitamin C and hence it has a huge demand in the market. This species requires less care for growing, hence it is found almost everywhere in the country. As of today, a total eight varieties of pummelo are available in Bangladesh, among which six were developed and released by BARI, and the other two were released by BAU (Azad et al., 2020; Mia, 2016d; Rahim et al., 2011).

FIGURE 10.19 BARI Batabilebu-1.

10.6.4.1 BARI Batabilebu-1

BARI Batabilebu-1 was developed from local germplasms that were collected from different districts of Bangladesh. It is the first ever released variety of pummelo in this country. It is a regular fruit bearer. Leaves are long and deep green coloured. The fruit is round and medium. Only 45% of the fruit is edible. Each fruit weighs 900–1,100 gm. The pulp of the fruit is a little bitter in taste as well as soft and juicy. The fruit retains a yellowish colour upon ripening. A plant can hold up to 45–55 fruit during the fruiting season and the yield is around 14–16 tonnes per ha-1. The most interesting part is that it can be grown anywhere around the country with little care (Azad et al., 2020; Mia, 2016d) (Figure 10.19).

10.6.4.2 BARI Batabilebu-2

This variety was also developed through screening of the local germplasm collected by BARI from different districts of Bangladesh. This is a modern variety of pummelo and was released at nearly the same time BARI Batabilebu-1 was released. The plant of the variety resembles an umbrella. Leaves are round, winged, and deep green. Fruits are medium with an average weight of 750–780 gm. The rind is thick and the edible portion is 40% of the total fruit. The pulp is juicy and sweet and it turns into a yellow colour at ripening time. Each plant can bear 40–50 fruits during the fruiting season and yields 12–14 M. tonne of fruit in season. The variety grows everywhere in the country (Azad et al., 2020; Mia, 2016d) (Figure 10.20).

10.6.4.3 BARI Batabilebu-3

This modern pummelo variety was developed by evaluating the germplasm collected by BARI from the northern territories of Bangladesh. It is a regular fruit-bearing plant having long deep green-winged leaves and semicircular medium-sized fruits. Each fruit weighs around 1,000–1,150 gm. The rind is yellow coloured and can be separated from the pulp easily. The pulp is sweet and juicy and constructs 55–60% of edible parts. Single plants of this variety can bear 70–75 fruits and yield 25–30 M. tonne ha-1 during the fruiting season. It is also cultivable all over the country (Azad et al., 2020; Mia, 2016d) (Figure 10.21).

10.6.4.4 BARI Batabilebu-4

The variety was released in 2004 by evaluating the germplasm collected from different districts of the country. The plant resembles to the shape of an umbrella. Leaves are deep green circular and

FIGURE 10.20 BARI Batabilebu-2.

FIGURE 10.21 BARI Batabilebu-3.

winged. Fruit-bearing habit is regular. The fruit is round in shape and medium in size. A fruit can weigh 850–1,100 gm. The pulp is white and sweet without any bitter taste. It is a late variety of pummelo and each plant can produce 40–50 fruit during the season and yield 15–20 M. tonnes per ha. It is also suitable for all over the country (Azad et al., 2020; Mia, 2016d) (Figure 10.22).

10.6.4.5 BARI Batabilebu-5

This is a regular fruit-bearing and high-yielding variety of pummelo. The plant is medium-sized with long leaves. The bright yellow-coloured fruits are roundish and bigger in size. A single fruit can be weighed up to 850 gm. The fruits appear in solitary. An eight-year plant of the variety can give an average of 18 fruit per plant and the yield is 10.03 M. tonnes per ha. in a year (Azad et al., 2020; Mia, 2016d) (Figure 10.23).

10.6.4.6 BARI Batabilebu-6

BARI Batabilebu-6 is a late pummelo variety recently released by BARI through the selection from local germplasm. The plants of this variety are medium-sized and wide in architecture. The leaves are long, deep green, and winged and the petioles are wide. The fruit is semi-spherical and coloured greenish yellow. The pulp is juicy, soft, and red in colour without bitterness and the pulp consists of 57% of the fruit (Azad et al., 2020; Mia, 2016d) (Figure 10.24).

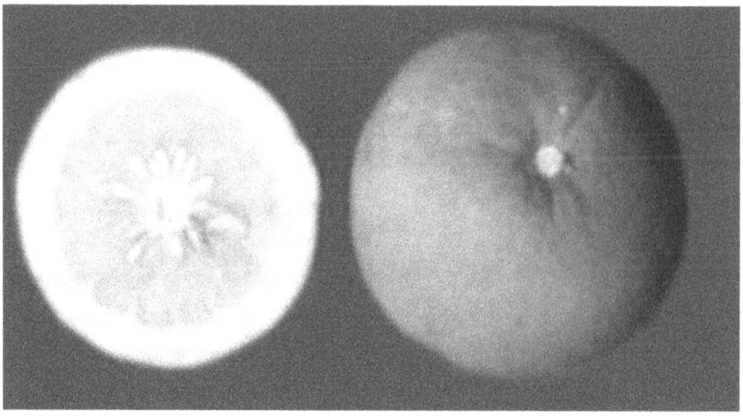

FIGURE 10.22 BARI Batabilebu-4.

FIGURE 10.23 BARI Batabilebu-5.

10.6.4.7 BAU Jambura-1

This variety was developed at Bangladesh Agricultural University through the selection of the local germplasm. The variety has a medium height. The variety is also high-yielding as well as regular-fruit-bearing variety. The fruit is big enough and round in shape. A single fruit can weigh up to 2,000–5,000 gm. The pulp is white and juicy with a lesser bitter taste. Around 42.7% of the total fruit is edible. This variety gives fruits year-round and can produce 20–25 M. tonne ha-1 in a year (Mia, 2016d; Rahim et al., 2011)

10.6.4.8 BAU Jambura-2

This variety was also developed by BAU through the selection of local germplasm. It is a year-round pummelo variety. The fruit is medium-sized with an average weight of 1,000–1,500 gm per fruit. The pulp of the fruit is red coloured juicy and very sweet in taste. The edible portion of total fruit is around 40–52%. It produces 20–25 tonnes of fruits per ha in a year (Mia, 2016d; Rahim et al., 2011).

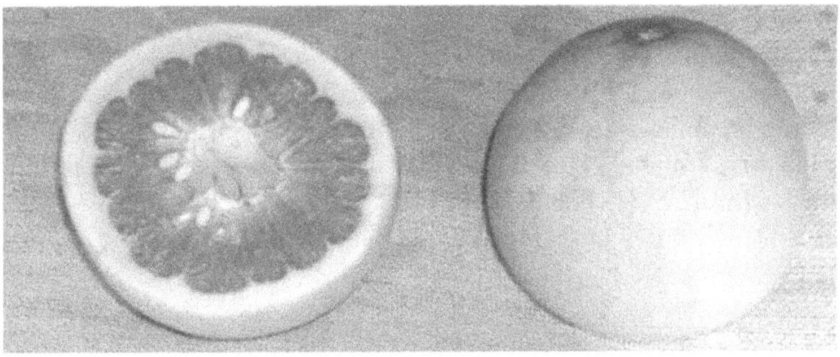

FIGURE 10.24 BARI Batabilebu-6.

FIGURE 10.25 BARI Kagzilebu-1.

10.6.5 Lime

Lime (*Citrus aurantifolia*) is a very common type of citrus fruit that is found all over Bangladesh. Locally it is called "Kagzilebu" and is very popular among the common people. Similar to lemon it is also used in making lemonade and hence it has a high demand among the mass people. It is also useful for culinary purposes. This species is enriched with nutrition and medicinal value. The fruit has a very strong taste and attractive fragrance. As of today, there are two released varieties of lime in the country namely BARI Kagzilebu-1 and BAU Kagozi Cebu-1 (Azad et al., 2020; Mia, 2016c; Rahim et al., 2011).

10.6.5.1 BARI Kagzilebu-1

This variety was developed by BARI through the selection and evaluation of local germplasm collected from different parts of the country and released as a variety in 2018. This is a high-yielding regular-fruit-bearing variety. The plant is bushy type with semicircular leaves having a dark green colour. The fruit is elliptical in shape and relatively bigger in size (82-g). However, the fruit is bright green coloured and appears in clusters. It is rich in vitamin C (65 ml/100 gm). Its edible part is 57% of the total fruit. It is tolerant to almost all common diseases and pests of citrus and has resistance to canker and gummosis diseases of citrus (Azad et al., 2020; Mia, 2016c; Rahim et al., 2011) (Figure 10.25).

10.6.5.2 BAU Kagozi lebu-1

BAU released the variety in 2006 and it was the first locally released lime variety in Bangladesh. The variety was developed from locally available germplasm. It is also a high-yielding variety and fruits year-round. The fruit is semicircular in shape weighing 80–120 g each fruit. It starts to produce fruit just a year after planting, from that view it is an early-bearing variety. The rind is thin and the edible part is 28% of the total fruit. Every branch of the plant bears 2–5 fruits and 15–30-tonne produce can be harvested from one hectare of land (Mia, 2016c; Rahim et al., 2011).

10.6.6 Sweet Lime

Sweet lime (*Citrus liberticides*) is a sweet pulped citrus fruit just like the mandarins, sweet oranges, and pummelo. This fruit has a high demand in the fruit juice processing industry as well as in households for beverage preparation. Nowadays its demand is increasing in Bangladesh and it is cultivated commercially in some parts of the country. BARI Mistilebu-1 is the only available variety of sweet lime in Bangladesh (Azad et al., 2020).

10.6.6.1 BARI Mistilebu-1

This variety was released in Bangladesh by BARI in 2013 to address the increasing demand for sweet lime. It is a high-yielding and regular fruit-bearing variety. The plant is dwarf and bushy. The fruits are round and medium-sized and possess an attractive green colour, having thin and tight skin. The fruit pulp is light yellow juicy and very sweet (brix value 7.5%). The edible part of the fruit is 55–60% of fruit. Besides, the fruit is enriched with calcium, phosphorus, and vitamin C. During the fruiting season, a single plant gives around 300–580 fruits per plant and 35–40 tonnes of fruits per ha (Azad et al., 2020) (Figure 10.26).

FIGURE 10.26 BARI Mistilebu.

10.6.7 Citron

Citron (*Citrus medica*) is a minor citrus of Bangladesh. This is a primitive type of citrus species that is highly appreciated for its fragrance and rind. The rind can be consumed too. To date, only one citron variety BARI Jaralebu-1 is released in Bangladesh (Azad et al., 2020).

10.6.7.1 BARI Jaralebu-1

This variety was developed from a selection of local germplasm and released by BARI as a variety in 2018. The plant of the variety is rapidly growing with moderate or highly branching habits. Leaves are long and deep green. The fruits are large enough and weigh 882 gm per fruit of which 70% is edible. The rind is thick, uneven, scented, and also consumable. It is a regular-bearing fruit variety and gives fruit year-round. A plant can bear 60–80 fruit per fruiting cycle (Azad et al., 2020) (Figure 10.27).

10.6.8 Wild Orang or Satkara

Satkora or wild orange (*Citrus macropetra*) is a minor citrus fruit and is very popular in the Sylhet region. Like other citrus its pulp is not consumed rather its rind is used in curry due to its attractive scent to make the curry more attractive and more palatable. It is also used to make pickles. In Bangladesh, the only released variety of satkora is BARI Satkara-1 (Azad et al., 2020).

10.6.8.1 BARI Satkara-1

It was developed from local germplasm and released as a variety in 2004. It is a high-yielding and regular fruit-bearing variety. The plant has medium height and is wide to moderately bushy in architecture. The fruit is yellow in colour, medium in size, and tapered in shape like mandarin orange. The weight of each fruit is 300 g and the variety yields 10 M. tonne per hectare (Azad et al., 2020) (Figure 10.28).

FIGURE 10.27 BARI Jaralebu-1.

FIGURE 10.28 BARI Satkara-1.

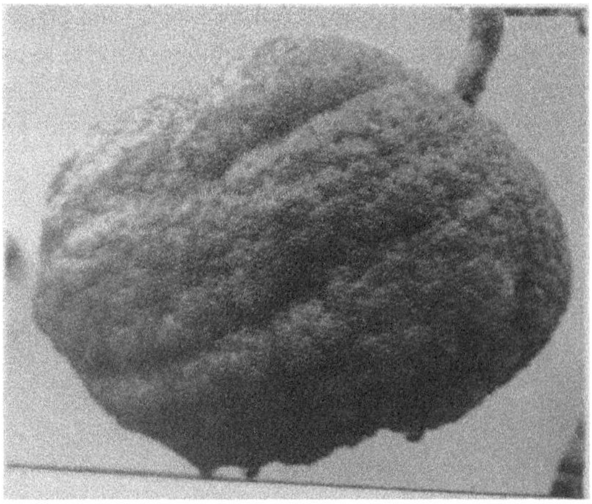

FIGURE 10.29 Ada lebu.

10.6.9 Ada Lebu

Ada Cebu or ada jamir (*Citrus assamensis*) is a citrus fruit that mainly grows in the eastern parts of Bangladesh and is cultivated on a limited scale. The species is an indigenous species extensively found in the greater Assam and Sylhet regions. Its fruit and leaves contain a strong ginger-like fragrance and flavour for which it is popular. Leaves are long, and fruits as roundish. Fruit has thick smooth to semi-smooth skin. The pulp contains highly acidic juice. This is a vigorous late-planting species (Rahim et al., 2011). To date, there is no released variety of ada lebu in Bangladesh but the locally available germplasm is multiplied and cultivated extensively in its growing areas (Figure 10.29) Ada lebu

10.7 CULTIVATION PROCESS AND SIGNIFICANT TECHNOLOGIES IN CITRUS FARMING IN BANGLADESH

The cultivation process of every citrus fruit requires more or less the same steps which start with land selection and end with the harvesting of the produce. So, what differs in cultivation technique from species to species is the technique and the protocols to carry out those steps. The whole process can be divided into two parts, viz. pre-harvest operation or the growing part and the harvesting and post-harvest operations. The pre-harvest operation includes the field operations whereas the harvest and post-harvest operation includes the processing of harvested fruits.

10.7.1 Pre-Harvest or Crop Growing Operations

This part of the cultivation includes the selection of land and planting material, processing of land, planting of the seedling, rearing of the seedling, fertilization, pest disease management, cultural operations, and other operations that are related to the rearing of fruit plants for fruit production. This phase contains the crucial steps of the production process and affects citrus production volume directly. These steps are almost similar in different citrus species and what differs here is the spacing, fertilizer dose, pest management process etcetera. The cultivation process, fertilizer doses, and pest and disease management techniques of citrus cultivation are summarized in Table 10.4, Table 10.5, and Table 10.6 respectively.

10.7.2 Harvesting and Post-Harvest Operations of Citrus Production

- **Mandarin orange:** Fruit should be harvested after proper ripening and proper care should be taken during the process to avoid any injury. Fruit can be preserved in cold storage for up to 2 months at 10°C with 80-90% relative humidity whereas it could thrive for 3 months if the temperature is reduced to 5.6°C. On the other hand, dipping the fresh fruit in 13% liquid wax just after harvest enables the shelf life of the fruit to 25 days at normal temperature (Azad et al., 2020).
- **Sweet orange:** After ripening fruit needs to be harvested on time. Generally, sweet oranges are harvested from September to October. Mature fruits are collected by hand or net and sorted after harvesting to remove the damaged and injured fruits. The fruits are finally graded according to their shape and size and wiped with soft cloths. Finally, the fruits are placed in wooden or plastic carts and kept in a cool and dry place (Kundu et al., 2020).
- **Lemon:** Lemons are harvested year-round. Lemons are harvested before ripening. After harvest fruits are graded and sorted. Finally, they are placed in wooden or bamboo carts and sent off to market (Strano et al., 2022).
- **Lime:** Limes are also harvested year-round but are mainly harvested during May and September. Like lemons, it is also harvested before ripening. After harvest fruits are graded and sorted. Finally, they are placed in wooden or bamboo carts and sent off to market (Azad et al., 2020).
- **Sweet lime:** Sweet limes are generally harvested from September to October. Ripen fruits are harvested by hand or net. After harvest fruits are graded and sorted. Finally, they are placed in wooden or plastic carts for transportation (Azad et al., 2020).
- **Pummelo:** Pummelo is harvested year-round but major harvesting is done from September to November. The fruit is harvested just after it attains a slight yellow colour. Harvested fruits are graded, sorted, and wiped with a soft cloth. Finally, they are placed in wooden or plastic carts for transportation or sent off to market. Here it should be mentioned that pummelo can be kept on the tree for a long time even after ripening and it is called "Tree storage" (Azad et al., 2020; Kundu et al., 2020; Strano et al., 2022).

TABLE 10.4
Field Operations for Production of Different Citruses

Citrus species	Soil	Planting material	Land preparation	Pit size and spacing	Pit fertilizer	Planting time	Planting method	Cultural operation	Reference
Mandarin Orange	Well drained, high, acidic	Graft	Criss cross ploughing, laddering/ spading for hills	75×75×75cm / 4–5m	Cow dung-15kg Wood ash- 3-5kg, TSP-250gm, MP-250gm, Lime-250gm	May–June/year-round with irrigation facility	Hexagonal/ contour	Irrigation, drainage, pruning	Ollitrault & Luro, 2001; Abbas & Fares, 2009; Curk et al., 2022; Rao et al., 2011
Sweet Orange	Well drained, high to medium high, acidic	Graft	Ploughing and cleaning on plain land, only cleaning in hilly land	75×75×75cm / 4m	Cow dung-15kg Wood ash- 3–5kg, TSP-250gm, MP-250gm	May–June/year-round with irrigation facility	Hexagonal/ contour	Irrigation, drainage, pruning	Abbas & Fares, 2009; Curk et al., 2022; Azad et al., 2020
Lemon	Plain loamy soil, moderate acidic	Graft	Ploughing and cleaning on plain land, only cleaning in hilly land	60×60×60cm / 2.5–3m	Biofertilizer-15kg, TSP-200gm, MP-200gm	May–October/ year-round with irrigation	Line	Irrigation, drainage, pruning	Abbas & Fares, 2009; Curk et al., 2022
Lime	Well drained loamy, moderate acidic	Graft	Ploughing and cleaning on plain land, only cleaning in hilly land	80×80×80cm / 4–5m	Cow dung-20kg, TSP-300gm, MP-250gm, gypsum-250gm, boric acid- 30 gm, dolomite-1.5kg (for acidic soil)	June– August/ year-round with irrigation	Line, Square/ vertical line	Irrigation, drainage, pruning	Ollitrault & Luro, 2001; Abbas & Fares, 2009; Azad et al., 2020
Sweet Lime	Well drained loamy	Graft	Ploughing and laddering, pruning of nearby plants	80×80×80cm / 3m	Cow dung-20kg, TSP-300gm, MP-250gm, gypsum-250gm, boric acid- 30gm	May–August/ year-round with irrigation	Square, hexagonal/ contour	Irrigation, drainage, pruning	Curk et al., 2022; Azad et al., 2020
Pomelo	Well drained loamy, moderate acidic	Graft	Ploughing and laddering	60×60×50cm / 6m	Cow dung-20kg, TSP-300gm, MP-300gm, gypsum-200gm	June–September/ year-round with irrigation	Line	Irrigation, drainage, pruning	Abbas & Fares, 2009; Curk et al., 2022
Satkara	Well drained high land, moderate acidic	Graft	Criss cross ploughing, laddering/ spading for hills	75×75×75cm / 6m	Cow dung-15kg Wood ash- 3–5kg, TSP-250gm, MP-250gm, Lime-250gm	April–May, August–September	Line	Irrigation, drainage, pruning	Azad et al., 2020
Ada lebu	Well drained loamy, moderate acidic	Graft	Ploughing and laddering	90×90×90cm / 3.5–4.5m	Cow dung-15kg Wood ash- 3-5kg, TSP-250gm, MP-250gm	May–August	Line	Irrigation, drainage, pruning	Rahim et al., 2011

TABLE 10.5
Requirement of Fertilizers

Age of plant (yr.)	Species	Well rotten cow dung (Kg)	Urea (gm)	TSP (gm)	MoP (gm)	Gypsum (gm)	Zinc sulphate (gm)	Boric acid (gm)	Reference
1–2	Mandarin orange	10	200	100	150	-	-	-	Abbas & Fares, 2009; Azad et al., 2020
	Sweet orange	10–12	200–300	100–150	100–150	-	10	5	Srivastava et al., 2002; Azad et al., 2020
	Lemon	25	250	250	250	100	25	50	Srivastava et al., 2002; Azad et al., 2020
	Lime	15	200	200	200	20	5	5	Abbas & Fares, 2009; Azad et al., 2020
	Sweet lime	10–12	200–300	100–150	100–150	-	10	10	Srivastava et al., 2002; Azad et al., 2020
	Pomelo	7–10	175–225	80–90	140–160	-	-	-	Srivastava et al., 2002; Azad et al., 2020
	Satkora	7–10	175–225	80–90	140–160	-	-	-	Azad et al., 2020
3–5	Mandarin orange	15	300	150	200				Abbas & Fares, 2009; Azad et al., 2020
	Sweet orange	12–15	300–450	150–200	150–200	-	15	8	Srivastava et al., 2002; Azad et al., 2020
	Lemon	30	500	400	250	150	30	60	Srivastava et al., 2002; Azad et al., 2020
	Lime	20	400	300	300	50	8	10	Abbas & Fares, 2009; Azad et al., 2020
	Sweet lime	12–15	300–450	150–200	150–200	-	15	12	Srivastava et al., 2002; Azad et al., 2020
	Pomelo	10–15	270–300	140–170	400–500	-	-	-	Srivastava et al., 2002; Azad et al., 2020
	Satkora	10–15	270–300	140–170	400–500	-	-	-	Azad et al., 2020
6–7	Mandarin orange	20	500	400	300				Abbas & Fares, 2009; Azad et al., 2020
	Sweet orange	15–18	450–600	200–300	200–250	-	20	10	Srivastava et al., 2002; Azad et al., 2020
	Lemon	40	500	400	400	200	40	75	Srivastava et al., 2002; Azad et al., 2020
	Lime	25	500	400	400	70	10	15	Abbas & Fares, 2009; Azad et al., 2020
	Sweet lime	15–18	450–600	200–300	200–250	-	20	15	Srivastava et al., 2002; Azad et al., 2020
	Pomelo	20–25	500–600	400–450	500–550	-	-	-	Srivastava et al., 2002; Azad et al., 2020
	Satkora	20–25	500–600	400–450	500–550	-	-	-	Azad et al., 2020
8–10	Mandarin orange	20	500	400	300				Abbas & Fares, 2009; Azad et al., 2020
	Sweet orange	18–20	600–700	300–450	250–300	-	25	12	Srivastava et al., 2002; Azad et al., 2020
	Lemon	-	-	-	-	-	-	-	Srivastava et al., 2002; Azad et al., 2020
	Lime	-	-	-	-	-	-	-	Abbas & Fares, 2009; Azad et al., 2020
	Sweet lime	18–20	600–700	300–450	250–300	-	25	18	Srivastava et al., 2002; Azad et al., 2020
	Pomelo	20–25	500–600	400–450	500-550	-	-	-	Srivastava et al., 2002; Azad et al., 2020
	Satkora	20–25	500–600	400–450	500–550	-	-	-	Azad et al., 2020

(continued)

TABLE 10.5 (Continued)
Requirement of fertilizers

Age of plant (yr.)	Species	Well rotten cow dung (Kg)	Urea (gm)	TSP (gm)	MoP (gm)	Gypsum (gm)	Zinc sulphate (gm)	Boric acid (gm)	Reference
Above 10	Mandarin orange	30	650	500	500				Abbas & Fares, 2009; Azad et al., 2020
	Sweet orange	20–25	750	500	450	-	30	15	Srivastava et al., 2002; Azad et al., 2020
	Lemon	-	-	-	-	-	-	-	Srivastava et al., 2002; Azad et al., 2020
	Lime	-	-	-	-	-	-	-	Abbas & Fares, 2009; Azad et al., 2020
	Sweet lime	20-25	750	500	450	-	30	22	Srivastava et al., 2002; Azad et al., 2020
	Pomelo	25–30	600–700	400–450	600–680	-	-	-	Srivastava et al., 2002; Azad et al., 2020
	Satkora	25–30	600–700	400–450	600–680	-	-	-	Azad et al., 2020

TABLE 10.6
Pests and Diseases in Citrus

Leaf miner	Pest	All *Citrus* species	Twisted and discoloured leaf	Burning of infected leaves, application of imidacloprid @ 2.5ml/litre or quinalphos @ 2ml/litre
Fruit fly	Pest	Mandarin, Sweet orange, Pomelo	Rotten fruit	Burning of infected fruit, harvesting of fruit just after maturity, application of pheromone trap
Lemon butterfly	Pest	Lemon	Damaged leaf	Burning of infected leaves, application of the combination of organophosphate and pyrethroid @ 2ml/litre
Psyllid bug	Pest	Mandarin and Sweet orange	Damaged leaf and stem followed by gummosis	Destruction of damaged plant parts, application of malathion to prevent the bug and Bordeaux mixture to prevent gummosis
Mites	Pest	Lemon	Damaged leaf and fruit	Destruction of affected plant parts, application of malathion or nicotinic sulphate
Damping off	Disease	All *citrus* species	Death of seedling	Avoiding excessive irrigation ensuring proper drainage, low density planting on seedbed
Greening	Bacterial disease by psyllid bug	Mandarin and Sweet orange	Reduced leaf, yellowing of leaf and reduced fruit number	Destruction of affected plant, application of the combination of organophosphate and pyrethroid once in a month during May–October
Gummosis	Fungal disease	Mandarin, Sweet orange, lemon, lime	Secretion of gummy exudates from cracked stem	Using resistant variety as root stock, avoiding excessive irrigation, use of Bordeaux mixture
Canker	Bacterial disease by leaf miner	All *citrus* species	Yellowing of leaves, cracked fruit	Spraying with Bordeaux mixture or copper oxychloride in every month during rainy season
Dieback	Fungal Disease	All *citrus* species	Gradual death of plant	Pruning of dead tip of branches along with live part, spraying with copper oxychloride @ 2gm/litre

- **Satkara:** The fruits are harvested just after they start to ripen from September to November. At ripening it turned a light green colour and the fruit skin became smooth. Mature fruits are collected by hand or net and sorted after harvesting to remove the damaged and injured fruits. The fruits are finally graded and kept in wooden or plastic carts for transportation (Azad et al., 2020).

10.7.2.1 Additional Practices

For citrus fruits, there is no additional practices are followed but in recent times fruit bagging become popular for mandarin, sweet orange, and sweet lime. In this practice, the fruits are covered by breathable paper bags at an early stage of fruit development which saves the fruit from invasion by insects and pests besides the fruit also remains free from pesticides (Azad et al., 2020).

10.8 PROCESSED CITRUS PRODUCTS

In Bangladesh, citrus fruits are consumed mainly in fresh condition. Besides that, some processed products are also available. Majority of the processed products are not produced on a large scale rather they are produced as homemade food. Among the processed products only orange jelly and orange drinks are popular and produced on a large scale. Besides, Satkora is also processed in a small quantity as a canned product and used exclusively for export purposes. In Bangladesh, few companies deal with the processing of citrus which are the following (EPB, 2024).

- Pran Foods Limited: Manufacturer and Exporter of orange jelly and orange drinks.
- Square Food and Beverage Limited: Manufacturer and Exporter of Orange Jelly.
- BD Foods: Manufacturer and Exporter of orange jelly.
- Sajeeb Foods Limited: Manufacturer and Exporter of orange jelly and orange drinks.
- Prome Agro Foods: Manufacturer and Exporter of Orange Drinks.
- Ahmed Food Products Limited: Manufacturer and Exporter of orange jelly.
- Cocola Food Products Limited: Manufacturer of orange jelly.

10.9 CHALLENGES AND STRATEGIES OF CITRUS PRODUCTION IN BANGLADESH

Bangladesh is a small South Asian country with a high population density with 1,301 people per square kilometre (World Bank, 2024). The ever-increasing population is negatively impacting the land area. Every year 1% of land is decreasing due to housing and other needs of this vast population. On the other hand, due to climate change soil salinity, drought, heat waves, cold waves, flash floods, hail storms, and other negative environmental events are frequently happening leaving challenges to crop productivity. Hence, it is a great challenge for the scientific community to feed this enormous population keeping sustainable productivity of any crop.

The major challenges for citrus fruit production are soil fertility, disease, and insect infestation as well as abiotic stress due to climate change. Citrus fruits require a significant amount of nutrients every year, but farmers are not aware of proper fertilization in their orchards. The farmers apply organic nutrients in their orchards once a year, mostly in the winter season, and apply chemical fertilizers before and after the rainy season, which is not sufficient for the plants. As a result, plants starve from nutrient deficiency. Teaching the farmers to apply proper nutrients based on soil tests is a challenge for the extension workers. Moreover, soil test facilities are not everywhere, although the Soil Resource Development Institute (SRDI) is trying to motivate farmers to use fertilizers on a soil test basis.

Climatic vulnerability is another challenge for sustainable citrus fruit production. Citrus fruits are mainly grown in the north-eastern part of Bangladesh, where the rainfall is high leaving poor fertility of the soil. Sudden rain in the flowering season causes poor fruit set; contrary prolonged dry season at the fruit maturity stage hampered fruit quality. Moreover, soil pH is also a negative factor for a sustainable yield and fruit quality. Some new technologies adopted to mitigate these problems but still we need more research on these aspects. Soil salinity is another problem for growing citrus in the southern part. To overcome this problem salt tolerant rootstock trial is carried out and using local pummelo rootstock sweet orange is now growing in the southern part of Bangladesh. Drought in winter is a very critical problem in Bangladesh. Researchers are now mitigating this problem using drought-tolerant rootstock and drip irrigation. In flush flood prune areas still new technologies are required to grow sustainable citrus production, although some water-logging tolerant lemon cultivars have already been identified.

The most important challenge for citrus growing all over Bangladesh is the infestation of bacterial diseases viz. citrus canker and citrus greening. In 2008 citrus export from Bangladesh stopped due to citrus canker. Afterward, a significant number of research works established the management strategies for citrus cankers, and in 2013 the export started again. But in the case of citrus greening, still there is no management technology. Research works are ongoing including disease-free planting material production, holistic approaches to orchard management, etc. to address the disease. Some breeding works are also in the pipeline for preparing greening tolerant hybrids.

Post-harvest loss is another problem for the farmers. In the picking season, the lemon process dropped to BDT 1.00 per piece of lemon which is very disappointing for the farmers. Therefore, some new technologies are disseminated to farmers for producing year-round lemons by lowering pick season fruit load. In this aspect, some research activities are required for the post-harvest processing of lemon and lemon juice for industrial purposes.

Manpower development is also important for sustainable citrus production in Bangladesh. Although DAE and the research organizations are providing hands-on training to farmers still there is a lack of knowledge of farmers regarding management practices and disease and insect management, which should be mitigated through extensive practical trading. Moreover, other stakeholders like nursery personnel, and post-harvest processors especially women should be trained. Some other persons like the agricultural machinery maintenance workers, irrigation service workers, electrical workers, and transporters should be trained to smoothen their jobs so that farmers can get the service timely and efficiently.

Finally, the socio-economic conditions of the farmers should be taken into consideration. More female members should be involved in the production and processing of citrus fruits. As a result, the income of the family will be increased which will ultimately uplift the social security as well as the production of citrus fruits.

10.10 CONCLUSION

The citrus industry is an emerging agro-based industry in Bangladesh and has a great opportunity ahead. In spite of being a new industry, it has developed a lot within a few years and it is still growing. For the development of this industry, proper synchronization between the government and the private sector is necessary. With flexible and supportive policies, the government can encourage the private sector to expand its industry and production. This will create a chance to reduce the import dependency since a majority of the market share depends on imported citrus fruits. Besides, collaborative work between different research organizations and proper extension channels is necessary to add some more to the total production. Through R&D, high-yielding varieties can be developed, and at the same time, more land can be added to citrus cultivation through the development of adaptive good citrus varieties. There is also an opportunity to reduce production loss by developing resistant varieties and by inventing efficient post-harvest technologies. It is also necessary to ensure the proper price of produce which will encourage the growers. There is also a chance to expand the processing industry and add more items to the list because there is a potential export destination of Bangladesh existing in Kuwait Qatar and the UK for these processed items. By summing up everything it can be concluded that with proper planning and execution, the citrus industry can be turned into a valuable sector for the national economy of Bangladesh.

10.11 ACKNOWLEDGEMENTS

We would like to thank Dr. Mohammad Quamrul Islam Martin and Md. Moshiur Rahman, Bangladesh Agricultural Research Institute (BARI), Gazipur, Bangladesh for providing us with information on citrus cultivation and production in Bangladesh.

10.12 AUTHOR CONTRIBUTIONS

The outline was prepared by JR and MZR. The manuscript was written by MZR, MHMBUB and JR. Final drafts of the manuscript were read and approved by all authors.

REFERENCES

Abbas, F., & Fares, A. (2009). Best Management Practices In Citrus Production. *Tree and Forestry Science and Biotechnology, 3*(1): 1–11.

Alamgir, Md., Islam, M., Hossain, N., Kibria, M.G., & Rahman, M. (2015). Assessment of Heavy Metal Contamination in Urban Soils of Chittagong City, Bangladesh. *International Journal of Plant & Soil Science, 7*: 362–372. 10.9734/IJPSS/2015/18424.

Ara, N., Bashar, M. K., Uddin, M. K., & Khalequzzaman, K. M. (2008). Evaluation of Pummelo, (Citrus Grandis L) Cultivars In Northern Area Of Bangladesh. *Journal of Agricultural Research, 46*(1): 65–76.

Azad, A. K. et.al. (2020). *Krishi Projukti Hatboi (Handbook on Agro-Technology).* Rahman, M. H. H. (Ed.). Bangladesh Agricultural Research Institute, Gazipur, Bangladesh. 9. pp. 254–273.

Banglapedia (2021) Chittagong Hill tracts, Banglapedia. Available at: https://en.banglapedia.org/index.php/Chittagong_Hill_Tracts (Accessed: 23 April 2024).

BARI (Bangladesh Agricultural Research Institute) (2024). *Annual Report (2022–2023).* BARI, Gazipur. pp. 1346.

BBS (Bangladesh Bureau of Statistics). (2018). *Yearbook of Agricultural Statistics 2017.* Bangladesh Bureau of Statistics. Statistics Division, Ministry Planning, Govt. Peoples Republic of Bangladesh, Dhaka. pp. 268.

BBS (Bangladesh Bureau of Statistics). (2019). *Yearbook of Agricultural Statistics 2018.* Bangladesh Bureau of Statistics. Statistics Division, Ministry Planning, Govt. Peoples Republic of Bangladesh, Dhaka. pp. 276.

BBS (Bangladesh Bureau of Statistics). (2020a). *Bangladesh Environment Statistics. Bangladesh Bureau of Statistics.* Statistics Division, Ministry Planning, Govt. Peoples Republic of Bangladesh, Dhaka. pp. 140.

BBS (Bangladesh Bureau of Statistics). (2020b). *Yearbook of Agricultural Statistics 2019.* Bangladesh Bureau of Statistics. Statistics Division, Ministry Planning, Govt. Peoples Republic of Bangladesh, Dhaka. pp. 269.

BBS (Bangladesh Bureau of Statistics). (2021). *Yearbook of Agricultural Statistics 2020.* Bangladesh Bureau of Statistics. Statistics Division, Ministry Planning, Govt. Peoples Republic of Bangladesh, Dhaka. pp. 277.

BBS (Bangladesh Bureau of Statistics). (2022). *Yearbook of Agricultural Statistics 2021.* Bangladesh Bureau of Statistics. Statistics Division, Ministry Planning, Govt. Peoples Republic of Bangladesh, Dhaka. pp. 265.

BBS (Bangladesh Bureau of Statistics). (2023). *Statistical Yearbook Bangladesh 2022.* Bangladesh Bureau of Statistics. Statistics Division, Ministry Planning, Govt. Peoples Republic of Bangladesh, Dhaka. pp. 180, 252, 268.

Berk, Z. (2016). *Citrus Fruit Processing.* Oxford, UK. Academic Press.

Curk, F., Luro, F., Hussain, S., & Ollitrault, P. (2022). Citrus Production. In S. Hussain, M. F. Khalid, M. A. Ali, N. Ahmed, M. Hasanuzzaman, & S. Ahmad (Eds.), *Citrus Production* (1st ed., Vol. 1, pp. 1–21). CRC Press. https://doi.org/10.1201/9781003119852

DAE (Department of Agricultural Extension). (2024). Information on fruit production 2022–2023, Horticulture Wing, DAE, Farm gate, Dhaka. Retrieved from http://hort.dae.gov.bd/sites/default/files/files/hort.dae.gov.bd/files/5600c78b_7377_466b_adc3_a867482a5c76/2024-04-13-09-30-783cbb0312ab158eec0f2e73988f9d12.pdf dated: 25.04.2023.

EPB (Export Promotion Bureau). (2024). Annual report 2022-23. Export Promotion Bureau, Ministry of Commerce, Govt. Peoples Republic of Bangladesh, Dhaka. pp. 330.

FAO. (2020). *Bangladesh. FRA 2000 – Forest resources of Bangladesh – Country report.* www.fao.org/3/ad104e/AD104E03.htm

Fuller, D. Q., Castillo, C. K. B., Eleanor, Q. L., & Weisskopf, A. (2017). Charred Pummelo Peel, Historical Linguistics, and Other Tree Crops: Approaches to Framing the Historical Context of Early Citrus Cultivation in East, South, and Southeast Asia. In V. Zech-Matterne & G. Fiorentino (Eds.), AGRUMED: *Archaeology and History of Citrus Fruit in the Mediterranean* (pp. 29–48). Publications du Centre Jean Bérard. doi:10.4000/books.pcjb.2017.

Guerra, M. (2009). Chromosomal Variability and Origin of Citrus. In C. L. Mahony & D. A. Springer (Eds.), *Genetic Diversity* (1st ed., Vol. 1, pp. 51–68). Nova Science Publisher.

Hossain, M. A., Khatun, M., Matin, M. A., & Dewan, M. F. (2017). Growth rates of fruit cultivation in hilly areas of Bangladesh. *Bangladesh Journal of Agricultural Research, 42*(2): 387–392.

Huq, S. M. I., Shoaib, M., & Uddin, J. (2013a). Climate. In S. M. I. Haq, J. Uddin, & M. Shoaib (Eds.), *The Soils of Bangladesh* (1st ed., pp. 15–30). Springer. https://doi.org/10.1007/978-94-007-1128-0_3

Huq, S. M. I., Shoaib, Md., & Uddin, J. (2013b). Major Soil Types. In S. M. Imamul Huq, Jalal Uddin, & Md. Shoaib (Eds.), *The Soils of Bangladesh* (1st ed., pp. 41–49). Springer. https://doi.org/10.1007/978-94-007-1128-0_5

Islam, Md., Hasan, Md., & Farukh, M. (2017). Application of GIS in General Soil Mapping of Bangladesh. *Journal of Geographic Information System,* 09, 604–621. 10.4236/jgis.2017.95038.

Kundu, D., Das, M., Mahle, R., Biswas, P., Karmakar, S., & Banerjee, R. (2020). Citrus fruits. In C. M. Galankis (Ed.), *Valorization of Fruit Processing By-Products* (1st ed., Vol. 1, pp. 145–166). Elsevier. https://doi.org/10.1016/B978-0-12-817106-6.00007-1

Mia, Md. A. B. (2016a, July 16). *Orange. Digital Herbarium of Crop Plants.* https://dhcrop.bsmrau.net/2815-2/

Mia, Md. A. B. (2016b, July 16). *Malta – digital herbarium of crop plants.* Digital Herbarium of Crop Plants. http://dhcrop.bsmrau.net/malta/

Mia, Md. A. B. (2016c, April 26). *Lemon.* Digital Herbarium of Crop Plants. https://dhcrop.bsmrau.net/lemon/?doing_wp_cron=1752363811.9725470542907714843750

Mia, Md. A. B. (2016d, July 16). *Pummelo.* Digital Herbarium of Crop Plants. https://dhcrop.bsmrau.net/pummelo/

Nazirul, M., Sarker, I., Chandra Barman, S., Islam, M., & Chakma, A. S. (2017). Role of lemon (Citrus limon) production on livelihoods of rural people in Bangladesh. *Journal of Agricultural Economics and Rural Development, 2*(1), 053–063. www.premierpublishers.org.

Nicolosi, E., Deng, Z. N., Gentile, A., La Malfa, S., Continella, G., and Tribulation, E. (2000). Citrus phylogeny and genetic origin of important species as investigated by molecular markers. *Theoretical and Applied Genetics, 100,* 1155–1166.

OEC (Center for Collective Learning) (2024). Citrus in Bangladesh. *The Observatory of Economic Complexity.* https://oec.world/en/profile/bilateral-product/citrus/reporter/bgd

Ollitrault, P. F. L., and Luro, F. (2001). *Tropical Plant Breeding* (1st ed.). CRC Press. https://doi.org/10.1201/9781482279863

Paul, D. K., & Saha, R. K. (2004). Nutrients, Vitamins, and Minerals Content in Common Citrus Fruits in the Northern Region of Bangladesh. *Pakistan Journal of Biological Sciences, 7*(2), 238–242.

Rahim, M. A., Alam, A. K. M. A., Alam, M.S., & Anwar, M. M. (2011). Underutilized Fruits in Bangladesh. In M. A. Rahim, M. A. Hashem, M. S. H. Choudhury, T. R. Kwon, M. H. Rahman, & M. M. Hossain (Eds.), *BAU-GPC, Bangladesh Agricultural University, Mymensingh* (Vol. 1, pp. 148–168). Bangladesh; Bioversity International Malaysia and Rural Development Administration (RDA), Korea.

Rao, M. N., Soneji, J. R., and Sahijram, L. (2011). Citrus. In C. Cole (Ed.), *Wild Crop Relatives: Genomic and Breeding Resources* (1st ed., Vol. 1, pp. 43–59). Springer Berlin Heidelberg. https://doi.org/10.1007/978-3-642-20447-0_3

Sarker, J. C., Bhuyan, M. H. M. B., and Rahman, S. M. L. (2019). Evaluation of Citron (Citrus medica) Accessions under the North Eastern Region of Bangladesh. *International Journal of Current Microbiology and Applied Sciences, 8*(04), 2891–2898. https://doi.org/10.20546/ijcmas.2019.804.338

Scora, R. W. (1975). On the History and Origin of Citrus. *Bulletin of the Torrey Botanical Club, 102*(6), 369. https://doi.org/10.2307/2484763

Shahid, S. (2010). Recent Trends in the Climate of Bangladesh. *Climate Research, 42*(3), 185–193. https://doi.org/10.3354/cr00889

Sherozzaman, M. (2012). "Panchagarh District". In S. Islam & A. A. Jamal (Ed.), *Banglapedia: National Encyclopedia of Bangladesh* (2nd ed., Vol. 2, pp. 212–218). Asiatic Society of Bangladesh.

Srivastava, A. K., Singh, S., & Marathe, R. A. (2002). Organic Citrus: Soil Fertility and Plant Nutrition. *Journal of Sustainable Agriculture, 19*(3), 5–29. https://doi.org/10.1300/J064v19n03_03.

Strano, M. C., Altieri, G., Allegra, M., Di Renzo, G. C., Paterna, G., Matera, A., & Genovese, F. (2022). Postharvest Technologies of Fresh Citrus Fruit: Advances and Recent Developments for the Loss Reduction during Handling and Storage. *Horticulturae, 8*(7), 612. https://doi.org/10.3390/horticulturae8070612.

Talon, M., Wu, G. A., Gmitter, F. G., & Rokhsar, D. S. (2020). The Origin of Citrus. In T. Manuel, C. Marco, & F. G. Gmitter (Eds.), *The Genus Citrus* (1st ed., Vol. 1, pp. 9–31). Elsevier. https://doi.org/10.1016/B978-0-12-812163-4.00002-4.

Uddin, M. J., Dey, S., & Taslim, T. (2016). Trend and output growth analysis of major fruits in the Chittagong region of Bangladesh. *Bangladesh Journal of Agricultural Research, 41*(1), 137–150. https://doi.org/10.3329/bjar.v41i1.27679

Uddin, M.N. (1987). History of Horticultural Research Station, Jaintapur, Sylhet, BARI, Jaintapur, Sylhet-3156

Wikipedia contributors. (2024a). Citrus. In *Wikipedia, The Free Encyclopedia.* Retrieved 14:22, March 13, 2024, from https://en.wikipedia.org/w/index.php?title=Citrus&oldid=1213788988

Wikipedia contributors. (2024b). Rangpur Division. In *Wikipedia, The Free Encyclopedia.* Retrieved 11:02, April 23, 2024, from https://en.wikipedia.org/w/index.php?title=Rangpur_Division&oldid=1220065059

Wikipedia contributors. (2024c). Rangpur (fruit). In *Wikipedia, The Free Encyclopedia.* Retrieved 08:17, April 13, 2024, from https://en.wikipedia.org/w/index.php?title=Rangpur_(fruit)&oldid=1218143673

World Bank. (2024). The World Bank open data. Retrieved from https://data.worldbank.org/indicator/EN.POP.DNST?locations=BD dated: 27.04.2024

Wu, G.A., Terol, J., Ibanez, V., López-García, A., Pérez-Román, E. et al. (2018). Genomics of the origin and evolution of Citrus. *Nature, 554*(7692): 311–316.

Zhong, G., & Nicolosi, E. (2020). Citrus Origin, Diffusion, and Economic Importance. In A. Gentile, S. La Malfa, & Z. Deng (Eds.), *The Citrus Genome* (1st ed., Vol. 1, pp. 5–21). Springer International Publishing. https://doi.org/10.1007/978-3-030-15308-3_2

11 Citrus Industry of Vietnam

Present Status and Future Opportunities

Nguyen Van Hoa, Dang Thi Kim Uyen,
Nguyen Thi Kim Thoa, and Vo Huu Thoai

11.1 INTRODUCTION

According to the Ministry of Agriculture and Rural development of Viet Nam (MARD) in 2022, the total area for citrus was 259,967 ha with the total production of 3.67 million tones. Of them, in the North it was over 106,125 ha, the central North over 29,630 ha, the South-central coast and central high land of 11,813ha and the South was over 91,702 ha (Figure 11.1). Citrus in Vietnam (orange, pummelo, mandarin, lemon and lime) is occupied a biggest horticultural crops area in Vietnam, it is around 24%. Of them, the orange area is of 41%, next is pummelo (33%), lemon and lime (16%) and the rest is mandarin 10% (Figure 11.2). Citrus places not a big contribution for farmers' income since the price is low and many other factors effected.

11.2 MAJOR CULTIVARS OF CITRUS CULTIVATED IN VIETNAM

According to Tran Van Hau and Tran SyHieu, 2023, there were several citrus varieties are being grown in Vietnam, such as a group of orange eg. Cam mat, cam ngot (*Citrus sinensis* (L.)), cam xoan, cam tron (Hamlin, Valencia), cam Navel, cam huyet; Group of mandarin eg. Satsuma, DiaTrungHai, Hong (Tieu), King mandarin, Duong, Goup of Lemon, eg. Chanh tau (*Citrus limonia*); Group of lime as Chanhgiay (*C. aurantifolia*), chanhkhong hat (*C. latifolia*); Group of Pummelo (*Citrus maxima*) such as buoi Nam Roi, buoi Da xanh, buoi Long Co Co, buoi Thanhkieu, buoi Duong La Cam, buoi ThanhTra, buoi PhucTrach, buoi Dien; and the Calamondin (Table 11.1).

11.3 MAJOR CITRUS VARIETIES GROWN IN VIETNAM FOR COMMERCIALIZATION

11.3.1 Sweet Orange – Cam Mat

Citrus sinensis is a popular citrus variety grown in Vietnam, particularly in the Mekong river Delta, Southern Vietnam, which provides more Vitamin C to customers. If you are not a VN citizen, please look forward to have and enjoy the Sweet orange from VN, this variety is suitable to the tropical conditions of temperature of 28 to 38°C and high humidity, it can be grafted on its own rootstock or marcotting (Le Van Duc *et al.*, 2023a).

11.3.2 Xoan Orange – Cam Xoan (Diamond)

This is the same family as sweet orange; it is growing more popular in the Mekong river delta region, it becomes famous in Lai Vung district, Dong Thap province. This orange fruit has a thin,

DOI: 10.1201/9781003584599-11

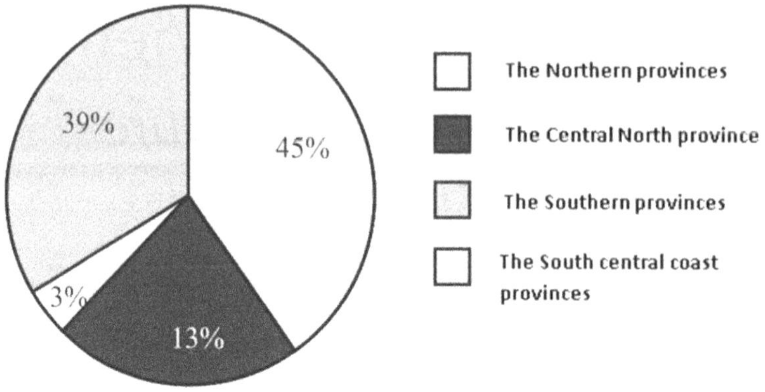

FIGURE 11.1 The concentrated citrus fruit growing area of Vietnam.

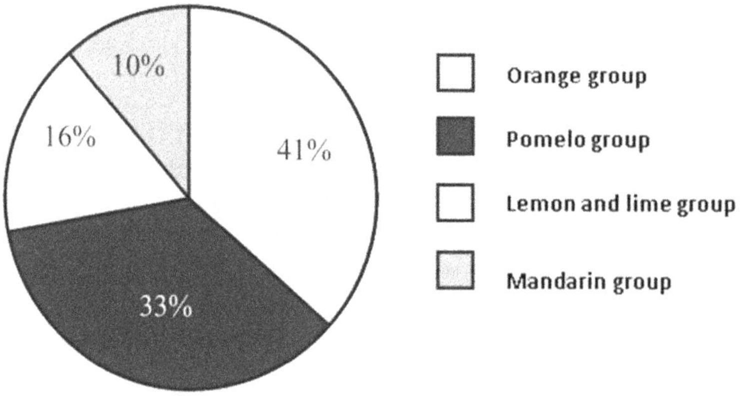

FIGURE 11.2 The concentrated citrus fruit growing area of Vietnam.

succulent skin, yellow flesh, very few seeds, a sharp sweet taste and a characteristic aroma. The average weight of each orange is 250–300 grams. However, the smaller the orange, the sharper the sweetness. Delicious fruits usually do not weigh more than 250grams. Orange fruits usually have a balanced shape, shiny and slightly tanned skin. Diamond oranges are also called two-coins oranges because the bottom of the fruit and the stem end are round, coin-shaped, slightly concave. The peel of the orange fruit is green and mixed with a little yellow. Diamond oranges are best at that level. When the fruit peel turns completely yellow, the amount of sugar in the fruit will decrease, causing the orange to no longer be sweet but gradually turn sour (Tran Van Hau and Tran Sy Hieu, 2023).

11.3.3 King Mandarin – Cam Sanh

Citrus nobilis is about 3.30 m high, the canopy is 4.40m, the normal harvesting is around Lunar New Year in Vietnam, however, the Mekong river delta farmers can induce off-season fruit whole around the year for commercialization, especially it if shipping to the Northern parts of the country in the dry season for juice. The fruit is round to heart shape of average weight of 275g/fruit and 3x2 cm size with juicy of a little source and sweet, at young stage the fruit skin is dark green and it will be orange to yellow skin at ripening time (Tran Van Hau and Tran Sy Hieu, 2023), and in the North, it is grown in Ha Giang province for northern market (Le Van Duc *et al.*, 2023)

TABLE 11.1
List of Citrus Germplasm in SOFRI, VietNam (2024)

Sr. No	Variety name/VN name	Scientific name	Source of origin	Collection time
		Pummelo group		
1	Buoi Nam Roi	*C. maxima (Burm)Merr*	***TienGiang***	1998
2	Buoi Nam RoiBinh Dai	*C. maxima (Burm)Merr*	Ben Tre	1998
3	Buoi Bang Luan	*C. maxima (Burm)Merr*	Doan Hung, PhuTho	2011
4	Buoi Chua	*Citrus maxima*	Nghe An	2011
5	Buoi Chua Ha Tri	"	Ha Tri, Cao Bang	2011
6	Buoi Chua PhucTrach	"	PhucTrach, Ha Tinh	2011
7	Buoi Chum/Grapefruit	"	Ha Noi	2011
8	Buoi Da Xanh	"	Cai Be -TienGiang	2011
9	Buoi Dang	"	Ha Noi	2011
10	Buoi Dao HuongKhe	"	Huongkhe, Ha Tinh	2011
11	Buoi Dao Ngot	"	Ha Tinh	2011
12	Buoi Do	"	Ha Noi	2011
13	Buoi Hong	"	ThanhHoa	2011
14	Bu BuoioiPhucHoai	"	Cao Bang	2011
15	Buoi Rung	"	Cao Bang	2011
16	BuoiSuu	"	Doan Hung, PhuTho	2011
17	BuoiTru	"	BinhDinh	2012
18	BuoiTru Dai Binh	"	Quang Nam	2013
19	Buoi Duong La Cam	"	Binh Duong	2017
20	Buoi Long Co Co	"	TienGiang	2015
21	Buoi Luan Van	"	FAVRI, Ha Noi	2023
22	Buoi Duong BanhXe	"	Binh Duong	2023
23	Buoi Long Hong	"	TienGiang	2023
24	Buoi Ruby	"	Vinh Long	2023
25	Buoi Thank Kieu	"	Dong Thap	2023
26	Buoi Nam Du	"	TienGiang	2023
		Orange group		
27	Cam Bo Ha	*Citrus reticulate*	Nam Dinh	2011
28	Cam Bu Ha Tinh	*Citrus reticulate*	Ha Tinh	2011
29	Cam ChanhHaiHau	*Citrus sinensis x C. grandis*	Hai Duong, QuangBinh	2011
30	Cam mat	*Citrus sinensis*	Can Tho	2011
31	Cam Duong Canh	*Citrus reticulate*	Vinh Long	2011
32	Cam Long Vang	*Citrus sinensis*	HoaBinh	2011
33	Cam Song Con	*Citrus sinensis*	Nghe An	2011
34	Cam Thien	*Citrus sinensis*	QuangBinh	2011
35	Cam TrungVuong	*citrus siensis*	Cao Bang	2011
36	Cam Van Du	*Citrus sinensis*	Nghe An	2011
37	Cam Voi	*Citrus sinensis*	QuangBinh	2011
38	Cam XaDoai	*Citrus sinensis*	Nghe An	2011
39	Cam Sanh (King mandarin)	*Citrus nobilis*	TienGiang	2011
40	Cam xoan	*Citrus sinesisOsbeck*	Ben Tre	2012
41	Cam Valencia My	*Citrus sinensis*	Ha Noi	2011
42	Washington Navel	*Orange*	Ha Noi	2011
		Mandarin group		
43	QuytBacGiang	*Citrus reticulate*	TuyenQuang	2011
44	Quyt Chum	*Citrus reticulate*	Ha Giang	2011

(continued)

TABLE 11.1 (Continued)
List of Citrus Germplasm in SOFRI, VietNam (2024)

Sr. No	Variety name/VN name	Scientific name	Source of origin	Collection time
45	Quyt Duong	*Citrus sp.*	TraVinh and HauGiang	2011
46	Quyt Ha Tri	*Citrus reticulate*	Cao Bang	2011
47	Quyt Hoi	*Atalantiaguillauminii*	NinhBinh	2011
48	Quyt Ly Nhan	*Citrus reticulate*	Ha Nam	2011
49	Quyt Nam Son	*Citrus reticulate*	HoaBinh	2011
50	QuytTichGiang	*Citrus reticulate*	Hưng Yen	2011
51	QuytVangBac Son	*Citrus reticulate*	Lang Son	2011
52	Quyt Hong (Tieu)	*Citrus reticulate*	Dong Thap	2011
		Lemon and Lime group		
53	ChanhBuoi	*Citrus medica*	Nghe An	2011
45	Chanh Do	*Citrus limon*	Ha Noi	2011
55	Chanh San	*Citrus jambhiri*	Hung Yen	2011
56	Chanh Tau	*Citrus limon*	Dong Thap	2014
57	Chuc La Nho	*Citrus histrix*	-	2011
58	Chap	*Citrus sinensis x limon*	Ha Tinh	2011
59	Chuc	*Citrus histrix*	Cambodia	2009
60	Chanhgiay	*C. aurantifolia*	TienGiang	2009
		Calamondin group		
61	Tac se	*Fotunella japonica*	Ben Tre	2019
62	Tat HuongKhe	*Citrus sp.*	HuongKhe, Ha Tinh	2011
		Other relation to citrus		
63	Fortune	*Orange*	France	1995
64	Murcott	*Orange*	France	1995
65	Orlando	*Orange x Mandarin*	France	1995
66	Osceola	*Mandarin*	France	1995
67	Acolemon		Australia	1998
68	CaRi		Ha Noi	2011
69	Cleopat	*Citrus sp.*	Ha Noi	2011
70	Com Ruou	*Glycomis sp.*	PhuQuoc, KienGiang	2011
71	Da Tu Bien	*Limnocitruslittoralis*	NhaTrang	2011
72	Ky Da HaiHau	*Citrus maxima*	HaiHau, Nam dinh	2011
73	Mac Run	*Citrus lime*	Lang Son	2011
74	Thanh Yen	*Citrus medicavarethroge*	Dien Bien	2011
75	TienChot	*Seveinabuxifolia*	ThanhHoa	2011

11.3.4 Hong (Tieu) Mandarin – Quyt Hong (Tieu)

Citrus reticulata is being grown mostly in south-western provinces (the Mekong river delta), especially in Lai Vung district, Dong Thap province; it has a spherical shape, the thin orange-pink shell mixed with some green spots when fruit is ripe, look like Satsuma mandarin from Japan, but it is only the mandarin variety has such colour in the tropical conditions like Vietnam, with orange to red flesh, few seeds and sweet taste (Tran Van Hau and Tran Sy Hieu, 2023).

11.3.5 Quat (Tac Or Hanh)

Fotunella japonica is a small citrus tree of 1.5 m high with very good canopy distribution, people use it as an ornamental bush tree during Lunar New Year with beautiful tree bearing many fruits of small size of 2–3 cm each, in addition, each fruit of this citrus provides pectin 10%, vitamin C 0.13–0.24 mg, Fe 5.1 mg, Cu 0.8 mg, sugar, organic acids etc. That is why it provides very good juice during dry summer days.

11.3.6 Da Xanhpummelo – Buoi Da Xanh

Citrus grandis/C. maxima originates from the Ben Tre province and is widely grown in the Mekong river delta region with the green skin. That is why it is named Da xanh (green skin), Da Xanhpomelo seedlings grow healthily and are less susceptible to pests and diseases. The variety has round fruit with smooth, shiny green skin. The pulp when ripe is light pink in colour, with an average weight of 1,500 to 2,000 grams. The fruit's flesh is pinkish in colour, succulent, crunchy, sweet and not bitter. Harvest time is from September to early November; the fruit can be stored under normal conditions for 1 to 2 months (Tran Van Hau and Tran Sy Hieu, 2023).

11.3.7 Dienpummelo

It is a famous one in the North Vietnam (Ha Noi), it grows well, has less pest and disease invasion; this variety has round fruit. When ripe, the skin of the fruit is smooth, yellow in colour, the average weight of the fruit is 800–1,000g, the segments are easy to separate, the flesh is light yellow, crispy and sweet. When ripe during the Lunar New Year, the fruit can be stored under normal conditions for 3 months (Le Van Duc *et al.*, 2023b). See Table 11.2.

In Vietnam, we do not grow graft fruit since in Vietnamese the graft fruit (buoi chum) is easy to be misunderstood with pummelo (buoi) and it can't be sold in the market.

11.4 AGRO-CLIMATE ZONES IN VIETNAM WHERE CITRUS IS CULTIVATED

In Vietnam, with diversity of the climate from tropical to subtropical conditions of the long country and diversified of plain delta to hilly regions, which is very suitable to citrus growing at different stages of the trees like vegetative growth, fruiting and repining (Le Van Duc *et al.*, 2023a & 2023b) e.g. Temperature of 23 to 39°C; while if the temperature is less than 12°C and over 40°C citrus trees will not be developed well and then affect to all tree's activities making lower yield and worse quality. If the air temperature is high, it also effects the activities of the root system, which makes root uptake water and nutrient more or less and then leading to less or over development of the tree, the citrus leaves will be over respiratory. The total temperature ranges from 25,00–35,00°C can grow citrus trees.

Citrus tree can be grown in different of soil types. However, if it is planted in not good soil, the more investment must be inputted, then the income will be less. The good soil for citrus must have a soil layer over a metre, rich in humus (humus content in soil from 2–2.5% or more), the content of nutrients N, P, K, Ca, Mg ... , reaches average or higher (N: 0.1% –0.15%; easily digestible P_2O_5: 5–7 mg/100 g of soil; easily digestible K_2O:7–10 mg/100 g of soil; Ca, Mg: 3–4 mg/100 g soil). The suitable pH is of 5.5 to 6.5. Especially good drainage is required; mechanical composition includes sandy soil, riverine alluvial soil, or light loam (coarse sand to light loam soil accounts for 65–70%); Slope from 3–8 degrees.

Citrus is a tree that loves moisture but cannot tolerate waterlogging because the tree's roots are fungal roots. Therefore, if the soil is flooded with lack of oxygen, the roots will function poorly. If submerged for a little long time, the roots will rot, causing leaves and young fruits to fall. This explains why growing citrus on land with trees does not last as long as growing on sloping land. The periods when plants need water are: the period when the plant is basically building, the period when the plant is growing, especially the period when the plant differentiates flower buds, bursts flower

TABLE 11.2
Some Major Citrus Varieties Grown in Vietnam

Sweet Orange (Cam Mat – *Citrus sinensis*)	
Xoan orange (Cam Xoan)	
King Mandarin (Cam sanh)	

(Continued)

TABLE 11.2 (Continued)
Some Major Citrus Varieties Grown in Vietnam

Chanh tau Lemon (*Citrus limonia*)	
Hong (Tieu) mandarin (quyt Hong (Tieu)	
Quat (*Fotunella japonica*) (Hanh, Tac))	

(*continued*)

TABLE 11.2 (Continued)
Some Major Citrus Varieties Grown in Vietnam

Da xanh pummelo (*Citrus maxima*) (Buoi Da Xanh)	
Nam Roi pummel (Buoi Nam Roi)	
Fruit and tree of Dien pummelo	

(*continued*)

TABLE 11.2 (Continued)
Some Major Citrus Varieties Grown in Vietnam

buds, flowers and develops fruit. The annual amount of water needed for 1 hectare of citrus ranges from 9,000–12,000 m^3, equivalent to 900–1,200 mm/year of rainfall. In general, rainfall in agricultural production areas of our country ranges from 1,400–2,500mm/year. In terms of total quantity, it is enough or even more than the citrus tree's needs. However, rainfall is unevenly distributed between months of the year, causing excess water or water shortage at different growth stages. Therefore, it is necessary to have technical measures to maintain moisture and supplement water during dry periods; good drainage during prolonged and localized rain.

Citrus does not like strong light, preferring scattered light with an intensity of 10,000–15,000 Lux, corresponding to 0.6 cal/cm^2 and corresponding to light at 8 am and 4–5 PM on clear summer days. This is because light intensity directly affects CO_2 assimilation. Strong light intensity reduces CO_2 assimilation because of increased radiation on the leaf surface. The optimal temperature on the leaf surface for CO_2 assimilation ranges from 28–30°C. Temperatures below the optimum also reduce CO_2 assimilation. To have scattered light, it is necessary to arrange reasonable tree density and regularly prune properly.

Particularly with these conditions, farmers can induce flowers anytime for year-round harvesting. These conditions are also suited to pests and diseases development.

In Viet Nam, important citrus growing area concentrate at provinces of Nghe An, Ha Tinh, Ha Giang, Ha Noi, PhuTho, Lang Son, Yen Bai in the Northern part, and in the South, it is grown more at Vinh Long, Ben Tre, TienGiang, Can Tho, Dong Thap, TraVinh, Dong Nai, Binh Duong provinces (Figure 11.3).

11.5 SIGNIFICANT TECHNOLOGIES AND PACKAGE OF PRACTICES FOLLOWED IN CITRUS FARMING

Over 133 research studies have been carried out by SOFRI in 30 years of establishment. The main packages of new advanced technologies have been developed and used by citrus farmers/growers and cooperatives in Vietnam for citrus farming.

FIGURE 11.3 Map showing area of domain citrus growing in Vietnam.

The significant technologies and package of practices used in Vietnam for citrus farming are:

- Less seeds Duong La cam Pummelo LD 4 (Figure 11.4) variety has been developed by SOFRI using gamma radiation and got recognized by the Vietnam Ministry of Agriculture and Rural development (MARD) in the year 2011 (under decision number 242/QĐ-TT-CCN, on May 19th, 2011). The average weight of fruit is from 0.8–1.2 kg/fruit, low pear shape, smooth surface, pale yellow and tightly packed flesh, very few seeds (5–6 seeds/fruit compared to >50 seeds of mass-produced varieties), Brix degree average > 10%. Average vitamin C content > 29mg/100g. Less susceptible to stem gummosis, pink fungus disease, scab, aphids, fruit borers and green stink bugs. Suitable on alluvial soil along rivers or in the Southeast region.
- Seedless King Mandarin LĐ 6: This seedless LD 6 (Figure 11.4) had been developed from gamma mutation and got recognized by the MARD in the year 2011 (under same decision number 242/QĐ-TT-CCN, on May 19th, 2011). Number of seeds/fruit is from 0–3, thin and smooth skin with orange flesh and juicy, sweet to light soure with characteristic fragrance.
- Successful development of the shoot tip grafting (STG) technology for viral free seedlings in sweet orange, mandarin, pummel, king mandarin, etc. for huge amount of free-diseased seedlings multiplication since 2000 to now. All the seedlings must be kept in double doors green house before release to growers.

Seedless King mandarin LD 6

Less seeds Duong La Cam Pummelo LD 4

FIGURE 11.4 Some new citrus varieties developed by SOFRI in Vietnam.

- Only use of the free-diseased seedlings through STG for cultivation, regularly monitoring citrus orchards for any abnormally growing plant appear, then the tree should be treated, removed and/or destroyed completely.
- Grow citrus at suitable density (average of 400–500 citrus trees/ha depending on type of citrus and soil types, conditions).

- Orchards sanitation regularly, especially after each harvest to remove any source of inoculum, plant debris to avoid any infection/infestation.
- Orchards management and canopy pruning at early stage of the citrus trees growth, which make orchards more ventilation and getting more sun light for better biosynthesis, in addition, it provides less chance of pests and diseases development. Besides that, growers can control the tree height for better distribution of branches and canopy management.
- Balance the nutrition by using proper suitable fertilization, organic matters and antagonist organism (not using foliage fertilizer regularly).
- Good weed management: except the intercropping plant on the orchard, the entire weed must be cleaned; especially at near the trunk base to avoid pests and diseases invasion to citrus trees.
- Good drainage and irrigation system to avoid high humidity which is suited to soil borne disease development or drought in high land during dry season. And suit to induce off-season flowers.
- Using suitable rootstock for each type of commercial citrus varieties/types at different regions of different conditions like drought (Chuc, Volkameriana variety) in high, low (sweet orange, mandarin) lands, acidity and salinity soil type and conditions (sweet orange, mandarin, Da xanhpummelo).
- Determination best time of harvesting of different types of citrus, like Nam roipummelo is 162 to 169 days after fruit set, but Hong mandarin is 38 to 39 weeks, which provides the best quality and long storage time.
- Minimally processed Nam roipummelo products were prepared and handled to maintain their fresh nature for providing convenience to the users. Fresh Nam roipummelo after preparing segments and dipping in acid ascorbic 1.5% and the addition of antimicrobials with Natrimetasiphite 0.06% or Potassium sorbate 2%, they were packed in polystyrene trays covered with PVC film. These packages were stored at 10–12^{0}C, the products remained good qualities in 20–25 days.
- In SOFRI, we have the Transfer technology centre with all facilities to train TOT staff from local provinces and direct to farmers/growers, cooperatives' members, especially at the Mekong River Delta provinces, with the demonstration sites in SOFRI (0ver 60 ha at TienGiang province and over 200 ha in Ba Ria – Vung Tau province).
- In addition, in Vietnam we have the national extension centre, which can help to transfer technology to farmers, cooperative within the country too. Through this centre, SOFRI can help, support and collaboration for transferring citrus technologies.
- In Vietnam, the horticultural product should follow the GAP standards (VietGAP, GlobalGAP) for better providing safe products for local consumption or exportation to overseas customers. So far, SOFRI's staff have consultancy to growers, cooperatives, companies to produce citrus complain with GAP standards for many thousand hectares.

11.6 STATUS OF CITRUS INDUSTRY

Harvesting Window of major cultivars of citrus: For citrus, the main harvesting time in Vietnam is around October to February or March of next year; it is suitable for customers' consumption and New Year celebration, Lunar New Year decoration. Some beautiful colour citrus varieties which people keep as bonsai in front of the house like Hong (Tieu) mandarin, Dienpomelo, etc. In the South of Vietnam, growers/farmers can induce off-season flower (Tran Van Hau and Tran SyHieu, 2023 and Le Van Duc *et al.*, 2023a & 2023b) for harvesting fruits whole year around, which make diversity for consumption by fresh eating or juice during the sun light days in the dry season with high temperature period of time.

Marketing network of citrus: Regarding citrus fruits (oranges, mandarin and pummelo,) Vietnam always has a high position in the world compared to other fruits (1.4 million tonnes/year; accounting for 0.9% of the world's citrus fruits).

However, in Vietnam, the main market of citrus such as orange, mandarin, lime and lemon is domestic markets; since we have a long country with diverse climate conditions, so the consumption of citrus fruit is also diversity. Vietnam only exports pomelo to foreign countries since it can be stored for longer time over 2 months by sea freight container shipping.

11.6.1 Potential, Advantages, and Some Major Limitations For Oranges

11.6.1.1 Orange

Belongs to a group of fruits with high and continuously increasing demand. Export value on the world market from 2009–2013: fresh fruit from 4.4 billion USD to 5.5 billion USD; Orange juice reached over 1.16 billion USD/year from 2009–2010 and reached over 3 billion USD/year from 2011- two thousand and thirteen.

Vietnam belongs to the group of countries with the largest orange harvest area and output in the world (1.06% and 0.74% respectively); Ranked 2nd in Southeast Asia (after Indonesia). Our country's fresh orange products are mainly used for domestic needs and do not have much export competitiveness due to unattractive designs and varieties with many seeds. Vietnam's orange (fresh fruit) export value is almost insignificant: 15 thousand USD in 2015 and 43 thousand USD in 2016.

11.6.1.2 Pummelo

In the ASEAN region, in recent years, the growth rate of fruit and vegetable exports has been fast and strong, an average of 32.2% per year in the period 2011–2016, and in 2017 it increased sharply to 42.4% compared to the previous year 2016. In 2018, fruit and vegetable export turnover reached 3.81 billion USD, an increase of 8.8% over the year 2017, in which fruit export turnover alone reached 3.13 billion USD, an increase of 4.8% over the year 2017 and accounted for 82.05% of total fruit and vegetable exports.

Regarding citrus fruits (oranges, pummelo,), Vietnam always has a high position in the world compared to other fruits (1.4 million tonnes/year; accounting for 0.9% of the world's citrus fruits).

11.6.1.3 Status of Export and Import of Citrus

The exported value of pummelo in 2020 was 4.93 million USD (Doan Huu Tien *et al.*, 2020). And the main market of Vietnamese Da xanhpummelo was China, EU, Russia, Canada. According to data from the Ministry of Industry and Trade, Vietnam, pummelo exports in 2021, 2022, and 2023 was reached 9.77 million USD, 18.78 million USD and 43.3 million USD, respectively.

Production of Nam roi and Da xanhpomelo has been helped farmers and agents of pomelo supply chain to get income. On the supply chain of Nam roipomelo from Vinh Long to HoChiMinh City (HCMC), the profit of farmer, collector, wholesaler and retailer was 350VND/kg, 240VND/kg, 210VND/kg and 1.127VND/kg, respectively. On the supply chain of Da xanhpomelo from Ben Tre to HCMC, the profit of farmer, collector, wholesaler and retailer was 4,429VND/kg, 690VND/kg, 645VND/kg và 4,284VND/kg respectively. On the supply chain of Nam roipomelo from Vinh Long to HCMC the price of pomelo was increasing 2.2 times. On the supply chain of Da xanhpomelo from Vinh Long province to HCM City the price of pomelo was increasing 1.7 times. The export supply chain gave farmer higher profit than domestic supply chain.

11.7 STATUS OF CITRUS PROCESSING

In Vietnam, the processing status is still low, only some citrus juice making; the processed products from citrus need to be developed for future diversity products beside the fresh consumption.

The external colour of citrus fruits normally relates to climatic conditions more than to maturity. The treatment of 18 ppm in 37 hours was found effective in degreening of Cam sanh orange fruits (King mandarin).

Minimally processed Nam roipummelo products were prepared and handled to maintain their fresh nature for providing convenience to the users. Fresh Nam roipummelo, after preparing segments, and dipping in acid ascorbic 1.5% and the addition of antimicrobials with Natrimetasiphite 0.06% or Potassium sorbate 2%, were packed in polystyrene trays covered with PVC film. These packages were stored at 10–12^{0}C, the products remained in good quality in 20–25 days.

11.8 STATUS OF CITRUS RESEARCH AND EXTENSION

Many research projects both national and international have been carried out in Vietnam both at the North (Fruit and Vegetable Research Institute – FAVRI), National Plant Protection Institute (PPRI) and in the South (Southern Horticultural Research Institute, SOFRI) for both breeding, cultivation techniques, root stock for different types of soils and adverse conditions, pests and diseases etc. We have over 130 studies on citrus tree both in nation and international collaboration. The international organizations are CIRADFHLOR and INRA of France; ACIAR and Sydney University of Australia; Taiwan National University of Taiwan; JIRCAS, JICA and Aichi Steel Company of Japan; and USDA and Florida University of the USA, etc.

As mentioned above, we have the Transfer technology centre, which helps to do training and demonstration to local staff and growers from the South Vietnam.

In addition, under the plant wise system, we have established the plant doctor system in collaboration with Agribusiness' company to run online or onside plant clinics to make consultancy to farmers any time they needed.

11.9 CHALLENGES AND STRATEGIES

In Vietnam, the major problem for citrus cultivation is pests and diseases, since our country is at tropical conditions. In the following sections, we will list all these and their management in a friendly environmental way.

The Southern Horticultural Research Institute has been studying citrus for the subjects on breeding, cultivation techniques and plant protection. SOFRI has committed various international projects for citrus culture, especially plant protection, with foreign organizations mentioned above. National projects were also carried out to screen for tolerant citrus varieties. In particular, the following diseases and arthropod pests have been focused on in these projects (Table 11.3). The main plant diseases are citrus greening disease (HLB), citrus Tristeza virus (CTV), canker, root rot and gummosis diseases, while arthropods are Asian citrus psyllid (*Diaphorinacitri*), aphids, mealy bugs, mites and thrips. In this chapter, we describe the characteristics of some important pests and diseases and explain how to manage them with integrating compatibly different measures. See Table 11.4.

TABLE 11.3
Important Citrus Diseases and Pests in Vietnam

Disease/Pests	Causal agent/Scientific name	Time of infection/ infestation	Damaged part	Incidence severity
Greening disease (Huanglongbing)	*Candidatus* Liberibacterasiaticus	Year around	Whole tree	+++
Citrus Tristeza	*Clostero virus*	Year around	Leaf, branch, fruit	++
Citrus Canker	*Xanthomonas Axonopodis*pv. *citri*	Rainy season	leaf, fruit	++
Citrus Root rot	*Fusariumsolani, Phytophthora* sp., *Pythium* sp.	Rainy season	Root, whole tree	+++
Citrus Scab	*Elsinoefawcettii*	Rainy season	leaf, fruit	++
Greasy spot	*Mycossphaerellacitri*	Year around	Leaf	++
Sooty Mould	*Capnodium*sp.	Rainy season	Twig, flower, leaf	+++
Phytophthorarootrot and Gummosis	*Phytophthoranicotianae / P. citrophthora*	Rainy season	Trunk and root	++
Anthracnose	*Colletotrichum Acutatum*	Rainy season	leaf, fruit	++
Quick decline – wilt syndrome	*Pseudococcus*sp. plus *Clitocybetabescen*	Rainy season	Root	+
Slow decline	*Tylenchulus Semipenetrans*	Rainy season	Root	+
Root lesions	*Pratylenchuscoffeae*and/or *P. citri, Tylenchulus* spp.	Rainy season	Root	+
Thrips	*Scirtothripsdorsalis Thripshawaiiensis*	Dry and hot weather	flower, young leaf, fruit	+++
Mealy bug	*Planococcus citri Planococcus* sp.	Dry and hot weather	whole tree	+++
Fruit borer	*Citripestissagittiferella*	Year around	fruit	++
Leaf miner	*Phyllocnistiscitrella*	Year around	young leaf, fruit	++
Psyllid	*Diaphorinacitri*	Year around	young leaf, shoot	+++
Aphid	*Toxopteracitricidus Toxopteraaurantii*	Year around	young leaf	++
White moth plant hopper	*Lawana conspersa*	Year around	whole tree	+
Spine fruit bug	*Rhynchocoris poseidon*	Year around	fruit	+
Citrus red mite	*Panonychuscitri*	Year around	leaf, fruit	++
Citrus rust mite	*Phyllocoptrutaoleivora*	May–July and October–November	leaf, fruit	++
Broad mite	*Polyphagotarsonemuslatus*	Year around	Young leaf	+

TABLE 11.4
Some Important Citrus Diseases and Pests in Vietnam

Causal agent	Typical symptom	Transmission	Integrated management
	GREENING DISEASE (HUANGLONGBING – HLB)		
Candidatus *Liberi-bacterasiaticus*	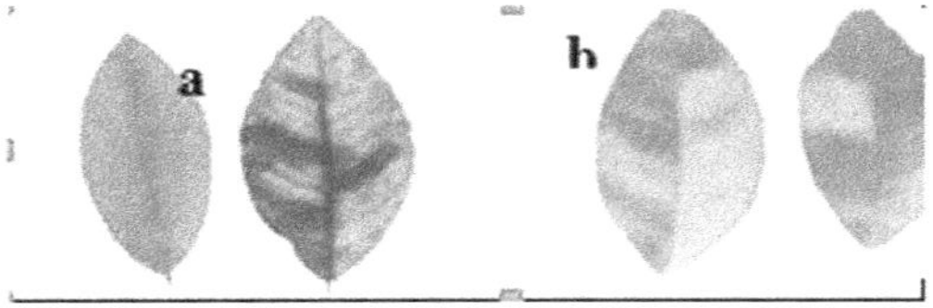 Healthy leaf (a) and yellow, mottling leaves due to HLB (b) 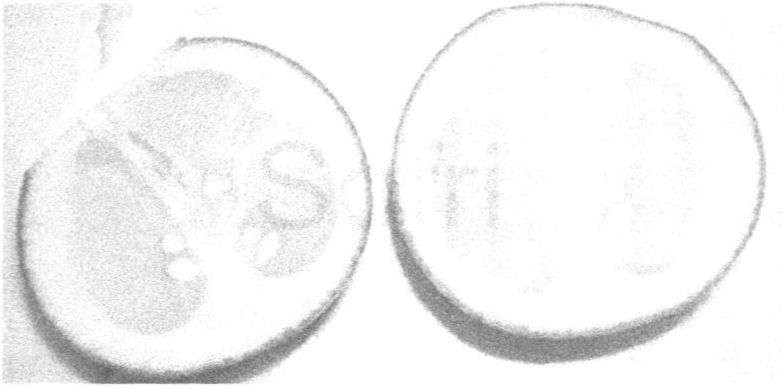 Eccentric fruit symptoms on Da xanh pomelo	• Sap transmission • Horticultural tool • Dodder *Cuscutacampestris* • psyllid *Diaphorinacitri* 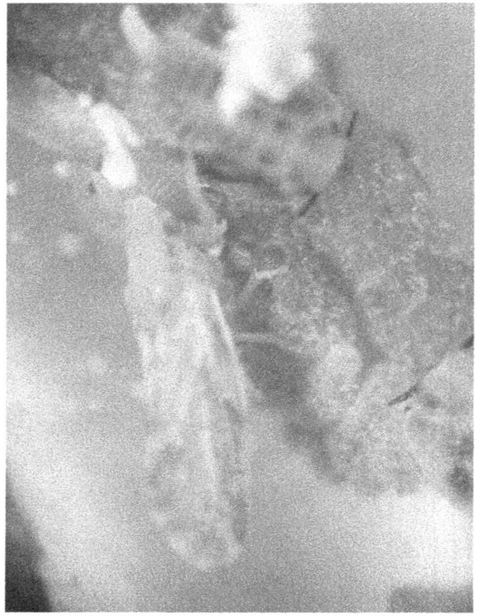 *Diaphorina citri* the HLB vector	• The use of free disease seedlings (STG) • Keep orchard isolated from infection source/ • Systemic Chemical use for its vector • Horticultural and/or mineral oil for vector at each new flesh/shoots • Guava inter-planting one by one at early stage of citrus – Guava should grow at 6 months in advance. • Use of *Tamarixi radiate*, *Diaphorencyrtusaligarhesis*. *Isaria*sp and *Oecophyllasmaragdina*. • Proper nutrients contribute greatly to increasing tolerance of the citrus trees to the disease, utilizing a variety of micronutrients as part of disease management solution prolongs the life of citrus trees (Fe, Zn, Mn, etc), especial increasing the use of organic compounds/fertilizer. • Scientists are going for breeding the tolerant/ resistance citrus variety.

	2. CITRUS TRISTEZA VIRUS (CTV)		
It is caused by Closterovirus with long filamentous string of 11 x 2000 nm.	 Vein clearing symptom on Mexican lime 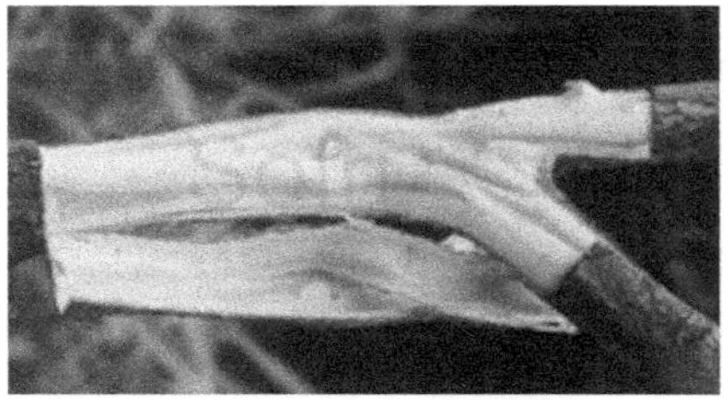 Symptom of stem pitting on the twig of Lemon  Symptom of yellow end of sweet mandarin	• Sap transmission • Horticultural tool • Aphids such as *Aphis gossypii*, *A. spiraecola, Myzuspersicae.* 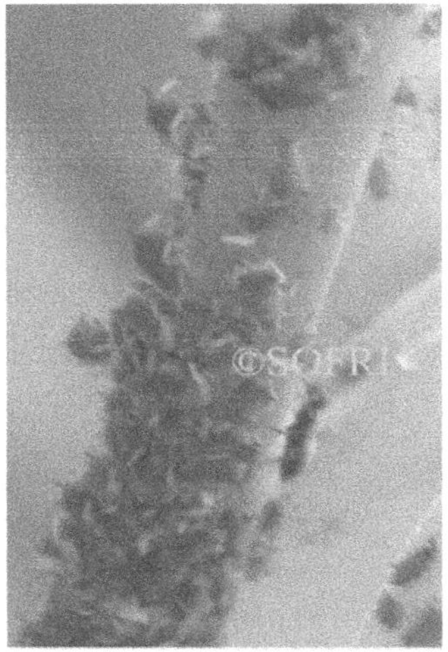 A vector aphid of CTV	• Using Mild strain for cross protection (Vaccination), In Vietnam, we can use vein clearing as a mild strain for cross protection on other citrus varieties. • The use of free disease seedlings (STG) • Keep orchard isolate from infection source/ • Systemic Chemical use for its vector • Horticultural and/or mineral oil for vector at each new flesh/shoots. • Use of *Tamarixi radiate*, *Diaphorencyrtusaligarhesis*. *Isaria* sp and *Oecophyllasmaragdina*. • Proper nutrients contribute greatly to increasing tolerance of the citrus trees to the disease, especial increasing the use of organic compounds/fertilizer.

(continued)

TABLE 11.4 (Continued)
Some important citrus diseases and pests in Vietnam

Causal agent	Typical symptom	Transmission	Integrated management
3. CITRUS CANKER			
A rod-shaped gram -negative bacterium, *Xanthomonas-axonopodis*pv. *citri* (*Xac*)	   Symptoms on leaf (a), young stem (b) and fruit (c)	• Leaf miner (*Phyllocnistiscitrella)* makes wound on the leaf then the bactria can be easy to enter to the young leaf, shoot and then infection occurs.	i) Using of the citrus disease-free seedlings through STG; ii) Immerse tools in Javel water of 1,500 ppm for 5 to 10 min after used on one citrus tree before going to use for the next tree; iii) Keep the orchards under ventilate well with sufficient insolation; iv) Apply limestone powder every year and paste lime on the trunk base to prevent the bacteria inoculum colonization; v) Observer the orchard regularly for any signal of the infection of canker or Infestation of leaf miner. vi) Remove infected parts and keep the remnant away from the citrus orchard; vii) Avoid sprinkler-irrigation system and use bactericides such as copper oxychloride (Isacop), streptomycin orneem cake suspension for spraying on young leaves and shoots. viii) Good management of the leaf miner in field especially at the young stage of new shoots, leaves

4. CITRUS ROOT ROT

The causal organism is *Fusariumsolani, Phytophthora* sp., *Pythium* sp.

Symptoms on leaves and root

- Tolerant variety for rootstock: such as Buoi Do (*C.maxima*), Long CoCopummelo (*C. maxima*) for pomelo commercial variety; Sweet orange for orange and King mandarin; Lemon (tau) for lemon and Mexican lime.
- *Cultural measures:* as those for HLB disease.
- *Mechanical measures:* After every harvesting, citrus trees should be pruned; remove infected/affected parts; the copper fungicide should be passed over the wound.
- *Biological and Chemical control measures*: Symptomatic trees should be immediately treated with dripping of dimethomorph (Insuran) around the trunk on the soil surface every seven to ten days; The inclusion of antagonistic biocontrol agents, like *Trichoderma*, will synergistically regulate the development of the disease; Plant growth regulators for root such as Rootwell, DS Gold, Rhizomyx, promise better root grow and help plant recovery from the disease.

(*continued*)

TABLE 11.4 (Continued)
Some important citrus diseases and pests in Vietnam

Causal agent	Typical symptom	Transmission	Integrated management
	5. *PHYTOPHTHORA* ROOT ROT AND GUMMOSIS (TRUNK CANKER)		
Phytophthoranicotianae, P. citrophthora, P. citropthora, P. citricola and P. palmivora	 Gummosis, also as trunk canker		• The rootstock from the flowing varieties Troyer, Carrizo, Trifoliata and Cleopatra are good candidates for the region heavily infested by the disease. • The soil and/or compost for nursery should be treated with copper fungicide before sowing seeds. • Orchards and nurseries should be kept away from inoculum of *Phytophthora* diseases. • The irrigating water coming from infected areas should be avoided. • Trees need to be planted in suitable densities, which would allow better ventilation and less overlapping of canopies each other. • In the fruiting season, bactericides should be applied every 10 to15 days. No applications should be performed after 20 to 15 days before harvesting. • Application of more organic matters can facilitate antagonistic biocontrol agents like *Trichoderma* to be maintained more and longer.

		• Lime powder is recommended to apply every year in order to increase to soil pH while lowering the survival of the pathogen. Pasting lime on the trunk 50 cm above the ground surface may be considered for the elimination of the pathogen remained near the trunk, especially during the rainy season.
6. THRIPS		
***Scirtothripsdorsalis* and *Thripshawaiiensis* (Thripidae: Thysanoptera)**		• Do not stress the tree. • Make sure the plant is well watered, especially during dry periods. • Spray water with a high-pressure nozzle onto the canopy. • When the tree has just budded/flowered, spray 2-3 times with insecticide.

(continued)

TABLE 11.4 (Continued)
Some important citrus diseases and pests in Vietnam

Causal agent	Typical symptom	Transmission	Integrated management
	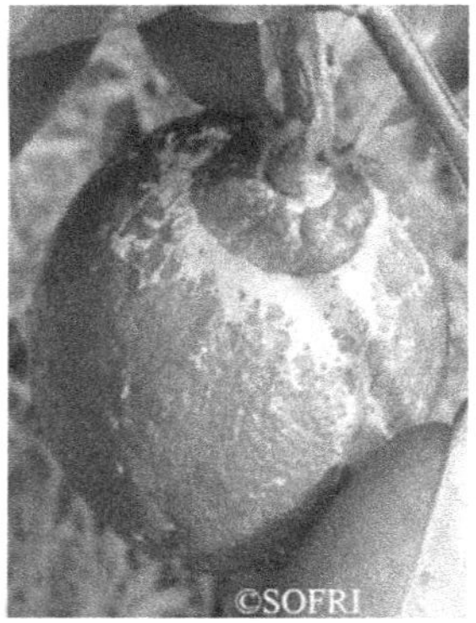 Thrips, its symptoms on flower and fruit		
		7. MEALY BUG	
Planococcuscitri (Pseudococcidae: Hemiptera)	 Mealy bug and its symptoms on fruit		• Prune off excess branches, canopy branches, etc. to reduce conditions for mealy bug to develop. • Remove and destroy fruits, leaves and branches infected with mealy bug to avoid spreading. • Regular cleaning on the garden to avoid creating favourable conditions for ants to shelter and spread bugs.

		• Rearing and releasing ladybird *Menochilussexmaculatus, Coccinellatransversalis, Cryptolaemusmontrouzieri* and lacewing *Suarius* sp., *Chrysoperlacarnea* to control mealybug population. • Chemical control should part of an Integrated Pest Management.
	8. APHIDS	
Toxopteracitricida and Toxopteraaurantii (Aphididae: Hemiptera)		• Prune branches, fertilize properly, and control the tree to produce concentrated buds to limit the continuous infection throughout the year. • The most common are ladybird beetles (adults and larvae), lacewing larvae, and tiny parasitic wasps that lay their eggs in the adult aphids. • Raising weaver ants *Oecophyllasmaragdina* in the garden • Chemical control should part of an Integrated Pest Management.

(continued)

TABLE 11.4 (Continued)
Some important citrus diseases and pests in Vietnam

Causal agent	Typical symptom	Transmission	Integrated management
	 Aphids and its symptoms on young leaves		
9. LEAF MINER			
Phyllocnistiscitrella (Gracillaridae: Lepidoptera)	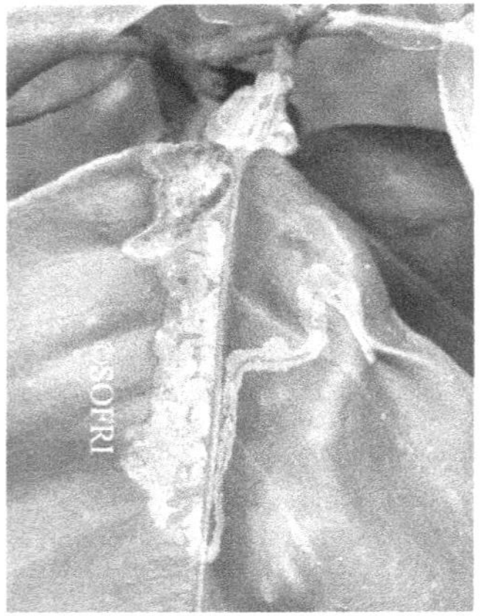 		• Where citrus leafminer is a problem, remove water shoots that might act as a site for the moths to lay eggs (oviposition). All the shoots originating from the rootstock below the graft union, should always be pruned because they produce more vegetative parts and do not usually produce desirable fruit.

	 Leaf miner and its symptoms on leaves	• Nitrogen fertilizer should NOT be applied at times of the year when leafminer populations are high. The presence of Weawearant Oecophyllasmaragdina should bepreserved in citrus orchard due to its deterrent effect to the adult leaf miner • Chemical control should part of an Integrated Pest Management.
	10. PSYLLA	
Diaphorinacitri (Psyllidae: Homoptera)	 Psylla and its present on young leaves	• Infected trees should be removed to prevent them from being reservoirs of the pathogen. • Clean and healthy plants should be transported. • Use yellow traps to track psyllids. • Raising weaver ants *Oecophyllasmaragdina* in the garden • Chemical control should part of an Integrated Pest Management.

(continued)

TABLE 11.4 (Continued)
Some important citrus diseases and pests in Vietnam

Causal agent	Typical symptom	Transmission	Integrated management
11. FRUIT BORER			
Citripestissagittiferella (Pyralidae: Lepidoptera)	 Fruit borer and its symptoms fruit		• Collect and destroy infested fruits (both the fallen fruits and remaining fruits on tree). • Monitoring of the nocturnal activity of adults by using light traps (White light, Compact 15 W lamp, lighting time from 6–9 pm). For an optimal monitoring of adult population, it is recommended to place ten light traps per orchard (1 ha). After 7 days of the adults' appearance, it is advised to start the control programmes including pesticide application or fruit bagging. • Bagging of the fruits at 14–21 days after fruit set reduces damage. Pesticide should be sprayed before bagging the fruits. • Spray 4–5 times/season (depending on the number of fruit borers present in the garden) with some active ingredient pesticides.

The other challenge for citrus in Vietnam, especially in the Mekong River Delta, is saline water coming with the climate changes. We had carried the experiment for salt tolerance, and the result has shown the classifiable same citrus rootstocks into four groups: Excellent salt tolerant genotype was *Citrus nobilis* ($S_{5.1}$); good salt tolerant genotypes were *Citrus reticulata* ($QT_{3.2}$), *Citrus aurantifolia* ($CT_{5.2}$), *Citrus sinensis* ($CM_{6.1}$); medium salt tolerant genotypes were *Citrus maxima* ($BLH_{2.1}$), *Citrus maxima* ($BX_{2.0}$), *Citrus maxima* ($BC_{2.4}$), *Citrus maxima* ($BC_{3.1}$); salt sensitive genotypes were *Citrus maxima* ($BC_{2.3}$), *Citrus maxima* (BC_{1}), *Citrus maxima* ($BTT_{7.0}$) (To Viet Diem Ca *et al.*, 2007).

11.10 CONCLUSION

In Vietnam, the citrus industry has been established for a very long time, and many technological packages have been developed and applied by growers. However, there are still gaps that need to be filled, especially on citrus diseases. Of them, the HLB is the biggest problem which has not fully been understood and not found a way to manage. At the moment, we are collaborating with Japanese scientists to work on this; it is opening another way for this problem, but it still needs more hands to join. Besides that, in Vietnam and Asia, each household has very small land of citrus plantation; the growers need to join with each other for larger producing areas and follow the same safe protocols (GAP, organic, eco-friendly production standards) for uniform good quality products and larger quantities at different times of the year, and more processed products should be there instead of only fresh products as in Vietnam. In addition, for the cultivating techniques, the motorization needs to be developed and used in big production areas.

REFERENCES

Doan Huu Tien, Dai Thu Tuyet Lan, Vo Huu Thoai. 2020. *Present status of Da xanhpummelo production and consumption in the value chain study*. Project report, Hochiminh National University Publisher.

Le Van Duc, Cao Van Chi, Doan Van Lu, Tran Thi My Hanh, Nguyen Quang Huy, Nguyen Quoc Manh, Dao Quang Nghi, Nguyen Van Nghiem, Vo Huu Thoai. 2023a. *Handbook guidline on Orange Cultivation technology follow VIETGAP standards*. Asian Agritrade and GIZ, 104.

Le Van Duc, Cao Van Chi, Doan Van Lu, Tran Thi My Hanh, Nguyen Quang Huy, Nguyen Quoc Manh, Dao Quang Nghi, Nguyen Van Nghiem, Vo Huu Thoai. 2023b. *Handbook guidline on Pomelo Cultivation technology follow VIETGAP standards*. Asian Agritrade and GIZ, 106.

Nguyen Minh Chau, Nguyen Van Hoa, Yusa Kazuyoshi, Koshida ryu, Do Hong Tuan, Vo Huu Thoai, Le Quoc Dien, Nguyen Thi Ngoc Truc, Nguyen Van Son, Nguyen Thanh Hieu, Dang Thuy Linh, Huynh Thanh Loc, Nguyen Huy Cuong, Doan Van Bang. 2013. *Handbook on cultivation and care of King Mandarin*. Agriculture Publisher, 145.

Nguyen Thi Thu Cuc. 2015. *Insect, mite pest of fruit trees in Vietnam and their natural enemies*. Can Tho University Publisher, 1–95.

Nguyen Thi Thu Cuc, Pham Hoang Oanh. 2009. *Citrus pest and their management in Vietnam*. Agriculture Publisher, 151.

Nguyen Van Hoa, Nguyen Ngoc Anh Thu, Dang Thi Kim Uyen, Dang ThuyLinh, Nguyen HuyCuong. 2007. *Hand book on some important diseases on fruit crops in Ben Tre province*. Phuong Dong Publisher, 10–54, 75–85.

Pham Van Du, Nguyen Thi Hong Thuy, Nguyen Van Hoa, Nguyen Thi Thanh Mai, Le Thanh Tung, Nguyen Thi Bich Van. 2014. *Handbook on VietGAP, GlobalGAP Certification Procedures and Certificating bodies*. Agriculture Publisher (ISBN 978-604-60-1735-6).

To Viet Diem Ca, Nguyen ThanhBinh, Le ThiKhoe , VoHuu Thoai. 2007. Reliable and easy screening technique for salt tolerance of commercial citrus rootstocks under *in-vitro*. (Research result on horticultural research from 2005 to 2006. Agriculture Publisher, 20–27).

Tran Van Hau, Tran SyHieu. 2023. *Off season induction technology for Citrus spp*. Agriculture Publisher, 360.

12 Citrus Industry of Iran

Seyed Mehdi Bani Hashemian, Abuzar Hashempour, and Mehdi Azadvar

12.1 INTRODUCTION

Due to its nutritional value and pleasant flavour, citrus is one of the most popular fruits in the world. These fruits are a rich source of vitamins, organic acids, mineral salts, fibres, and antioxidants that effectively strengthen the body's health systems (Lim 2012). Oranges, mandarins and tangerines, limes and lemons, and grapefruits and pomelos are the main horticultural groups of citrus fruits consumed as fresh fruit and processed products. The citrus fruits were the second most produced fruit worldwide in 2021, accounting for 161.8 million tonnes produced in more than 10.2 million hectares (FAOSTAT 2023). Citrus trees are cultivated in the tropical and subtropical regions of the world. They are grown in more than 150 countries from five continents with the highest share (51.7 percent) in Asia (Figure 12.1). China, India, Turkey, Iran, Vietnam, Pakistan, Indonesia, Thailand, Syria, and Japan are the top ten Asian citrus producer countries in 2021 (FAOSTAT 2023) (Figure 12.2). According to the latest FAO statistics, China is the largest citrus producer in Asia and the world. Islamic Republic of Iran with about 3.9 million tonnes has a share of 5 percent in citrus production in the Asia (FAOSTAT 2023) (Figure 12.1).

Iran has 137.5 thousand hectares of cultivated area bearing citrus orchards, of which oranges, mandarins and tangerines, limes and lemons, and grapefruits and pomelos have respectively 69.6, 12, 27.7, and 3.9 thousand hectares. The 10-year trend of citrus harvested area of Iran has been decreasing except in 2021 (Table 12.1). Accordingly, Iran's citrus fruit production fluctuated in recent years and according to FAO statistics in 2021 (FAOSTAT 2023), of 3.86 million tonnes of citrus fruit production estimated (Table 12.2), oranges, mandarins, and tangerines, limes and lemons grapefruits and pomelos have a share of about 55, 14, 13 and 2 percent, respectively (Figure 12.2).

12.2 AGRO-CLIMATIC ZONES IN IRAN WHERE CITRUS IS CULTIVATED

Iran is an arid and semi-arid country with a subtropical region along the Caspian coast on the northern edge of the territory. Citrus is widely distributed in different regions of the country. Citrus cultivation areas of Iran can be divided into two northern and southern regions (Bani Hashemian et al. 2013a). The northern region with three provinces of Gilan, Mazandaran, and Golestan is located between the Caspian Sea and the Elburz mountain range. This region has a semi-Mediterranean climate with an average annual rainfall of about 1,000 mm and periodic frosts every four to five years. The southern region includes an inland belt near the Zagros mountain range, the coastal belt of the Persian Gulf and Oman Sea, and provinces of Fars, Kerman, Ilam, Khuzestan, Kohkiluieh, and Boyerahmad, Hormozgan and Bushehr. The region has a tropical and subtropical climate with low

DOI: 10.1201/9781003584599-12

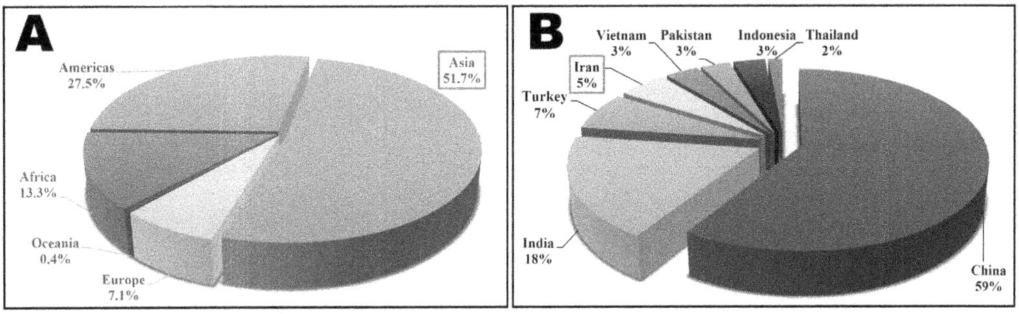

FIGURE 12.1 Citrus production shares in five continents of the world (A), and the top citrus producing countries in Asia (B) (FAOSTAT 2023.)

TABLE 12.1
Citrus Harvested Area in Iran During 2012–2021 (Thousand Hectares)

	2012	2013	2014	2015	2016	2017	2018	2019	2020	2021
Oranges	142.1	147.4	100.7	67.2	61.9	60.0	69.1	62.1	59.2	69.6
Mandarins & Tangerines	34.9	35.7	36.3	24.9	12.6	9.5	12.2	16.6	14.1	12.0
Limes & Lemons	25.5	25.7	26.0	31.2	27.7	27.8	28.6	28.0	27.9	27.7
Grapefruits and Pomelos	4.0	3.4	3.6	3.9	3.7	3.9	3.9	3.8	3.9	3.9
Other citrus species	25.7	26.0	25.5	25.4	24.8	24.4	24.6	24.1	24.4	24.4
Total area harvested	232.2	238.4	192.2	152.6	130.7	125.6	138.3	134.7	129.4	137.5

TABLE 12.2
Citrus Production in Iran During 2012–2021 (Thousand Tonnes)

	2012	2013	2014	2015	2016	2017	2018	2019	2020	2021
Oranges	2456.1	2547.2	1530.4	1530.4	1561.1	1755.6	2184.7	2308.7	2214.2	2139.9
Mandarins & Tangerines	615.5	650.2	693.8	559.9	348.4	369.1	534.8	586.6	572.2	527.0
Limes & Lemons	406.3	427.7	481.1	512.7	473.9	489.2	478.8	476.6	477.8	479.0
Grapefruits and Pomelos	82.4	86.9	93.7	88.5	89.7	88.7	90.3	94.0	91.0	91.8
Other citrus species	547.7	581.0	599.1	617.1	621.8	626.1	625.3	622.9	625.2	627.4
Total production	4108.0	4293.0	3398.2	3308.7	3094.9	3328.7	3913.8	4088.9	3980.4	3865.1

annual rainfall and high temperatures. The excessive heat is so that for example in Khuzestan province citrus is grown at temperatures close to 50°C in summer (Figure 12.2).

The provinces of Mazandaran (51.4%), Fars (25.4%), Hormozgan (7.7%), Kerman (5.9%), Gilan (4.2%), Khuzestan (1.8%), Golestan (1.6%) and Bushehr (0.8%), account for more than 98% of Iran's citrus production. Citrus is one of the most important fruit crops in Iran in terms of cultivated area and production share. According to the latest report, the production of horticultural products in Iran has been about 22.4 million tonnes. Among them, the highest share of production is related to apples, citrus, grapes, greenhouse cucumbers, and date palm with respectively 18%, 14.7%, 12.4%, 6.8%, and 6.7%. 6.3% of 2.5 million economic and bearing orchards of fruit trees belong to citrus (Anonymous 2021).

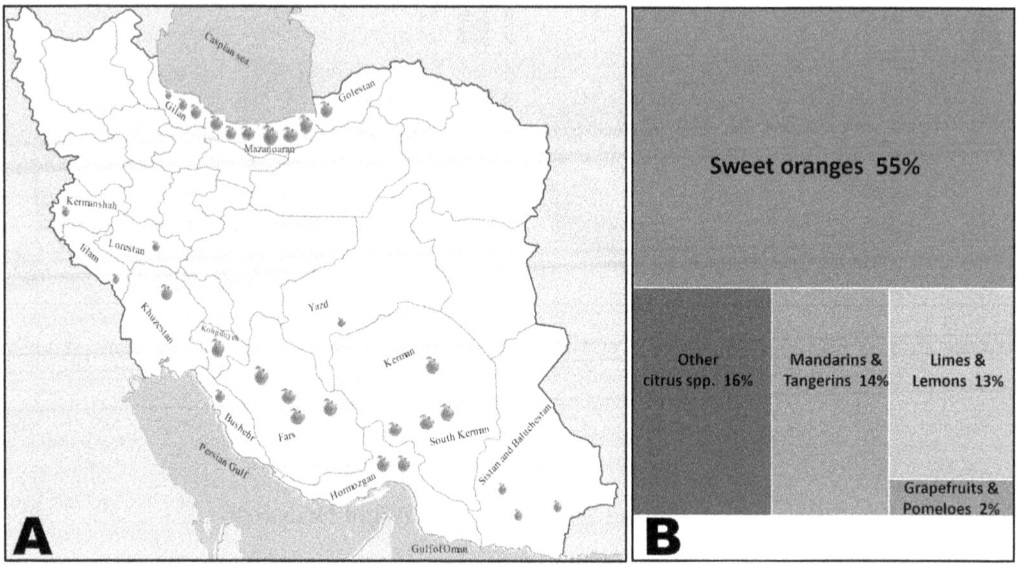

FIGURE 12.2 The northern and southern citrus cultivation areas (A), and the citrus production shares of different varieties (B) in Iran.

12.3 COMMERCIAL USED ROOTSTOCKS

1918 is considered a turning point in the citrus industry of Iran. Until then all local citrus varieties were propagated by seed and because of the losses caused by the *Phytophthora disease*, gardeners began to replace the seedling trees with graft-propagated trees on sour orange rootstock (Bové 1995). Some of these seedling trees can still be found in the northern and southern citrus cultivation areas of the country. Nowadays citrus is graft-propagated in Iran on several rootstocks with sour orange (*C. aurantium*) and Mexican lime (*C. aurantifolia*) as the dominant rootstocks (Bani Hashemian et al. 2013a). During the years, the seeds of different rootstocks were introduced into the country and were planted along with the local cultivars. Sour orange is still the major rootstock of the northern region of Iran. However, with the natural spread of *Citrus Tristeza Virus* (CTV) in this region (Rahimian et al. 2000), it is being replaced by other CTV-tolerant rootstocks such as Trifoliate orange (*Poncirus trifoliata*), Troyer citrange (*P. trifoliata* X *C. sinensis*) and Swingle Citromelo (*P. trifoliata* X *C. paradisi*). Citranges C35 and C32, sour oranges Smooth Flat Seville (SFS) and Gou Tou, Forner alcaide (*P. trifoliata* X *C. reshni*), Cleopatra mandarin (*C. reshni*), Yuzu (*C. junos*) and the local Shel-mohalleh (*citrus* sp.) rootstock are also used at lower levels in this area. Mexican lime and local Bakraee (*citrus* sp.), both drought tolerant, are the major rootstocks in the southern citriculture areas of Iran (Bani Hashemian et al. 2013a). Bakraee is an unknown natural biotype that is being used widely as rootstock as well as edible fruit in this part of the country. Based on morphological features, it was thought that it is a natural hybrid between mandarin and lime (Bové 1995) but further phylogenic studies among "Bakraee" and some commercially important citrus varieties by SSR and PCR-RFLP (Golein et al. 2012) showed a close relationship between this biotype and Rough lemon (*C. jambhiri*) (Figure 12.3). In addition to Mexican lime and Bakraee, Sour orange, Volkamer lemon (*C. volkameriana*), Rough lemon, Rangpur lime (*C. limonia*) and Cleopatra mandarin are other popular rootstocks in the south of Iran.

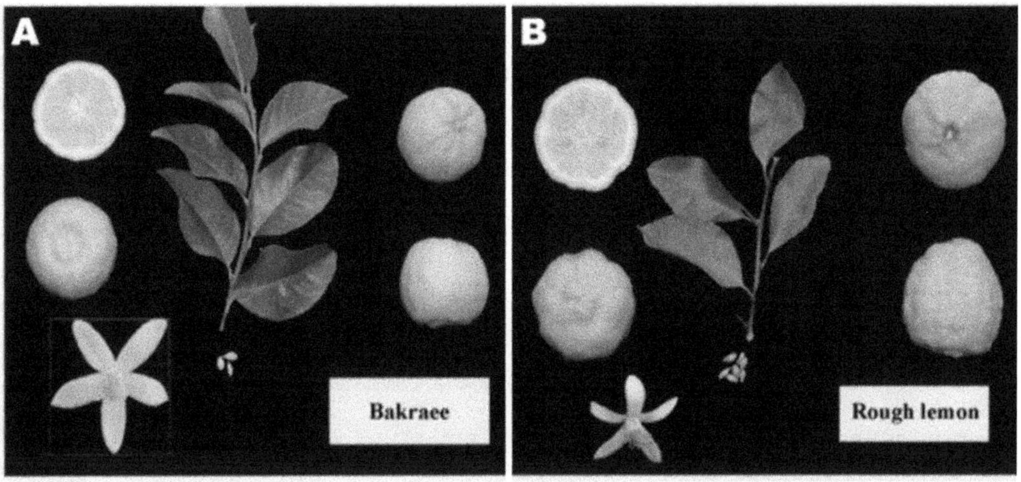

FIGURE 12.3 Morphological comparison of leaf, flower, fruit and seed of Bakraee (A), with rough lemon (B) (Golein et al. 2012.)

12.4 MAJOR CULTIVARS OF CITRUS

Citrus cultivation in Iran has a history of about 2,500 years. Citron (*C. medica*) was the first species cultivated during the Median empire. The attractive colour, taste, and aroma of citrus fruits have had a significant impact on their global expansion. Sailors, traders, emperors, and soldiers had an essential role in transferring fruits, seeds, or plants to different parts of the world. Citrus species such as sour orange, lemon, and lime were introduced to Iran by the Arabs in the tenth century. Much later, many other species reached the country. Since the Safavid era, when Portuguese ships were travelling the southern ports of Iran, the seeds of citrus fruits were available to the people of the region. According to historical evidence, Iran has been the gateway to the transfer of citrus fruits from Asia to other areas of the world. Specially with the opening of the direct shipping route between India and Portugal by Vasco da Gama, different citrus types entered Iran from India. Sweet orange, called "Portugal" in Iran and many neighbouring countries, reached the Persian Gulf ports by Portuguese in the same way. It is believed that about 300 years ago, citrus seeds brought from the south of the country were planted in the north of Iran. The former Union of Soviet Socialist Republics was also a source for citrus introductions to this part of the country (Golein and Aduli 2011).

Nowadays, citrus cultivars available in Iran are a combination of local and imported varieties. More than 150 citrus varieties were introduced into Iran from countries such as Turkey, Italy, Lebanon, Palestine, Morocco, USA, Japan, France, and Spain since 1933 with the aim of economic development due to frost damage (Table 12.3). These cultivars were usually cultivated in the national places and then became popular. However, there are varieties that have been illegally entered directly into the country by passengers. Among the different citrus species and cultivars available in Iran, limited varieties of sweet oranges, mandarins, tangerines, limes, lemons, and grapefruits are cultivated at the commercial level. Some cultivars like Thompson navel, Valencia, Moro, and several satsumas have been widely used by gardeners and have made a significant contribution to the country's citrus production.

In the Caspian Sea belt, citrus production is focused mainly on the early maturing cultivars because of the risk of periodic frost and snow damage. Thomson and Washington navels, Hamlin, Marrs and local sweet oranges, Moro and Sanguine blood oranges, Clementines, Satsumas, Page and Younesi (see Figure 12.8) mandarins and Eureka and Lisbon lemons are the most important cultivars of the region. Although recently introduced Spanish varieties such as Newhall and Navelina were accepted by the gardeners. Washington navel, Marrs, Moro blood, Valencia, and local sweet oranges; Kinnow,

TABLE 12.3
History of Importation of Citrus Cultivars into Iran

Cultivars	Origin	Import year
Sweet Oranges: Thompson navel, Bourouhain, Jaffa, Shamouti, Valencia, Italian, Sanguinella, Gross Sanguine, Moro **Mandarins:** Clementine, Willowleaf, Dwarf Chinotto, Owari satsuma **Grapefruits:** Marsh seedless, Duncan **Other citrus:** Variegated sour orange	Turkey Italy Lebanon Palestine	1933
Sweet Oranges: Washington navel, Frost navel, Gillette navel, Parent navel, Frost Valencia, Olinda Valencia, Campbell Valencia **Mandarins:** Algerian, Dancy, Frost Dancy, Fortune, Silverhill satsuma, Orlando tangelo, Minneola tangelo **Grapefruits:** Shambar, Redblush, Red **Limes and lemons:** Frost Eureka, Cook Eureka, Allen Eureka, Frost Lisbon	USA	1963–1964
Mandarins: Emperor, Ishikawa satsuma, Sugiyama satsuma, Miyagawa satsuma	Australia Japan	1968–1970
Sweet Oranges: Frost navel, Parent navel, Dalaman navel, Atwood early, Skaggs bonanza, Marrs early, Salustiana, Hamlin, Pineapple, Parson Brown, Cadenera, Frost Valencia, Olinda Valencia, Campbell Valencia, Moro, Tarocco **Mandarins:** Cadoux Clementine, Wase satsuma, Robinson, Osceola, Lee, Fortune, Fairchild, Fremont, Kinnow, Wilking, Kara, Batangas, Honey, Younesi (Ponkan), Dancy, Ortanique, Carvaihal, Murcott, Temple, tangelo, Orlando, Minneola, Seminole tangelo, Wekiwa tangelo **Grapefruits:** Frost Marsh, J.B.C. Marsh, Shambar, Redblush, Red **Limes and lemons:** Santarosa, Taroudant, Burneo, Frost Lisbon, Caver's Lisbon, Limoneira Lisbon, Cascade Eureka, Frost Eureka, Allen Eureka, South African sweet lime, Tunis sweet lime, Lyndden gurgh sweet lime, Sweet lemon, Thornless common lime, Millsweet limetta; Limonette de Marrakech, Tavares limequat, Eustio limequat **Other citrus:** Bergamot, Granito sour orange, Double-flowered Bouquetier sour orange, Grasse Bouquetier sour orange, Soft-fruited Bouquetier sour orange, large-flowered Bouquetier sour orange, Bouquet de fleurs sour orange, Nagami kumquat, Hairy Shaddock, No. 57 Shaddock, No. 103 Shaddock, Pink seedless Shaddock	USA Morocco	1971-1972
Sweet Oranges: Newhall, Ruby red valencia **Mandarins:** Marisol clementine, Nules clementine, Clousellina satsuma, Okitsu satsuma, Sugiama satsuma **Limes and lemons:** Tahiti lime, Persian lime, SRA lime	France	1999
Sweet Oranges: Navelate, Navelina, Washington navel, Fukumoto, Delta seedless, Midknight Valencia, Valencia late, Sanguinelli **Mandarins:** Okitsu satsuma, Hashimoto satsuma, Iwasaki satsuma, Marisol, Hernandina, Clemenules, Clemenrubi, Ortanique **Limes and lemons:** Fino, Bears lime, Limequat **Other citrus:** Calamandin, Fingered citron	Spain	2004
Sweet Oranges: Lane late, Navelate, Navelina, Foyos, Newhall, Cara Cara **Grapefruit:** Flame **Limes and lemons:** Verna, Mejicana	Spain	2010

Clementine, Orlando tangelo and local mandarins (see Figure 12.8); Marsh seedless, Red blush, Ruby red, and Ruby star grapefruits; Mexican lime, Eureka and Lisbon lemons, Palestine sweet lime and Persian lime are the dominant cultivars of the southern citriculture region of Iran. Some tropical areas of southern Iran in provinces of Hormozgan, Sistan Baluchestan, Kerman, Fars, and Bushehr are well suited to lime and lemon cultivation and related varieties represent the majority of production (Ebadi et al. 2017; Golein and Aduli 2011).

12.5 STATUS OF THE CITRUS INDUSTRY

12.5.1 Harvesting Window of Major Cultivars of Citrus

Most citrus varieties in Iran are ready for harvest between November and February, but some cultivars ripen at other times of year depending on the climatic conditions of the two regions. In fact, there are citrus fruits available for market almost all year round. Grapefruits, lemons, and the majority of oranges and mandarins are harvested in autumn and winter. In contrast, limes and some varieties such as Valencia sweet orange, some satsumas, and Yashar, a late ripening mandarin released in Iran (see Figure 12.7) are available in spring and summer (Ebadi et al. 2017; Golein and Aduli 2011) (Figure 12.4).

12.5.2 Marketing Network of Citrus

Citrus crop in Iran is generally harvested in plastic, wooden, or cardboard boxes through traditional and manual methods. The products of extensive orchards and agricultural complexes are directly offered for export, processing industries, the centre of provinces, and the capital (Figure 12.5). The fruits of small gardeners are first purchased by local retailers and then presented to wholesalers. Citrus fruits are exchanged between wholesalers in different regions of the country through trucks. Consumers provide their fruit directly or through greengrocers from wholesale centres. The price of the fruit increases by performing different marketing tasks of sorting, washing, disinfecting, waxing, packing, and storage.

12.5.3 Status of Export and Import of Citrus

Citrus fruit export from Iran tended to increase through the 2011–2021 period ending at 177,100 tonnes in 2021, worth about $66 million (Table 12.4). Iraq, China, Turkmenistan, and the Arab states of the Persian Gulf are the main export destinations of Iranian citrus fruits. Citrus exports from Iran are expected to increase in the following years (Anonymous 2021).

12.5.4 Status of Citrus Processing

Citrus juices constitute a significant portion of the food industry and are one of the most common fruit juices around the world. Lots of big and small citrus fruit processing industries have been established in Iran in recent years, whose products are exported to other countries, in addition to consumption inside the country. Usually, small fruits and extra harvested crops are sent to the factories. Fruit washing, grading, juice extraction, dehydrating, filtering, juice concentration, packaging and storing are carried out in the processing factories. Previously prepared or purchased concentrated juices are used for juice and syrup packaging after the harvest season. Citrus byproducts such as Peel oils, aromas and animal feeds are also manufactured along with juice extraction. Most lime and lemon juices are produced in the food processing industries of the south of Iran. Blossom jam is another product of citrus trees made homemade or industrially (Figure 12.5).

12.5.5 Status of Citrus Research and Extension

Synchronized with universities and colleges of agriculture, Agricultural Research, Education and Extension Organization (AREEO) based at the Ministry of Agriculture Jahad of Iran has been active

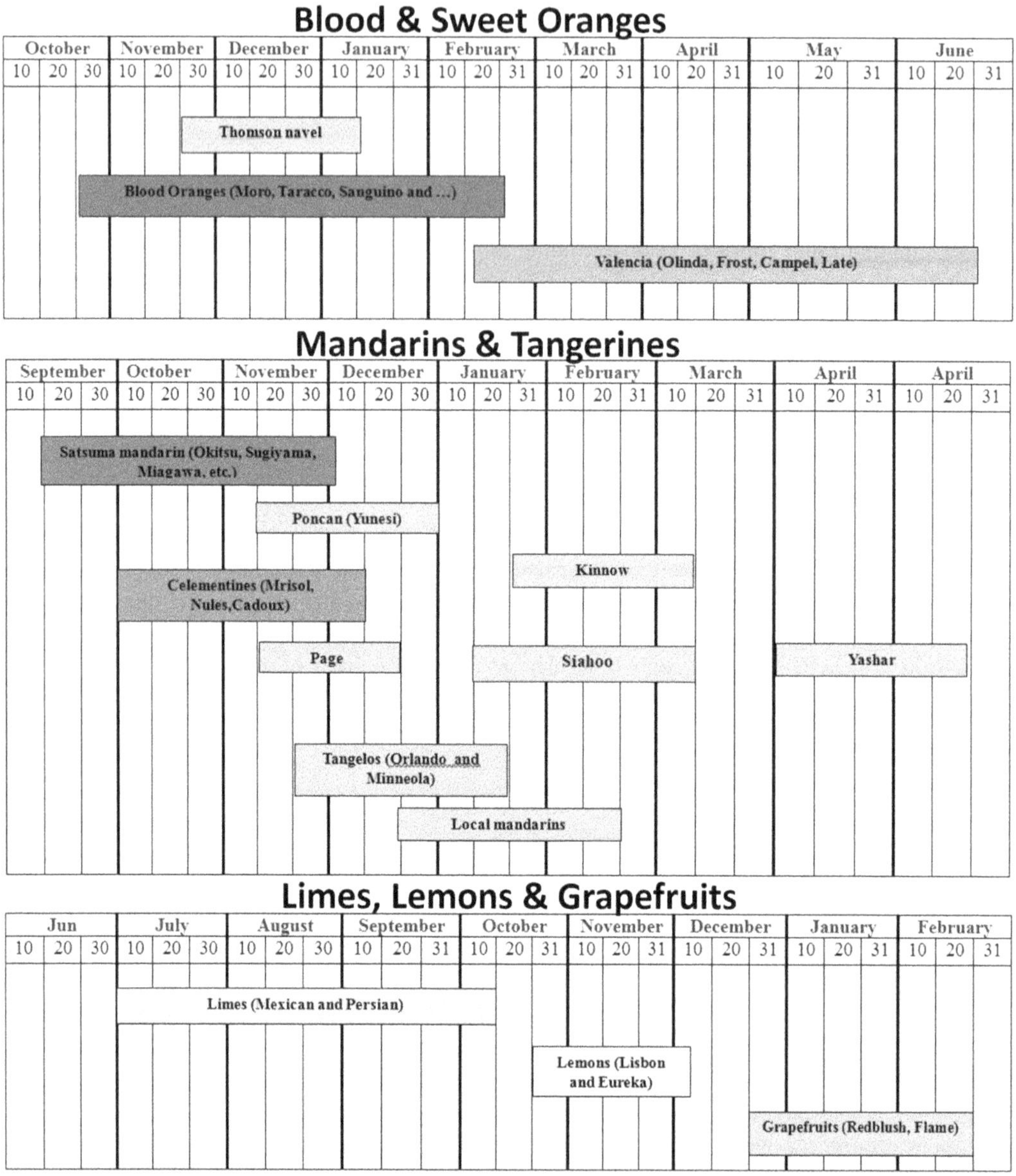

FIGURE 12.4 Harvesting window of the major citrus cultivars in Iran (Aduli et al. 2020.)

TABLE 12.4
Exports of Citrus Fruits from Iran from 2011 to 2021 (Thousand Tonnes)

	2011	2012	1013	2014	2015	2016	2017	2018	2019	2020	2021
Oranges	2.2	0.5	1.8	2.4	0.6	6.4	4.6	38	30	161.8	137.3
Mandarins & tangerines	7.4	4.1	6.5	3.8	3.2	6.8	6.1	11.5	11.5	-	-
Other citrus fruit	33.4	33.4	33.4	4.4	0.7	3.1	2.8	6.8	7.8	53.4	39.8
Total exports of citrus fruits	43	38	41.7	10.6	4.5	16.3	13.5	56.3	49.3	215.2	177.1

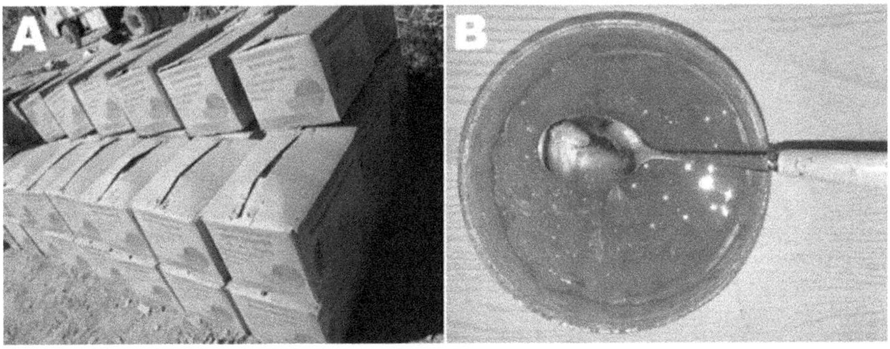

FIGURE 12.5 Field packing citrus fruits in a large orchard in Jiroft, Kerman province for direct transportation by truck to the capital, Tehran (A), and citrus blossom jam (B).

FIGURE 12.6 The central building of CSFRC located in Ramsar, Mazandaran province (A), and a memorial photo of citrus training course at Khorram Abad dependent station of CSFRC (B).

since 1959 and is responsible for agricultural research programmes within the country. Ramsar Agriculture Station in the north of Iran was established under the authority of this organization in 1960. In 1997 this centre was developed and began to work as Iran Citrus Research Institute (ICRI) and research programmes related to citrus trees were assigned to this institute. Research on Pomegranate, Fig, Olive, Kiwifruit and Persimmon were added to duties of ICRI in 2016. It was renamed to Citrus & Subtropical Fruits Research Centre (CSFRC) (Figure 12.6) and now it is governed by Horticultural Sciences Research Institute (HSRI) of AREEO.

The principal activity of CSFRC is to conduct research to improve production, productivity and quality of citrus crops and provide practical solutions to the problems confronting the citrus industry of Iran. Projects on breeding, pest and diseases, irrigation, nutrition, cultural practices and post-harvest of citrus are carried out by 25 scientific members including four associate professors, 15 assistant professors and six researchers at central unit and two dependent stations of CSFRC (Figure 12.6). Joint projects can also be executed by other institutes of AREEO and the universities. There are three citrus variety collections with respectively 135, 195, and 70 genotypes in Mazandaran, Khuzestan and Fars provinces. More than 300 research projects, 135 workshops and specialized training courses, 5,000 lectures to experts and gardeners, 60 technical and extension publications, ten books and 100 technical guide issues have been provided in CSFRC so far.

The development of citrus collections and the release of new varieties (Figure 12.7) by breeding methods are the research priorities of CSFRC. The establishment of a national certification programme was another requirement that was included in the citrus research programmes of the country

FIGURE 12.7 Some new citrus varieties released in CSFRC; Taban (A), Tabesh (B), Jahangir (C), Khorram (D), Zarin (E), Narin (F), Shahin (G), Nushin (H), and Yashar (I) (Aduli et al. 2020.)

because of the presence of infection to virus and virus-like agents in some available cultivars. In fact, to control diseases caused by graft-transmissible pathogens of citrus it is necessary to use healthy trees in the new plantings.

The first efforts in this field were started in 1996 through nucellar selection by Prof. Yunes Ebrahimi (1934–2020), the principal investigator, prominent expert, and Ex-director of ICRI (Figure 12.8). Production of the healthy mother trees was then continued by new methods. Rapid and complementary indexing of citrus varieties after the sanitation process is carried out for important virus and virus-like diseases. Rapid screening is necessary to select the candidates among the high number of plants produced by shoot-tip grafting (STG) and thermotherapy. After obtaining negative results from rapid indexing, plants will be entered into complementary indexing in which the critical pathogens of citrus will be analyzed by biological, serological, and molecular methods (Bani Hashemian et al. 2013b) (Figure 12.9).

12.6 SIGNIFICANT TECHNOLOGIES AND PACKAGE OF PRACTICES FOLLOWED IN CITRUS FARMING

Citriculture in Iran is a combination of tradition and modernity. Various technologies related to the citrus industry have been imported or localized to Iran. While old closely planted citrus groves still exist in two main citrus regions, the newly established orchards are cultivated usually 4x6 m apart with about 350–400 or less trees/ha. Despite the traditional system for citrus plant production, new

FIGURE 12.8 Siahoo, a local mandarin of Hormozgan province (A), and Younesi, a nucellar line of Ponkan mandarin obtained by Younes Ebrahimi (B).

FIGURE 12.9 Citrus health assessment process (A), and maintenance of the citrus primary sources in CSFRC (B).

nurseries have been established. Both container-grown and field-grown trees are produced in the nurseries (Figure 12.10).

Almost all orchards in two citrus regions of Iran are irrigated year round in different periods, except in some orchards in the north, which are rainfed. In fact, citrus cultivation and production without irrigation is impossible in more than 90% of the orchards in Iran. Irrigation is usually one of the most essential and expensive gardening operations in arid and semi-arid areas. Also, in humid semi-tropical areas, it is a crucial factor in keeping the yield and quality of the fruit. Climatic changes, multiple droughts, population growth, excessive extraction and harvesting of surface and

FIGURE 12.10 Some examples of traditional and modern nurseries for citrus production.

subsurface resources, and inappropriate temporal and spatial distribution of rainfall have all caused a lack of water resources and an increase in the actual cost of water and irrigation in most parts of Iran. The net water requirement for irrigation of citrus trees in the north and south of Iran is respectively about 3,200 and 10,800 cubic metres per hectare. Old flood or furrow irrigation methods are replaced in many orchards by drip irrigation to save water, especially in the south of Iran. More than 95% of citrus orchards in Fars and Hormozgan, 60% in Kerman, and 40% in Mazandaran provinces have drip irrigation nowadays (Ebadi 2017).

12.7 CHALLENGES AND STRATEGIES

12.7.1 MAJOR PESTS

Citrus pests are one of the main causes of quantitative and qualitative crop damage. Groups of arthropods, molluscus, and rodents are known currently as the most important pests of citrus trees in Iran. Because of favourable weather conditions, the diversity and density of the pests in citrus-growing areas of northern Iran are greater than in the south of the country. Due to severe damage and broad host range, the Mediterranean fruit fly (*Ceratitis capitata*) is one of the most economically important pest species of fruit trees in the world. It was first recorded from Ahwaz in southern Iran in 1958 (Salavatian and Sabzevari 1975), and then from north of Iran in 1980 (Sabzevari and Jafari 1991). Females lay their eggs on the underside of the fruit's skin. Larvae emerge from the eggs and feed on the fleshy part of the fruit (Figure 12.11). The wound created by the pest facilitates the entry of bacteria and fungi and causes fruit to drop and decay.

Large amounts of oils and insecticides in the form of summer and winter spraying are used every year against the scales and mites. Pulvinaria scale (*Pulvinaria aurantii*) and cottony cushion scale (*Icerya purchasi*) are the most important pests of citrus trees and can be found on different citrus cultivars. Pulvinaria scale is the dominant pest of citrus orchards in northern Iran. The cottony cushion scale is a widespread pest throughout the world, also in two citrus-growing areas of the country. Defoliation, fruit drop, shoot die back and decreasing tree vitality are the symptoms of the affecting trees. A decrease in crop marketability due to the accumulation of sooty mould is secondary damage that is created as a result of feeding and abundant honeydew secretion of the pest. Citrus brown scale (*Chrysomphalus dictyospermi*), Oriental yellow scale (*Aonidiella orientalis*), and Citrus mealybug (*Pseudococcus citri*) are the other species of scales in Iran. By feeding on fruits, leaves, and twigs, citrus mites decrease crop marketability and may cause tree defoliation. Citrus red mite (*Panonychus citri*) and citrus rust mite (*Phyllocoptruta oleivora*), with a worldwide

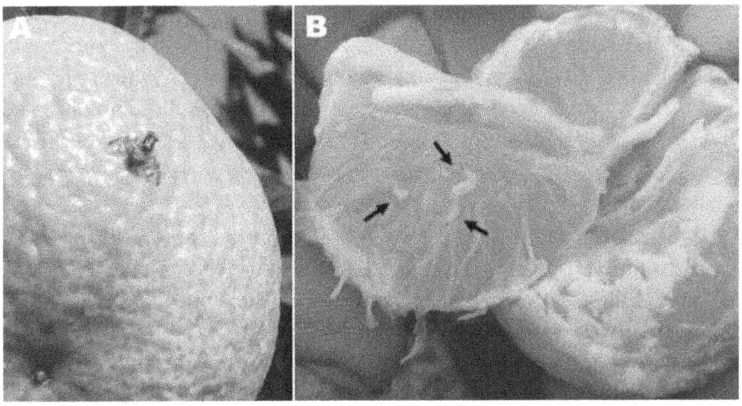

FIGURE 12.11 An adult Mediterranean fruit fly on the surface (A), and several larvae inside the citrus fruit (B).

distribution of citrus, are two important citrus pests in the north of Iran. In contrast, oriental mite (*Eutetranychus orientalis*) is the predominant mite species in the south of the country (Ebadi et al. 2017).

In addition to direct damage, some insect pests are vectors of dangerous viruses and virus-like agents. The Asian citrus psyllid (*Diaphorina citri*) has a wide distribution in Asia and other citrus-growing regions of the world. It is a sap-sucking pest of citrus and one of the vectors of a destructive disease, citrus Huanglongbing. The causal agent of the disease was first detected in two provinces in Southern Iran from their insect vectors in 2008 (Faghihi et al. 2009). Citrus green aphid (*Aphis spiraecola*), Citrus black aphid (*Toxoptera aurantii*), and Cotton aphid (*A. gossypii*) are the dominant species of aphids in Iran. The colonies of nymphs and adults usually appear in new growths of the trees. Due to sucking plant sap, they cause leaf curl and shoot-tip stunting. As they feed, they also produce sweet honeydew, which is colonized by black sooty mould fungi. Some aphid species are vectors of several citrus viruses. Among three species, *A. gossypii* was indicated as the vector of *Citrus Tristeza Virus* (CTV) in northern of Iran (Pourhamedi et al. 2020). *Citrus Vein Enation Virus* (CVEV) is another aphid-borne virus reported in the country (Nouri, 2015). Transmissibility of *Citrus Yellow Vein Clearing virus* (CYVCV) was confirmed under controlled conditions by the three dominant citrus aphids (Maghsudi 2023). CYVCV has been recently reported from Iran after being recognized in Pakistan, India, Turkey, and China (Bani Hashemian and Aghajanzadeh 2017).

12.7.2 Major Diseases

Several pathogens affect annually citrus trees in the north and south of Iran. In addition to common diseases, the emergence of citrus tristeza, citrus yellow vein clearing, citrus canker, Witches' broom, Huanglongbing, and decline during recent decades have caused significant damage to the citrus industry in Iran. Citrus tristeza disease caused by CTV, is the most important viral disease of citrus. Its severe damages have been reported from many citrus-growing areas worldwide. The typical tristeza disease due to infection by the decline isolates of CTV affects trees grafted on sour orange rootstock. CTV was first introduced in the north of Iran where sour orange is the major rootstock through the importation of infected materials from Japan in 1969 (Ebrahim-Nesbat and Nienhaus 1978). The presence of the virus was then confirmed in the southern provinces of Iran (Shafiee and Izadpanah 1996). *Aphis gossypii* has been known as the vector of CTV in the country (Rahimian et al. 2000). The main cause of the decline of citrus trees in the east of Mazandaran province, the primary focus of the disease, is related to tristeza infection (Bani Hashemian et al. 2013c) (Figure 12.12). The presence of viroids in some available cultivars in the CTV infected areas is another problem that limits the choice of alternative rootstocks to replace sour orange (Bani Hashemian et al. 2013a). *Citrus yellow vein clearing virus* (CYVCV) is a new emerging threat to the citrus industry reported from Iran after being recognized in Pakistan, India, Turkey, and China (Bani Hashemian and Aghajanzadeh 2017). Sour orange (*C. aurantium*) rootstock and lemon (*C. limon*) varieties are the main susceptible hosts of CYVCV in the country. Yellowing of the lateral veins and corresponding water-soaked areas on the lower surface of the leaves are the typical symptoms of infection (Figure 12.12). The distribution of the virus is currently restricted to the northern regions of Iran (Bani Hashemian and Aghajanzadeh 2020).

Citrus canker is a bacterial disease caused by the *Xanthomonas citri* subsp. *citric* that produces lesions on the leaves stems, and fruit of susceptible citrus trees (Figure 12.12). The disease was first reported on acid limes in the Kerman province of Iran (Alizadeh and Rahimian 1990). Subsequently, canker symptoms were observed in other lime-producing provinces of Kerman, Hormozgan, Ilam Sistan and Baluchistan. In recent years, following the reduction of early spring rains, the severity of the disease has decreased significantly in these areas. Citrus canker has not been reported so far from the northern citrus-growing region of the country. Witches'-broom disease of lime (WBDL) caused by "*Candidatus* Phytoplasma aurantifolia"-related strains is the most destructive disease

FIGURE 12.12 Symptoms of the emerging diseases of citrus in Iran: Tristeza decline of sweet orange on sour orange rootstock (A), CYVCV on lemon and sour orange (B), Citrus canker symptoms on acid lime (C), Witches'-broom disease of lime trees (D), HLB on grapefruit trees (E), and decline of citrus trees in south of Iran (F).

of acid limes in the southern citriculture provinces of Iran. The disease was first reported in Sistan and Baluchestan (Bové et al. 2000) and subsequently spread to Kerman and Hormozgan provinces. Disease symptoms include small chlorotic leaves, witches' broom and dieback (Figure 12.12) have been reported in acid lime (*C. aurantifolia*), barrage (*Citrus* sp.), grapefruit (*C. paradisi*), citron (*C. medica*), limequat (*C. aurantifolia* × *Fortunella* sp.), sweet orange (*C. sinensis*) and Orlando tangelo (*C. paradisi* × *C. reticulata*) (Azadvar et al. 2023). The casual phytoplasma is transmitted by leafhopper *Hishimonus phycitis* (Bagheri et al. 2009).

Huanglongbing (HLB) is a destructive disease limiting the production of citrus fruits in many areas of the world. The symptoms of HLB may differ depending on the strain of the pathogen, host plant, and the climatic conditions (Figure 12.12). Leaf "blotchy mottle" is the most characteristic symptom of HLB. Three liberibacter species of *Candidatus* Liberibacter asiaticus, *Ca.* Liberibacter africanus and *Ca.* Liberibacter americanus AND two psyllid species, *Diaphorina citri,* and *Trioza erytreae,* have been identified as respectively causal agents and vectors of the disease. In Iran, *Candidatus* Liberibacter asiaticus was first detected from its vector *D. citri* in Sistan Baluchestan and Hormozgan provinces (Faghihi et al. 2009). HLB was later reported from two other provinces of Kerman and Fars (Azadvar et al. 2019). A high correlation was found between the detection of HLB and the occurrence of citrus decline in southern Iran (Figure 12.12) (Passera et al. 2018; Azadvar et al. 2019; Alizadeh et al. 2017; Najafinia and Azadvar 2016; Bani Hashemian et al. 2013c).

12.7.3 Abiotic Stresses

Heat stress, frost, and soil salinity are limiting factors for citrus cultivation in Iran, which can affect the quantity and quality of the product. Higher temperatures of more than 40 to 45 degrees Celsius along with a decrease in air humidity by 20% or less will cause the leaves to fall and damage the tree. Most of Iran's citrus-growing areas in the southern provinces of Khuzestan, Hormozgan, Bushehr, Kerman Sistan, and Baluchistan are exposed to heat stress in spring and summer. As subtropical and evergreen plants, citrus trees are sensitive to cold and freezing stresses. Generally, when the temperature drops to -1.7 °C for 30 minutes or longer, some symptoms of frost damage occur in sensitive citrus trees. Heavy snow also breaks tree trunks and branches. During the last two decades, the orchards in the northern regions of Iran have faced at least three stresses of low temperatures. As a result of the snow and cold of November 2016, one-third of the orange crop of Mazandaran province was lost or had a quality change that made it impossible to eat. Soil salinity can also be a limiting parameter for citrus growth. Soil salinity has occurred in some citrus regions of Iran due to the decrease in rainfall, the increase in evaporation and transpiration, and the presence of salts in the irrigation water (Ebadi et al. 2017).

12.8 CONCLUSION

Iran has a diverse climate and an old history of citrus growing. Since the citrus research programmes started in the country, many attempts have been carried out and significant changes have occurred. New local and improved varieties were introduced and the technical citrus industry has given new faces to those involved. Climate change is a global problem that follows new challenges including drought and other abiotic stresses and the occurrence of new pests, among them graft-transmissible diseases of citrus. Increasing water efficiency and improving the existing orchards instead of groove establishment are novel strategies in the horticulture of the country. Fixing the defects of the national certification programme in quarantine, sanitation, and sanitary evaluation is one of the other necessities.

12.9 ACKNOWLEDGEMENTS

In part of the text, the previous information about the citrus industry of Iran (Ebrahimi 2001, 2002) has been updated. The authors thank T. Raisi, H. Ebadi and S. M. Mohammad Pour for providing some content in Persian and a photo (Figure 12.6B).

REFERENCES

Aduli, B., Khanjani, H., and Raheb, S. (2020). *Horticultural Features of Commercial Cultivars of Mandarin in Iran.* Horticultural Sciences Research Institute. (Booklet in Persian).

Alizadeh, H., Quaglino, F., Azadvar, M., Kumar, S., Alizadeh, A., Bolboli, F., Casati, P., and Bianco, P.A. (2017). First report of a new citrus decline disease (CDD) in association with double and single infection by '*Candidatus* Liberibacter asiaticus' and '*Candidatus* Phytoplasma aurantifolia' related strains in Iran. *Plant Dis* 101:2145.

Alizadeh, A., and Rahimian, H. (1990). Citrus canker in Kerman Province. *Iran J Plant Pathol* 26:1–4.

Anonymous (2021). *Agricultural Statistical*. Ministry of Agriculture, Department of Planning and Economy, Center for Information and Communication Technology.

Azadvar, M., Esmaeilzadeh, H.S.A., Salehi, M., Al-Subhi, A., Hemmati, C., Al-Ghaithi, A., and Faghihi, M.M. (2023). Updates on phytoplasma diseases associated with citrus plants in Asia. In: Tiwari, A.K., Caglayan, K., Hoat, T.X., Al Subhi, A., Nejat, N., and Reddy, G. (eds.). *Phytoplasma Diseases of Major Crops, Trees, and Weeds*, pp. 265–282. Elsevier, Academic Press. ISBN: 978-0-323-91897-8.

Azadvar, M., Golmohammadi, M., Safarnejad, M.R., Salari, M.R., Samavi, S., Faghihi, M.M., and Ghayeb, Z.M. (2019). Association of *Candidatus Liberibacter asiaticus* with citrus decline disease in southern provinces of Iran. *1st Iranian Plant Pathology Congress*. College of Agriculture and Natural Resources, University of Tehran.

Bagheri, AN., Salehi, M., Faghihi, M.M., Samavi, S., and Sadeghi, A. (2009). Transmission of *Candidatus* Phytoplasma aurantifolia to Mexican lime by the leafhopper *Hishimonous phycitis* in Iran. *J Plant Pathol* 91(4, suppl): S4.105.

Bani Hashemian, SM., and Aghajanzadeh, S. (2017). Occurrence of *Citrus yellow vein clearing virus* in citrus species in Iran. *J Plant Pathol* 99: 290.

Bani Hashemian, SM., and Aghajanzadeh, S (2020). Identification of *Citrus yellow vein clearing virus* in Mazandaran province. *Plant Protect (Sci J Agricult)* 43:49–61. DOI: 10.22055/PPR.2020.15997. (In Persian with English abstract).

Bani Hashemian, SM., Golmohammadi, M., and Duran, V.N. (2013b). Rapid screening of citrus plants from the sanitation process in the Iran citrus research institute. *IOCV-IX-Abstracts of Presentations at the 19th Conference of the International Organization of Citrus Virologists, South Africa. Journal of Citrus Pathology* 1 (2): 4. Retrieved from https://escholarship.org/uc/item/502200h0.

Bani Hashemian, SM., Golmohammadi, M., Mohammad, A.Y., Golein, B., and Moreno, P. (2013c). Decline of citrus trees in Iran. *IOCV-IX-Abstracts of Presentations at the 19th Conference of the International Organization of Citrus Virologists, South Africa. Journal of Citrus Pathology* 1 (2): 39. Retrieved from https://escholarship.org/uc/item/502200h0.

Bani Hashemian, SM., Taheri, H, Mohammad Alian., and Y, Bové, J.M, and Duran,V.N. (2013a). Complex mixtures of viroids were identified in the two main citrus-growing areas of Iran. *J Plant Pathol* 95:647–654.

Bové, J.M. (1995). *Virus and Virus-Like Diseases of Citrus in the Near East region*. FAO, p. 518.

Bové, J.M., Danet, J.L., Bananej, K., Hassanzadeh, N., Taghizadeh, M., Salehi, M., and Garnier, M. (2000). Witches'-broom disease of lime (WBDL) in Iran. *International Organization of Citrus Virologists Conference Proceedings (1957–2010)* 14 (14): 207–212. Retrieved from https://escholarship.org/uc/item/9x08w06f.

Ebadi, H. (2017). Improvement of citrus water use efficiency by reduction of incident radiation and partial root-zone drying. Doctoral dissertation, Sari Agricultural Sciences and Natural Resources University, Iran.

Ebadi H, Gholamian E, Fatahi Moghadam, J, Golein, B, Golmohammadi M, Moradi B (2017). *Guide to planting, growing, and harvesting citrus fruits*. (Book in Persian)

Ebrahimi, Y. (2001). Iran. *Options Méditerranéennes: Série B*, 33: 163–170.

Ebrahimi, Y. (2002). Citrus industry in Iran. *Options Méditerranéennes: Série B* 43: 31–35.

Ebrahim-Nesbat, F., and Nienhouse, F. (1978). Occurrence of *Citrus tristeza virus* in Iran. *Zeitschrift fur Pflanzenkrankheiten und Pflanzenschutz* 85:308–312.

Faghihi, M., Salehi, M., Bagheri, A., and Izadpanah, K. (2009). First report of citrus huanglongbing disease on orange in Iran. *Plant Pathol* 58:793–793.

FAOSTAT (2023). *World Food and Agriculture – Statistical Yearbook 2023*. Rome. https://doi.org/10.4060/cc8166en.

Golein, B. and Aduli, B. (2011). *Citrus Cultivation*. Novin Pouya publication. (Book in Persian).

Golein, B., Bigonah, M., Azadvar, M., and Golmohammadi, M. (2012). Analysis of genetic relationship between 'Bakraee' (*Citrus* sp.) and some known Citrus genotypes through SSR and PCR-RFLP markers. *Scientia Horticulturae* 148:147–153.

Lim, T.K. (2012). *Edible Medicinal and Non-Medicinal Plants*, Vol. 4. Dordrecht, Springer, pp. 656–687.

Maghsoudi, R., Nassrollahnejad S., Aghajanzadeh, S., and Bani, H.S.M. (2023). Transmissibility of *Citrus yellow vein clearing virus* by three dominant citrus aphids. *Iran J Plant Protec Sci* 54:101–113.

Najafinia, M., and Azadvar, M. (2016). Citrus sudden decline disease in Iran. *Indian Phytopathol* 69:41–13.

Nouri S., Dizadji, A., and Rahimian, H. (2015). First report of *Citrus vein enation virus* (CVEV) from Iran. *Iran J Plant Pathol* 51:1.

Passera, A., Alizadeh, H., Azadvar, M., Quaglino, F., Alizadeh, A., Casati, P., and Bianco, P.A. (2018). Studies of microbiota dynamics reveals association of *Candidatus* Liberibacter asiaticus infection with citrus (*Citrus sinensis*) decline in south of Iran. *Int J Mol Sci* 19:1817.

Pourhamedi, S., Barzegar, A., and Rahimian, H. (2020). Selective transmission of the seedling yellows-type genotypic variants of *Citrus tristeza virus* by *Aphis gossypii* in northern Iran. *Plant Pathol* 69:783–793.

Rahimian, H., Alavi, V., Shayegan, H. and Hadizadeh, AR. (2000). Spread of Citrus tristeza virus by *Aphis gossypii* in the north of Iran. *Iranian Journal of Plant Pathology* 36: 103.

Sabzevari, A., and Jafari, M.E. (1991). *Bio-Ecological Studies and Eradication of the Pest in Mazandaran «The last Infested Area in Iran»*. Bulletin of Plants Pest and Diseases Research Institute, p. 38. [In Persian].

Salavatian, M., and Sabzevari, A. (1975). *Mediterranean Fruit Fly, Ceratitis Capitata (Wiedemann)*. Bulletin of Plants Pest and Diseases Research Institute, p. 25. [In Persian].

Shafiee, V., and Izadpanah, K. (1996). Report of *Citrus tristeza virus* in southern Iran. *Iran J Plant Pathol* 32:191.

13 The Way Forward for Citrus Industry of Sri Lanka

Ishara Perera, H.M.U.A.G Jayalath Bandara, Himani Gammanpila, P.S.R Premarathne, I. Kalubowila, and R.M.N.T Amarsainghe

13.1 INTRODUCTION

Agriculture is one of the primary economic sectors contributing 7.5% to the Gross Domestic Product (GDP) in Sri Lanka (Central Bank Report, 2022) and a significant source of livelihood for 25% of the population (Department of Census and statistics, 2022). Fruit sector contributes 0.7% to the country's GDP in 2021 (Agstat, 2022), while citrus cultivation plays a significant role in the fruit industry of the country.

A wide variety of citrus fruits can be grown due to its favourable climate and the diverse agro-ecological zones (AEZ) of Sri Lanka (Dahanayake, 2015). A variety of citrus fruits including mandarin, orange, lime, tangerine, lemon, and pumelo are cultivated mainly focusing on the local markets. Orange and mandarins are priority crops of farmers due to their demand, medicinal and nutritional values. Apart from commercial cultivation, most of the farmers are willing to establish these in their home gardens.

Citrus has been cultivated in Sri Lanka since ancient times, due to its culinary, medicinal, and economic importance. With time, improved agricultural technologies have been involved in increasing yield and quality with pest and disease resistance. At present citrus cultivation contributes considerably to enhancing farmer income and also involves in export industry earning a foreign income to the country.

Citrus cultivation at present in Sri Lanka is practiced in many districts of the country. The production of lime and oranges from 2013 to 2022 are presented in Table 13.1. The lime cultivation area was around 9,500 ha while the production was 13,263 Metric tonnes (Mt) in 2022. Orange is cultivated in around 6,500 ha whereas, orange production slightly increased from 2013 and production in 2022 was7,607 Mt.

The agricultural practices, weather patterns and policy decisions affect the annual volume of the citrus industry in the country. Citrus fruits are widely consumed as fresh, and as value-added products such as juice, jam, chutney and so on. The fruit is rich in vitamins, and other important nutrients, which is important for the well-being of the people, though the consumption level is significantly low in the country. The per capita fruit consumption of Sri Lanka is 40 g per day (Bandara et al., 2021) but consumption of 200 g per day is recommended.

It provides extra income for the farmer families and with higher production targets it is a potential crop for the export market. Citrus cultivation as a major fruit crop is important as it supports in upgrading health of the population and enhancing the income level of the farmers contributing to upgrading the country's economy.

This chapter aims to provide the current status and future directions of the citrus industry in Sri Lanka with a comprehensive and forward-looking evaluation of the sector's potential growth,

DOI: 10.1201/9781003584599-13

TABLE 13.1
Production of Limes and Oranges (2013–2022)

Year	Lime production (Mt)	Orange production (Mt)
2013	12,447	5,730
2014	11,066	5,789
2015	10,792	6,792
2016	12,862	7,147
2017	12,383	7,049
2018	15,813	8,518
2019	16,727	8,882
2020	19,558	9,180
2021	16,651	8,797
2022	13,263	7,607

(Source: Agstat, DOA, Sri Lanka).

challenges, and strategies for sustainable development. The chapter integrates insights from experts, recent research, and industry trends to offer a holistic understanding of the citrus industry's status and its prospects.

13.2 PRESENT SCENARIO OF CITRUS CULTIVATION IN SRI LANKA

13.2.1 Major Cultivars and Varieties of Citrus Cultivated in Sri Lanka

Several cultivars and varieties of citrus species belonging to mandarins, orange, lime, lemon, and pumelo are cultivated in Sri Lanka. The Department of Agriculture (DOA) is the main government institute responsible for research and development of citrus crop. The DOA has released several varieties suitable for different AEZ of the country. Sri Lanka cultivates several citrus varieties, including sweet oranges (*Citrus sinensis*), mandarins (*Citrus reticulata*), lime (*Citrus aurantiifolia*), and pumello (*Citrus maxima*). Some popular local cultivars including Jaffa orange, the Malwana lime, and the Kagzi lime are also cultivated in the country.

The details of the released citrus varieties by the DOA are outlined below.

13.2.1.1 Mandarin Varieties

Mandarin can be cultivated in Wet, Dry, and Intermediate zones of Sri Lanka. However, dry cool climate with low humidity is most preferable for high-quality fruit production.

13.2.1.1.1 Madhu

This variety is well grown in dry, cool climates. The tree is medium size with 2–3 m height and an oval shape. The weight of the fruit is around 100–150 g, has sweet taste and the fruits are oval or round. The fruit is seedless or with 1–2 seeds and annually 2–3 harvests can be taken (Figure 13.1).

13.2.1.1.2 Rahangala Selection

The tree is 2–3 m in height and thornless. The weight of the fruit is around 70–90 g. When grown in a cool climate the colour of the fruit becomes yellowish green and in other areas fruit is green colour. The fruit is seedless and takes 4–5 months to maturity (Figure 13.2).

FIGURE 13.1 Madhu.

FIGURE 13.2 Rahangala selection.

13.2.1.1.3 Indu

The variety can be grown well in Wet and DZ. The tree is a spreading type and about 2 m height. The fruit weight is around 80–100 g and becomes yellowish green colour when mature. Few seeds can be seen in fruit (Figure 13.3).

13.2.1.1.4 Horana Juicy

The variety is recommended for the WZ of the country. The tree is 3–4 m in height and thornless. Fruits bear throughout the year and 600–1,000 fruits can be harvested in a year and around 60 kg of fruits can be harvested in a year. The fruit weight is around 100–130 g. Fruits become light yellowish when mature. The variety is more suitable for juice production (Figure 13.4).

13.2.1.1.5 Horana Ehime 1

The variety is suitable for dry and cool climates. The trees are spreading type and thornless. Fruits become orange when mature and seedless. Fruit surface is smooth and has sweet taste and high juice content. The fruit weight is around 80–100 g. Fruits can be harvested after 5–6 months of flowering (Figure 13.5).

FIGURE 13.3 Indu.

FIGURE 13.4 Horana Juicy.

13.2.1.1.6 Horana Ehime 2

The variety is suitable for dry and cool climates. The trees are spreading type and thornless. Fruits become orange when mature and seedless. Fruit surface is smooth and has sweet taste and high juice content. The fruit weight is around 80–90 g. Fruits can be harvested after 5–51/2 months of flowering (Figure 13.6).

13.2.1.1.7 Horana Ehimi 3

The variety is suitable for dry and cool climates. The trees are spreading type and thornless. Fruits become orange when mature and seedless. Fruit surface is smooth and has sweet taste and high juice content. The fruit weight is around 70–80 g. Fruits can be harvested after 41/2–5 months of flowering (Figure 13.7).

13.2.1.2 Orange Varieties

13.2.1.2.1 Arogya

The variety is suitable for cool climates in Up country and IZ. The tree grows up to 2–3 m and a spreading type. The fruit weight is around 400–500 g and contains higher juice content (Figure 13.8).

FIGURE 13.5 Horana Ehime 1.

FIGURE 13.6 Horana Ehime 2.

13.2.1.2.2 Sisila

The variety is suitable for Upcountry IZ of Sri Lanka. The tree height is around 2–3 m. The fruit is a round shape and contains higher juice content (Figure 13.9).

13.2.1.2.3 Bibile Sweet Orange

The variety is suitable for Dry and IZ of Sri Lanka. The tree height is around 3–5 m with a strong structure. The fruit is round and green colour and fruit weight is around 300–400 g and has sweet taste. The peel colour becomes yellowish when mature (Figure 13.10).

FIGURE 13.7 Horana Ehime 3.

FIGURE 13.8 Arogya.

13.2.1.2.4 Bibile Seedless Orange

The variety is suitable for Dry and IZ of Sri Lanka. The tree height is around 3–5 m with a strong structure. The fruit is round and green colour and fruit weight is around 300–400 g and has seedless. The peel colour becomes yellowish when mature (Figure 13.11).

13.2.1.2.5 Maduruketiya Selection

The variety is suitable for WZ and IZ of Sri Lanka. Few seeds can be seen in a fruit and the fruit peel is thick. Mature fruits are green, and the juice is orange colour (Figure 13.12).

13.2.1.3 Lime Varieties

13.2.1.3.1 Horana Lime 1

The tree is medium size and a spreading type. A year-round production can be taken (Figure 13.13).

FIGURE 13.9 Sisila.

FIGURE 13.10 Bibile Sweet.

13.2.1.4 Pumelo Varieties

13.2.1.4.1 Horana Pumelo 1

The weight of the fruit is around 1,900 g, and the flesh is pinkish white. Flowering occurs after three years of planting and fruiting will occur after 4.5–5 months after flowering (Figure 13.14).

13.2.1.4.2 Horana Pumelo 2

The weight of the fruit is around 1,325 g and the flesh colour is pinkish. Flowering occurs after three years of planting and fruiting will occur after 4.5–5 months after flowering (Figure 13.15).

13.3 AGRO-ECOLOGICAL ZONES AND DISTRIBUTION OF CITRUS IN SRI LANKA

There are three major climatic zones in Sri Lanka based on the rainfall distribution; “WZ”, “DZ” and the “Intermediate Zone (IZ)”. WZ includes the southwestern region and the central hills, DZ covers predominantly northern and eastern parts of the country and IZ lies between DZ and WZ. The WZ receives relatively high mean annual rainfall over 2,500 mm without pronounced dry periods. The DZ receives a mean annual rainfall of less than 1,750 mm and the IZ receives a mean annual rainfall between 1,750 to 2,500 mm. The Low country includes the land below 300 m in elevation

FIGURE 13.11 Bibile Seedless Orange.

FIGURE 13.12 Maduruketiya selection.

FIGURE 13.13 Horana Lime 1.

FIGURE 13.14 Horana Pumelo 1.

FIGURE 13.15 Horana Pumelo 2.

and the Mid country with elevation between 300–900 m while the Up country is the land above 900 m elevation. Both WZ and IZ spread across all three categories of elevation while the DZ is confined to the Low country resulting in seven agro-climatic zones covering the entire island. The country is divided into 46 agro-ecological regions (Figure 13.16) considering climatic factors, elevation, soil type, landform, land use, and relief (Punyawardena, 2008).

Mandarin variety Indu can be well grown in the Low country WZ and Madhu can be cultivated in Mid country. Rahangala selection is cultivated mainly in Mid country where dry and cool climate is prevailing. Ehimi varieties are well grown in Up country of Sri Lanka. The lower Uva area is regarded as the island's principal area of citrus fruits. Large extent of lime and sweet orange is being cultivated in Moneragala district (IZ) followed by Ratnapura (WZ) and Kurunegala districts. From the citrus cultivation lands, 50% area is covered under lime, followed by 30% under sweet orange and 20% under other citrus species (mandarins, grapefruits, pumelo, lemonime) in Sri Lanka. Further, different citrus varieties prefer different climatic and soil factors. As an example, the local orange variety called "Bibile Sweet" is performed well in Low Country IZ. The variety records good taste and brix value, when cultivated in Low Country IZ. Mandarin cultivation is well performed in Up Country IZ and Up Country WZ (above 900 m from the sea level). Though the "Valencea Orange" cultivar is grown and giving high-quality fruits and high yield in Low Country IZ, the variety "Washington navel" is not suitable for Low Country IZ or DZ as they will have low fruit quality. Washington navel gives good peel colour when it is grown in a dry and cool climate area in Up Country IZ.

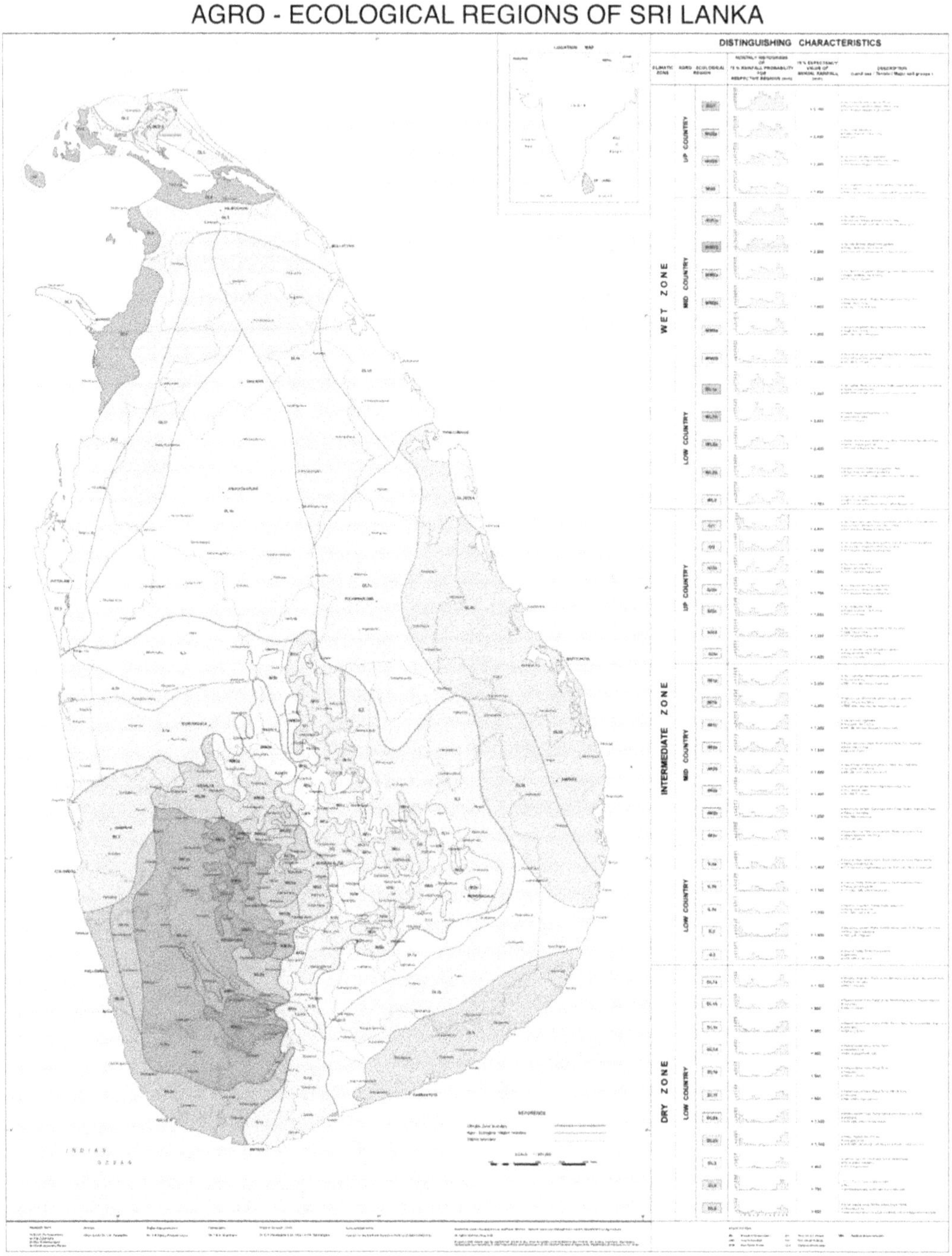

FIGURE 13.16 Agro-ecological zones of Sri Lanka (Natural Resources Management Centre, DOA, Sri Lanka).

13.3.1 Technologies and Package of Practices Followed in Citrus Cultivations

13.3.1.1 Climate and Soil

A well-drained soil with organic matter is preferable for the cultivation of citrus crops. Annual rainfall of 1,750–2,000 mm and a soil pH within the range of 5–6 is suitable for optimal crop growth. It is desirable to have a specific dry period and low atmospheric moisture for about 1–2 months per year to stimulate fruiting.

13.3.1.2 Field Preparation and Planting

For proper cultivation, it's crucial to implement effective soil conservation methods in the field and ensure adequate drainage. For cultivating oranges, creating pits sized at 60 x 60 x 60 cm with a spacing of 4 m x 5 m and for mandarins, pits at dimensions of 75 x 75 x 75 cm with a spacing of 3.5–4 m x 3.5–4 m is preferable. The pits should be filled with a mixture of topsoil and decomposed organic manure. Additionally, applying dolomite to each pit about two weeks before planting and using the recommended fertilizer 2–3 days prior to planting is advisable. It is recommended to plant a month before the rainy season begins or during the latter stages of the heavy rainy season. Moreover, the use of mulch during dry periods can be beneficial.

13.3.1.3 Crop Management Practices

In order to obtain a high yield and to develop a strong frame, citrus trees should be trained and pruned from the early stage. After making cuts, applying fungicide at the pruning site is recommended. After harvest, promptly remove all weak shoot ends. Additionally, eliminate any deceased plants, water shoots, and ensure the tree is thoroughly cleaned.

Soil moisture is important, especially after plant establishment and during the flowering stage of the crop. A weed-free area around the tree, extending to the branches' reach, is advised and care should be taken to prevent damages to the roots.

Both organic and chemical fertilizers should be applied as recommended by the DOA considering the AEZ. Applying approximately 10 kg of well-rotted compost around the tree twice a year can be beneficial.

13.3.1.4 Harvesting, Grading, and Storage

The grafted plants usually yield fruit within 2–3 years. It is advisable to harvest fruits that are fully ripened and cut the stalk close to the fruit. The timing of maturity varies with the specific variety. Thinning of the peel and peel colour becoming lighter are indicative of ripeness. Subsequently, sort the fruits based on their size and colour to create a consistent selection in grading. The fruits should be placed in boxes with good ventilation and should be handled carefully to avoid damages. It's beneficial to store fruits in cool conditions to extend the shelf life.

13.3.2 Status of the Citrus Industry in Sri Lanka

13.3.2.1 Harvesting Window of Major Cultivars of Citrus

The harvesting window for citrus fruits in Sri Lanka varies across different scenarios. For most farmers, citrus trees are found in home gardens (Rajapaksha et al., 2021) while a smaller number engage in commercial cultivations. The primary harvesting season of limes usually occurs from February to April in the country. However, certain citrus varieties can be harvested all year round and oranges and limes are available in the market throughout the entire year. The prevailing climate conditions play a pivotal role in determining variability of the harvest (Iizumi and Ramankutty, 2016).

13.3.2.2 Marketing Network of Citrus

The market network for citrus in Sri Lanka involves producers, distributors, and consumers across various stages around several key pathways. A majority of citrus produced in home gardens and commercial cultivations contribute to the national production.

Farmers play a key role in the market dynamics, actively connecting with consumers, retailers, and wholesalers. Wholesalers are important as they are involved in distributing citrus to retailers, ensuring a wider reach for the produce. They directly provide a significant portion of their citrus produce to local markets, which are often sold by roadside stalls and small grocery stores. These outlets cater to the needs of nearby communities, making fresh citrus readily available. Moreover, certain farmers engaged in large-scale cultivation also contribute to the supply chain for supermarkets, often through pre-established agreements that ensure a consistent flow of produce making citrus accessible to a broader consumer base.

Further, government institutes and farmer organizations often support the marketing network by organizing agricultural fairs during peak periods, enhancing direct interactions between producers and consumers. However, the climatic factors that affect production and the availability of citrus in the market. Adequate and efficient transport facilities and good storage conditions are crucial to maintain the quality of citrus. Additionally, the market benefits from food processing companies that create value-added products from citrus and distribute them within local markets, diversifying the range of offerings available to consumers. While the export of citrus is limited, a modest amount of lemons is gathered for overseas trade, contributing to the broader citrus market network.

13.3.2.3 Status of Export and Imports of Sri Lanka

As the domestic production of sweet orange and other citrus fruits are not sufficient to meet the national requirement, oranges and mandarin fruits are annually imported to the country expending a considerable amount of foreign exchange. The Table 13.2 provides data on the import of mandarin oranges and oranges (both fresh) into Sri Lanka from the years 2013 to 2022. In terms of mandarin, the quantities fluctuated over the years, recorded 5,728 Mt in 2022 while 2,407 Mt of fresh oranges were imported. The government has spent Rs. 496 million to import oranges in 2022 and Rs. 662 million for mandarin.

Sri Lanka imports different citrus products, including fresh and dried lemons, mandarins, and oranges. Fresh lemons are imported from France, Saudi Arabia and from the UAE, while dried lemons from India and the UAE. The mandarins in fresh form imported mainly China, Pakistan, and

TABLE 13.2
Imports of Mandarin and Oranges (2013–2022)

	Mandarin (Fresh)		Orange (Fresh)	
Year	Quantity (Mt)	Value (Rs. Mn)	Quantity (Mt)	Value (Rs. Mn)
2013	9,497	359.81	7,636	670.98
2014	6,552	245.79	6,370	596.05
2015	8,896	390.53	7,494	670.19
2016	10,566	504.15	8,130	863.81
2017	6,041	331.05	8,686	958.12
2018	10,961	603.70	8,356	939.63
2019	9,384	547.95	7,773	949.22
2020	7,036	429.53	4,696	670.42
2021	8,765	899.79	4,774	735.32
2022	5,728	662.95	2,407	496.24

(Source: Sri Lanka Customs).

TABLE 13.3
Exports of Lemon (2013–2022)

Year	Lemon (Fresh) Quantity (Mt)	Lemon (Fresh) Value (Rs. Mn)	Lemon (Dried) Quantity (Mt)	Lemon (Dried) Value (Rs. Mn)
2013	282	20	84	27
2014	470	25	174	59
2015	664	73	71	24
2016	131	16	7	3
2017	539	70	59	28
2018	289	69	135	52
2019	111	21	31	11
2020	87	10	130	58
2021	73	9	1	1
2022	42	9	13	21

(Source: Sri Lanka Customs).

South Africa. Oranges, both fresh and dried, bitter orange pulp and orange juice with various brix values are imported in 2022.

Factors such as market demand, trade dynamics, government policies on importation and domestic production might have contributed to these fluctuations in import quantities and values.

Table 13.3 provides export data on both fresh and dried lemons in terms of quantity and value from 2013 to 2022. Over this period, lemon exports exhibited varying trends. In 2013, 282 Mt of fresh lemons were exported, amounting to a value of 20 million Rupees. The figures for fresh lemon exports increased over the next few years, reaching a peak of 664 Mt in 2015, valued at 73 million Rupees. However, the quantities gradually decreased in subsequent years, with 2022 recording 42 Mt valued at 9 million Rupees.

The quantities of dried lemon exports were notably lower compared to fresh lemons and in 2022, 13 Mt were exported. Sri Lanka's major lemon export destinations include the Maldives, Italy, Qatar, the UAE, Saudi Arabia, the USA, and Bahrain (Export Development Board, 2022).

13.3.2.4 Status of Citrus Processing

Citrus processing and value addition encompass a diverse range of products that transform these fruits into various attractive and practical offerings. This includes the production of jams and marmalades, pickles as well as dried citrus products, such as peels and slices, which find uses in culinary and confectionery applications. Citrus-flavoured desserts and treats, including ice creams, snacks, and sweets are produced in the country. Furthermore, cosmetics infused with citrus extracts are produced widely as skincare products.

Few citrus species are already popular as fresh fruits and in the juice industry. However, many more species remain as underutilized species. Among them Pummelo (*Citrus maxima*), Grapefruit (*Citrus paradisi*), Sour orange (*Citrus aurantium*), Lemonime (Lemon x Lime), Nas naran (*Citrus japonica*), Heen naran (*Citrus madhurensis*), Rough lemon (*Citrus jambheri*), and Nattaran (*Citrus megalxyllocarpa*) have great potential in the processing industry and other industrial applications such as essential oils, pharmaceuticals, and therapeutic uses.

These products are manufactured by a multitude of producers, ranging from household-level initiatives to small and large-scale enterprises. Additionally, household cleaning liquids are also produced from citrus extracts as natural and eco-friendly formulations. Citrus processing and value addition shows a spectrum of options that cater to a wide array of consumer preferences and needs.

13.3.2.5 Status of Citrus Research and Extension

Citrus research and extension in Sri Lanka are primarily led by the DOA and play a vital role in advancing the citrus industry within the country. Research initiatives of DOA cover a wide range of areas aimed at improving the citrus sector comprehensively. One of their key focuses is on developing new citrus varieties that are better adapted to local conditions. At the same time, research is being carried out to introduce improved agronomic practices to enhance farmers' citrus cultivation. Several varieties and technologies have been developed and introduced to farmers during past decades.

Evaluating foreign germplasm of citrus is carried out to identify suitable types for cultivation in Sri Lanka. This not only enriches the diversity of citrus available but also ensures that the chosen varieties are well-suited to the country's specific environment. Mandarin varieties, Ehimi1, Ehimi2, and Ehimi3 were released, evaluating a Japanese germplasm and selecting the most suitable varieties for the up country of Sri Lanka. Researchers also explore ways to increase citrus productivity, looking into innovative methods and approaches that can make a positive impact on yield. Moreover, research is carried out to develop citrus varieties that are resilient against common pests and diseases, contributing to more sustainable cultivation practices and identifying the techniques for improving soil conditions and identification of optimal organic and inorganic formulations to enhance the quality of the harvest.

When pest and disease problems do arise, the research efforts focus on finding solutions and preventive measures to safeguard the citrus crops. Identifying outbreaks and studying potential solutions are conducted whenever necessary to effective disease management. Furthermore, the researchers engage in identifying value-added citrus products that cater to consumer preferences and market demand.

Citrus extension practices in Sri Lanka aimed at disseminating knowledge and empowering farmers in the citrus cultivation sector. Various projects initiated by the Ministry of Agriculture are carried out to facilitate dissemination of citrus crop directly to farmers island wide.

Pre-seasonal trainings are conducted to prepare farmers for upcoming agricultural activities, ensuring optimal yield outcomes. Additionally, technical working group meetings are organized by Provincial levels to foster collaborative problem-solving and address extension-related challenges arising for the crop. Monthly technical discussions are held bringing together extension officers and researchers to collectively find solutions to emerging issues of the crop.

The Department of Agriculture educates farmers through media announcements, including radio programmes and social media platforms. The DOA is carrying out toll free call centre (1920) to provide solutions to farmer problems over the phone without any delays. Further, printed publications, disseminated through newspapers and leaflets, contribute to knowledge dissemination, while exhibitions held at research centres showcase advancements and opportunities within the citrus industry. Notably, private organizations and farmers are supported through advisory services, and frequent field visits to cultivations by extension officers when necessary.

13.4 CHALLENGES AND STRATEGIES

The citrus industry in Sri Lanka faces several challenges that impact its growth. One of the main obstacles is low productivity, with sweet orange yielding 1.25 mt/ha and mandarin producing 2.9 mt/ha. This is significantly lower than the global average productivity ranging from 10 to 24 mt/ha. Most of the existing citrus cultivation comes from low-productive home gardens which the cultivations are not maintained according to required quality standards to achieve the maximum yield.

Moreover, the orchards in the country have a short economic lifespan of only 10 to 15 years, further worsening the issue of low productivity. The popular orange variety cultivated for making juice, called Bibile sweet, is grown in the Low Country IZ and DZ. The soil in most of these areas is alfisols, which are reddish-brown earth where a layer of gravel in the B horizon makes it harder for water to go deep into the soil. This often leads to water accumulating during the rainy season, a situation that's harmful for citrus cultivation creating various root and stem diseases caused by different types of fungi, such as *Phytophthora nicotianae* and *Phytophthora ciptoptera*. Moreover, because of difficulties in getting enough fertilizer and water, the yield from these trees is low, and sometimes the cultivation even starts to decline.

Another challenge is the lack of high-quality, productive citrus varieties with desirable traits such as good taste, fewer seeds, yellow peel colour, good appearance, and pest and disease resistance. Further, most of the table type mandarin varieties are well-suited for cultivation in the Up country as they develop a yellow colour peel. Such varieties with good quality and yellow peel colour suitable for other areas of the country are lacking at present.

The scarcity of commercial-scale farmers who can contribute to large-scale production, while a substantial number of low-productive home gardens is another significant issue. In cases where cultivation is not well managed, it has been observed that around 20–30% of trees experience decline during this period. This has emerged as a major obstacle to expanding the cultivation across larger areas.

Additionally, less adaptation of new farming technologies among growers hinders potential improvements in yield and quality. The lack of strong marketing channel to secure better prices for citrus products, along with a lack of value-added offerings, creates limitations within the industry. Considering lime cultivation, all trees are blooming at the same period of the year due to the weather pattern of Sri Lanka and producing massive yields in a short period of time. Hence considerable percentage of lime is wasted without use and the rest is sold at a cheaper price due to low demand. Furthermore, the development of a proper value chain is lacking, inhibiting efficient distribution and challenges related to inputs, labour, marketing and extension affects the farmers willing to produce for export market (Madanayake, 2016). These challenges collectively contribute to the limitations faced by the citrus industry in Sri Lanka.

To address these issues and enhance the citrus industry in the country, a series of forward-looking programmes will be implemented. One key aspect involves identifying the most suitable regions within diverse agro-ecological contexts to establish expansive commercial citrus gardens. These gardens will serve as hubs for large-scale cultivation, contributing to the growth of the industry.

Furthermore, establishment of mother plant orchards featuring recommended citrus varieties and producing high-quality planting materials to ensure a steady supply is carried out by the DOA. Programmes targeting off-seasonal fruit production to ensure a year-round availability of citrus products such as implementation of improved agricultural practices, and soil moisture management will help to optimize citrus cultivation. Programmes should be carried out to engage the younger generation in the industry by integrating modern technology practices targeting high quality and productivity.

As knowledge dissemination is essential, training programmes will support farmers to use a range of technologies, including irrigation and fertilization methods, pruning techniques, as well as strategies for pest and disease control. These packages will be delivered to farmers, especially home garden owners to enhance the quality and quantity of citrus yields.

Establishment of collection and distribution centres is needed facilitated by government coordination to develop the supply chain. These centres will offer valuable marketing support and preservation facilities to farmers of all sizes, leading to better communication and transportation of their products.

13.5 CONCLUSION

Sri Lanka's diverse agro-ecological zones and favourable climate enable the growth of various citrus fruits predominantly for local markets. Citrus cultivation is widespread revealing a notable growth in lime production over the years, while orange production exhibited certain fluctuations yet maintained a relatively steady course. While offering supplementary income for farmers, citrus cultivation also holds potential for the export market, contributing to public health and economic improvement. Agricultural practices, weather trends, and policy decisions outline the annual output of Sri Lanka's citrus industry, which is consumed both as fresh fruit and value-added products. Mainly the Department of Agriculture engages in research and extension activities for development and disseminating knowledge on citrus cultivation practices to obtain high-quality production. Further, private companies and organizations are also involved in disseminating and encouraging farmers to cultivate citrus in large-scale level. Importantly, the ongoing citrus research initiatives are driving the citrus industry forward by developing improved varieties, enhancing cultivation practices, ensuring crop health, and aligning products with consumer desires. Taken together, multifaceted extension programmes are set to transform the citrus industry by combining strategic planning, innovative techniques, and robust infrastructure development. These proactive initiatives will be important to make a significant impact in overcoming challenges and elevate the citrus industry in the country. With a focus on expansive cultivation through commercial gardens, and by improved practices and off-seasonal production, the industry's stability will improve further.

CONFLICT OF INTEREST

The authors declare that they have no conflict of interest.

REFERENCES

Agricultural Statistics (2022). Socio Economics and Planning Centre, Department of Agriculture, Sri Lanka.

Annual Report of Central Bank of Sri Lanka. (2022). Central Bank of Sri Lanka, Colombo 01.

Bandara, S., Kumara, T., Dharmadasa, S., and Samaraweera, R. (2021). Changes in food consumption patterns in Sri Lanka: Food security and sustainability: A review of literature. *Open Journal of Social Sciences* 9(10): 213–237. https://doi.org/10.4236/jss.2021.910016.

Dahanayake, N. (2015). Some neglected and underutilized fruit-crops in Sri Lanka. *International Journal of Scientific and Research Publications* 5(2): 1–7.

Export performance indicators. (2022). Sri Lanka Export Development Board, Colombo.

Iizumi, T., and Ramankutty, N. (2016). Changes in yield variability of major crops for 1981–2010 explained by climate change. *Research Letters* 11: 034003.

Madanayake, E. K. (2016). Challenges in vegetables and fruits exports in Sri Lanka. *Journal of Economics Students' Society, Economics Student Society* 11(1): 82–90.

MoFE. (1999). *Biodiversity Conservation in Sri Lanka: A Framework for Action.* Ministry of Forestry and Environment: Battaramulla, Sri Lanka. ISBN 955-9120-03-4.

Punyawardena, B. V. R. (2008). *Rainfall of Sri Lanka and Agroecological Zones.* Department of Agriculture: Peradeniaya, Sri Lanka.

Rajapaksha, L., Gunathilake, D. M. C. C., Pathirana, S. M., and Fernando, T. N. (2021). Reducing post-harvest losses in fruits and vegetables for ensuring food security – Case of Sri Lanka. *MOJ Food Process Technologies* 9(1): 7–16.

Sri Lanka Labor Force Statistics. (2022). *Bulletin: Sri Lanka Labour Force Survey 1st Quarter – 2022.* Department of Census and Statistics, Ministry of Economic Policies and Plan Implementation, Sri Lanka.

Venkatprahlad, S. N., and Wijeratnam, R. S. W. (2019). *Feasibility study for value addition in the fruit and vegetable sector of Sri Lanka* (STDF/PPG/576). International Trade Centre, Geneva, Switzerland.

14 Current Status and Prospect of Citrus Production, Research, and Development of Indonesia

Ir Hardiyanto, Nirmala Friyanti Devy, Kurniawan Budiarto, Chaireni Martasari, and Zainuri Hanif

14.1 INTRODUCTION

Indonesia is a tropical country located in the equatorial region at coordinates between 6° 08' NL – 11° 15'SL, and 95°45' – 141° 05'EL. It consists of 17,508 islands, and based on the plains, Indonesia's territory can be classified into highlands and lowlands with various agroecological conditions. Therefore, almost all horticultural crops, including subtropical crops, can be grown and well developed. Citrus is one of the major horticultural crops in Indonesia that it has been cultivated for over 55.000 ha in almost all provinces, with a total production and productivity of about 2.55 million tonnes and 35 tonnes/ha, respectively (BPS 2023). The main production centres were East Java, North Sumatera, West Sumatera, West Kalimantan, and Bali. Nevertheless, the contribution of citrus production to the total fruit production was only 9.8% as compared to banana (32,56%), Mango (11,52%), and Pineapple (11,01%) (Dirjen Hortikultura 2024). Many citrus species such as mandarin (*Citrus reticulata* Blanco), sweet orange (*Citrus sinensis* L.), Siam (*Citrus suhuiensis* hort. Ex Tanaka), pummelo (*Citrus maxima* Merr.), lime *(Citrus x hystrix* (Kieffer lime)), and lemon (*Citrus x limon* L. Burn F.) are commercially grown in Indonesia (Hardiyanto and Nirmala 2011). According to Anis et al. (2021) and Yulianti et al. (2020), approximately 255 varieties derived from more than 18 species have been collected, which the species is dominated by mandarins (43%), tangerines (4%), sweet oranges (23%), pummelos (15%), and others such as lime, and lemon. Mandarin citrus is the most economically important fruit crop in the highlands particularly the area around the volcano, such as in East Java (Malang District, Batu City), West Java (Garut District), and North Sumatra (Karo District, Brastagi and Simalungun) (Sutopo et al. 2021). In terms of major citrus cultivars, many commercial citrus cultivars have been established in Indonesia such as groups of "Keprok" as known as Mandarin (Batu 55, Garut, SOE, Madura, Kacang, Terigas, and Gayo); "Siam" as known as Tangerine (Pontianak, Banjar, Madu, and Gunung omeh); Sweet Oranges (Pacitan and Valencia); and Pummelo (Nambangan, Sri Nyonya, Magetan. and Pangkajene merah) (Agisimanto et al. 2007; Martasari et al. 2012).

Almost 90% of citrus in Indonesia in the 1980s was destroyed due to the attack of Huang Long Bing (HLB) caused by the bacterium *liberibacter asiaticum*, which can be transmitted through the vector *Diaphorina citri* Kuw. In 1987 FAO together with the Indonesian Government created the Citrus Rehabilitation in Indonesia Program, mainly to produce citrus seedlings free from systemic pathogens, including Citrus Vein Pholoem Degeneration (CVPD), Citrus Tristeza Virus (CTV),

DOI: 10.1201/9781003584599-14

Citrus Exocortis Viroid (CEV), Citrus Vein Enation Virus (CVEC), The success of citrus agribusiness began since the production of disease-free citrus seedlings through Shoot Tip Grafting (STG) technique, and the construction of Foundation Block (FB) and Budwood Multiplication Block (BMB) in 29 Provinces (Supriyanto and Setiono 2023). As a high-value crop, citrus farming offers farmers an opportunity to improve their income and living standards, because the citrus business activity is financially feasible as indicated by B/C ratio, IRR, and NPV could reach 2.86, 45.5%, and Rp. 188,363,727 or equivalent to US $ 12,558, respectively (Mutia et al. 2021). Indeed, fruit consumption tends to increase (4,233 kg/capita/year or equivalent to 1,33 million tonnes) (BPS 2023). It seems that the prospect of a sustainable citrus industry in Indonesia is promising. However, more than 90% of citrus is still imported, whereas export is less than 1%. Total value of imported citrus reaches US $ 234,092,000. This is a challenge and an opportunity for the Indonesian Government to substitute imported citrus with national citrus production. It means that synergies among national/local governments, research institutes, universities, extension workers, as well as stakeholders are very important. Research and development on citrus have been established particularly to produce superior varieties that are tolerant to biotic and/or abiotic stress, technologies for improving cultural practices under various locations, and qualified seed production. In the future, networking via research collaboration or scientific exchange and training as well is suggested.

This chapter presents an overview of the status and prospects of citrus production, research, and development in more detail. This paper aims to give up-to-date information and share the experiences in citrus production in Indonesia as well as feedback from readers on the citrus industry in Indonesia.

14.2 MAJOR CULTIVARS OF CITRUS CULTIVATED IN INDONESIA

In general, the proportion of citrus varieties grown by citrus growers is 80% of tangerine, 10% of mandarin, 6% of pummelos, and 4% other, like lemon, lime, and citrus hystrix. The major citrus species and cultivars commonly cultivated by farmers are as follows:

14.2.1 Tangerine

Tangerine or Siam cultivars (*Citrus nobilis* var. Siam Hort.), also known as "Jeruk Siam" are the most important group of citrus plants in Indonesia, which have the widest adaptability among the cultivated citrus cultivars. Siam cultivars are recognized by small-sized fruit with thin and easy peeling rinds and bright greenish-yellow skin colour when ripe. The fruit is sweeter than Mandarin (Martasari et al. 2012). There are many cultivars of Siam citrus locally grown in Indonesia, such as Siam Madu, Siam Pontianak, and Siam Banjar. Aside from the mentioned cultivars, there are also Indigenous cultivars endemically grown in various areas in Indonesia. The observed cultivars have similar leaves and stem characters, with slight variations in rind colour and thickness. Several Siam cultivars mostly grown in Indonesia are shortly illustrated in the following descriptions.

14.2.1.1 Tangerine cv. Siam Pontianak

Tangerine cv. Siam Pontianak was registered by the Provincial Government of West Kalimantan, Indonesia in 2003. The fruit is characterized by yellowish-green fruit rind when ripe with orange flesh (Figure 14.1). The uniform yellowish fruits would have commonly been observed when the plants were planted in highlands. The flesh has a smooth texture, sweet with a slight fragrance aroma. The fruit has a dimension of 7.1 cm in height, and 7.6 in diameter with thin rind (1–1.5 mm) (Anis et al. 2021).

FIGURE 14.1 The ripe fruit (a) and cross fruit section of tangerine cv. Siam Pontianak (b) (Chaireni Martasari with permission).

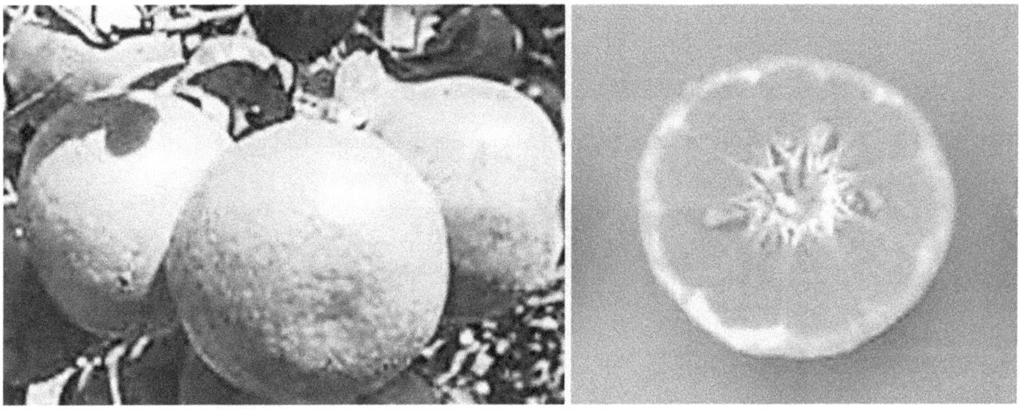

FIGURE 14.2 The ripe fruit (a)and cross fruit section of tangerine cv. Sinta Ponsoe (b) (Chaireni Martasari with permission).

14.2.1.2 Tangerine cv. Sinta Ponsoe

Tangerine cv. Sinta Ponsoe was released in 2019 by the Ministry of Agriculture, Republic of Indonesia. The cultivar is obtained from the crossing between Tangerine cv. Pontianak with Mandarin cv. SoE. Most fruit characteristics resemble to tangerine type, yet their thicker, rough rind (1.8–3 mm) (Figure 14.2) and profound pores are similar to mandarin type. The fruit has a size of 4.78–6.78 in height and 5.5–6.78 in diameter with longer fresh storage than other Siam cultivars (Anis et al. 2021).

14.2.1.3 Tangerine cv. Siam Madu

The cultivar was officially registered by Tanah Karo, North Sumatera, Indonesia in 1999. The fruit has spheroid to obloid shapes with a size of 5–7 cm in height, 6–8.5 cm in diameter, and rind thicknesses of 2–4 mm. The cultivars are adaptively grown from lowlands at 250 m asl to highlands at 850 m asl, but the best growth and fruit production are obtained at 600 to 700 m asl. The peel colour is yellowish green with orange flesh when ripe (Figure 14.3). At 6–9 years after planting with optimum maintenance, the productive plant can produce 750–900 fruits per plant per harvesting period (Anis et al. 2021).

FIGURE 14.3 The ripe fruit of tangerine cv. Siam Madu (Hardiyanto, private picture).

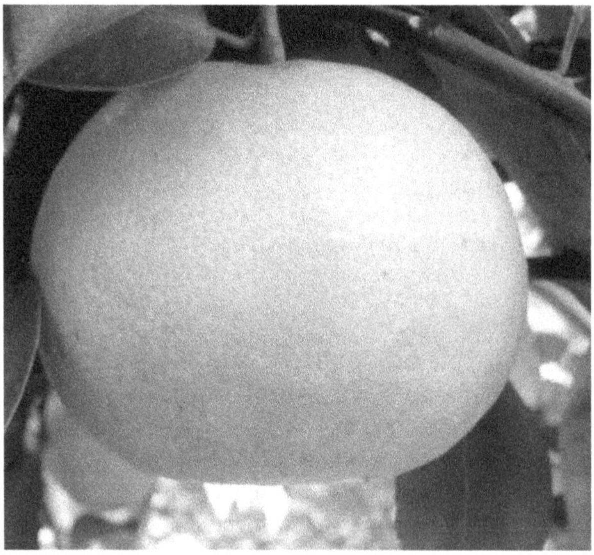

FIGURE 14.4 The ripe fruit of tangerine cv. Siam Banjar (Hardiyanto, private picture).

14.2.1.4 Tangerine cv. Siam Banjar

Tangerine cv. Siam Banjar is registered by city government of Banjar, South Kalimantan, Indonesia in 1998. The plant has an obloid fruit with a size of 6.5–7.5 in diameter and 1.3–1.7 mm rind thickness (Figure 14.4). Similar to cv. Siam Madu and Pontianak, the variety produces more uniform yellowish fruits at the physiologically ripening stage when planted in highland areas. At the productive stage under optimum maintenance, the plant can produce 500–600 fruits per plant in a harvesting period (Anis et al. 2021). The cultivar can be grown in indirectly inundated or catchment swamp areas with canopy modification (Rusmayadi et al. 2022).

14.2.2 Mandarin

Mandarin cultivars are commonly recognized for their fruit attractive rind and fruit characteristics. The rind has bright yellow-reddish orange with round, obloid, up to pyriform fruit shapes (Campos et al. 2005; Dorji & Yapwattanaphun 2011). In Indonesia, mandarins are merely officially registered than Siam cultivars i.e. Mandarin cv. Batu 55, Rimau Gerga Lebong (RGL), JRM 2012, Krisma Agrihorti, Monita Agrihorti, Orange Parahiyangan, SoE86 Agrihorti, DN Sambilulungan 1, Topazindo Agrihorti, Orinda Agrihorti, Kertaji, etc. Several mandarin cultivars mostly planted by Indonesian growers are presented in the following descriptions.

14.2.2.1 Mandarin cv. Batu 55

Mandarin cv. Batu 55 was officially registered in 2006 and is one of the most preferred by consumers and planted cultivars by growers. The average fruit size is 7.9 cm in height and 5.9 cm in diameter. The rind is yellow with a slight green tinge (Figure 14.5) and easy to peel when ripe. The cultivar has a yearly production cycle and can produce 36 kg of fruits per plant in a harvesting period under optimum maintenance (Sugiyatno et al. 2013)

14.2.2.2 Mandarin cv. Rimau Gerga Lebong (RGL)

Mandarin cv. RGL was registered by the city government of Lebong, the province of Bengkulu, in 2012. The fruit has orange rind and flesh, obloid shape (Figure 14.6) with a size of 4.9–5.9 cm in height, 6.7–9.6 cm in diameter and 173–347 g in weight per fruit. The rind is categorized as thick (4–5 mm) and adhesively attached to endocarp. The flesh contains 10.51–16°brix total soluble solid with 0.95% total acid representing of sweet flesh with little sour taste (Anis et al. 2021).

FIGURE 14.5 The ripe fruit of mandarin cv. Batu 55 (Hardiyanto, private documentation).

FIGURE 14.6 The ripe fruit of mandarin cv. RGL (Hardiyanto, private documentation).

FIGURE 14.7 The ripe fruit (a) and cross fruit section of mandarin cv. Krisma Agrihorti (b) (Chaireni Martasari with permission).

14.2.2.3 Mandarin cv. Krisma Agrihorti

Mandarin cv. Krisma Agrihorti is registered by Indonesian Citrus and Subtropical Fruit Research Institute (ICSFRI) in 2015. Unlike other mandarin cultivars, cv. Krisma Agrihorti has a pyriform fruit shape (Figure 14.7), yellow rind and a size of 7.36–8.77 cm in height and 7.42–8.46 in diameter. The cultivar also has less seed and sometime seedless fruits. The name "Krisma" represents its crispy and sweet endocarp/vesicle. Mandarin cv. Krisma Agrihorti can be adaptively planted in lowlands without significant alterations in fruit taste. In lowlands, the rind colours tend to be greener when ripe compared to those planted in highlands (Anis et al. 2021).

14.2.3 Pummelo

Up to 2021, 19 pummelo varieties have been officially registered by various institutions to the Ministry of Agriculture, Republic of Indonesia. These varieties include Pumelo cv. Nambangan, Kotaraja, Bageng Taji, Pamindo Agrihorti, MTR 19, and others.

FIGURE 14.8 The ripe fruit (a) and cross fruit section of Pummelo cv. Nambangan (b) (Chaireni Martasari with permission).

14.2.3.1 Pummelo cv. Nambangan

Pummeo cv. Nambangan was officially registered by the city government of Magetan, East Java, in 2000. The cultivar has obloid to spheroid fruit shapes (Figure 14.8) with rind thickness ranges from 1.7 to 2 cm. The rind colour was greenish yellow with pink to light endocarp when ripe, smooth texture with fewer trichomes. The single fruit is 1.2–2 kg in weight with a size of 15.3–17.1 in height and 20.1–20.8 cm in diameter (Anis et al. 2021).

14.2.4 Sweet Orange

Sweet oranges (*Citrus sinensis* L) are one type of fresh consumption oranges that has the character of a fruit skin that is close to the pulp. This type of orange is more consumed fresh in the form of cut fruits or juice. Until now, only five varieties have been registered as superior varieties and the most popular varieties are Pacitan sweet oranges.

14.2.4.1 Sweet Orange cv. Pacitan

One of the types of sweet oranges that is widely known to the public today, namely Pacitan sweet oranges or known as baby oranges. This orange was registered as a superior variety by the Regency Regional Government Pacitan, East Java in 2002 through the Decree of the Minister of Agriculture Number 412/Kpts/TP.240/7/2002. Pacitan sweet orange is a spherical fruit with a slightly flattened tip and a flat base. The fruit has a skin thickness of about 3.1–3.5 mm, weight 164.9–187.4 g, height 66–69 mm, and diameter 68–71 mm. When grown in the highlands, the skin of the fruit tends to be yellowish-green to yellow, and the flesh of the fruit is pale yellow to yellow when physiologically cooked. The advantage of the Pacitan sweet orange variety lies in the fruit taste of being almost acid-free with total dissolved solids of 8.4–11°Brix and total content acid is around 0.05–0.1%. Pacitan sweet oranges at the age of 6–10 years can produce up to 40–60 (Figure 14.9) (Anis et al. 2021).

14.2.4.2 Sweet Orange cv. Ortaji

In 2019, IAARD registered superior varieties of seedless sweet orange with the name Ortaji through the Decree of the Minister of Agriculture Number. 131/Kpts/SR.120/D.2.7/9/2019. Orange Ortaji sweet has a different flavour from Pacitan sweet orange. Ortaji citrus fruits are round, elongated (ellipsoid) with thick skin about 41.2–6.69 mm, weight per fruit 321–508 g, fruit height 69.7–98.9 mm, and a diameter of 84.6–13.33 mm. The colour of the skin of the fruit when ripe is yellow-orange (17A – RHS colour chart)) with a porous, rough surface texture

FIGURE 14.9 The ripe fruit of Sweet orange cv. Pacitan (Sri Andayani with permission).

FIGURE 14.10 The ripe fruit (a) and cross section of Sweet orange cv. Ortaji (b) (Sri Andayani with permission).

(Figure 14.10). The flesh of Ortaji sweet orange is bright orange-yellow (23A-RHS colour chart), with a slightly sour sweet taste with total solids dissolved 8.2–12°Brix and total acid 0.3–0.38%. At the age of 4 years, this orange can produce 24–33 kg per tree per year (Anis et al. 2021).

14.2.5 Other Citrus Types

In Indonesia, citrus is the second most popular fruit after banana and has been employed in many traditional ceremonies, as a food condiment, and for traditional therapeutics and pharmaceutical therapies (Kadowangko & Febriyanti 2023; Yulianti et al. 2016; Budiarto et al. 2017). As functional fruit, citrus fruits contain several phytochemical components that are beneficial to health, including vitamin C, vitamin B, potassium, calcium, phosphorus, magnesium, manganese, salt, carotenoids, and other phenolic chemicals (Codoner-Franch & Valls-Belles 2010). These citrus types include lemons, limau (*Citrus amblycarpa* (Hassk.), lime, *C. hystrix* and others. Several cultivars that have been registered to the Ministry of Agriculture, Republic of Indonesia are Lime cv. Borneo, Lime cv. Kalamansi FR, Lime cv. Nimas Agrihorti, Lemon cv. Montaji Agrihorti, *C. hystrix* cv. Puri Agrihorti, Limau (*C. amblycarpa*) cv. Sari Agrihorti, etc. Some of these cultivars are presented in the following descriptions.

14.2.5.1 Lemon cv. Montaji Agrihorti

Lemon cv. Montaji Agrihorti was registered by Indonesian Citrus and Subtropical Fruit Research Institute (ICSFRI) in 2018. This cultivar has very few seeds or even seedless in certain condition. In medium elevation, the flowers emerge simultaneously, thus enable the plant to bear fruits year-round. The fruit has greenish-yellow rind with the thickness of 2.3–2.4 mm and greenish white endocarp flesh when ripe (Figure 14.11). The plant has ellipsoid fruit and the size is recorded as 5.13–7.4 cm in height, 4.39–5.62 in diameter and 70–113 g in weight per fruit. Lemon cv. Montaji Agrihorti can be grown in the lowland to highland (Anis et al. 2021).

FIGURE 14.11 The ripe fruit of Lemon cv. Montaji Agrihorti (Sri Andayani with permission).

14.2.5.2 Lime cv. Borneo

Lime cv. Borneo was registered in 2011 by the provincial government of East Kalimantan. The fruit has a round shape with a rind thickness of 1–2 mm, 30–70 g of weight per single fruit and a dimension of 3.5–4.7 cm in height, 3.7–5.1 cm in diameter. The young fruit has green rind colour and turns into yellow when ripe. The cultivar has white endocarp flesh (Figure 14.12) and contains 53.66–55.58 mg/100g vitamin C, 10.2–10.8°brix total soluble solid and 11.86–15.53% total acid. Lime cv. Borneo is also recognized to have less seed with strong fragrance aroma and high fruit productivity. After 19 years with proper maintenance, the fruits can be harvested four times a year with total production of 220–223 kg (Anis et al. 2021).

14.2.5.3 Lime cv. Nimas Agrihorti

Lime cv. Nima Agrihorti was officially registered by Indonesian Citrus and Subtropical Fruit Research Institute (ICSFRI) in 2015. Unlike other lime cultivars, cv. Nimas Agrihorti has a sweet taste with an ellipsoid fruit shape (Figure 14.13). The fruit dimensions are 6–8 cm in height, 5–6 cm

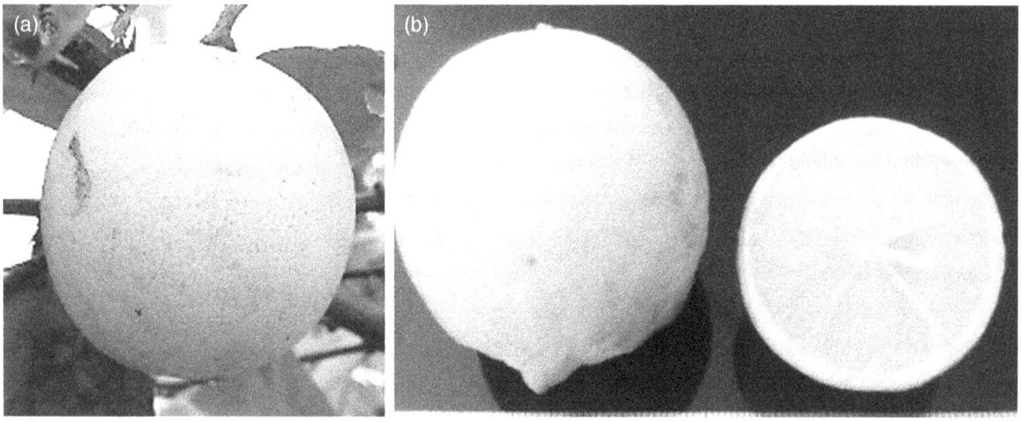

FIGURE 14.12 The ripe fruit (a) and the cross section of Lime cv. Borneo (b) (Sri Andayani with permission).

FIGURE 14.13 The ripe fruit (a) and cross section of Lime cv. Nimas Agrihorti (b) (Sri Andayani with permission).

in diameter with 72–80 g of single fruit weight. The rind is slightly thicker than that of cv. Borneo with yellow colour when ripe. The white endocarp contains 8.3°brix and 0.45% total acid, which are much lower than cv. Borneo. The cultivar is well adaptive in the lowland area (Anis et al. 2021).

14.2.5.4 *C. hystrix* cv. Puri Agrihorti

The leaves and fruit of *C. hystrix* have been long employed in the daily life of Indonesian people for favouring cuisine, biopharmacy, and biopesticide (Sophia et al. 2021; Maimunah et al. 2020). *C. hystrix* cv. Puri Agrihorti was registered by Indonesian Citrus and Subtropical Fruit Research Institute (ICSFRI) in 2015. The distinctive characteristic of the cultivar is the petiole size is the same or larger size than the ovate leaf blade (long petiole) (Figure 14.14). The whole leaf dimensions are recorded as 3–10 cm in length and 2–5 in width. The fruit bears a yellowish dark green rind when young and turns into yellowish bright green when physiologically mature with a thickness of 3–5 mm. The fruit is spheroid in shape with the dimensions of 2.4–4.8 cm in height, 3.3–4.9 cm in diameter, and 38–56 g in weight per fruit. The endocarp flesh is yellowish light green with 8–10°brix total soluble solid and 2.9% total acid containments (Andrini et al. 2021).

14.2.5.5 Limau (*C. amblycarpa* (Hassk.)) cv. Sari Agrihorti

Limau citrus (*C. amblycarpa*) is profoundly used as a food condiment in many Indonesian traditional foods (Humaira et al. 2022) and up to this moment, the plant has been extensively employed as biopharmacy and pharmaceutical properties (Nareswari & Kuncoro 2016; Cahyaningsih et al. 2016). Limau (*C. amblycarpa* (Hassk.)) cv. Sari Agrihorti was registered by Indonesian Citrus and Subtropical Fruit Research Institute (ICSFRI) in 2015. The distinctive features of Limau cv. Sari Agrihorti are an obloid fruit shape with smaller fruit size (2.56–3.59 cm in height and 3–3.8 cm in diameter) and lighter fruit weight (18–28.6 g per fruit) (Figure 14.15). The young fruit has a yellowish dark green rind and turns into a yellowish light green when ripe. The flesh has yellowish light green in colour and contains 26.5–29.5 mg/100 g vitamin C, 9–10°brix total soluble solid and 2.17% total acid. The fruits are usually harvested shortly before the fruit reaches the mature stage (Anis et al. 2021).

FIGURE 14.14 The ripe fruit, leaves and cross fruit section of *C. hystrix* cv. Puri Agrihorti (Sri Andayani with permission).

FIGURE 14.15 The fruit of Limau (*C. amblycarpa* (Hassk.)) cv. Sari Agrihorti (Sri Andayani with permission).

14.3 AGRO-CLIMATIC ZONES IN INDONESIA

Indonesia is the worlds's largest archipelago state, lies between 6^0 South Latitude - 11^0 South Latitude and 95^0 East Longitude-141^0 East Longitude. As passed by the equator, Indonesia has an almost entirely tropical climate. The climate is shaped by uniformly warm water that makes the country have little diurnal or seasonal fluctuation in temperature or day length from season to season. Freezes rarely occur except at the highest altitudes (Izza et al. 2022). Indonesia's climate is also characterized by three rainfall patterns, i.e. monsoonal, equatorial, and local patterns (Yassi et al. 2023). Franz Wilhem Junghuhn, a Germanist plant expert, divided Indonesia tropical climate into five regions based on continent elevations, i.e. tropical (0–700 m asl), characterized by yearly average temperature < 22°C, subtropical (700–1,500 masl), characterized by average temperature of 15–22°C, moderately cold (1,500–2,000 masl) characterized by average temperature of 11–15°C, cold (2,000–3,000 masl) characterized by average temperature of 11°C and tropical winter (>3,500) (Brillyansyah et al. 2023). The transitions between regions are not absolute but are part of a continuum. Moreover, there are distinct micro climates within each region that have unique rainfall, wind, or sunlight patterns that also impact on citrus growth, flowering and fruiting (Rokaya et al. 2016).

Pummelo or shaddocks can be naturally grown from tropical to subtropical regions (50–900 masl), but plant growth and fruit production would be best at 50–600 masl, less moist to dry and free from permanent inundation/submergence and flooding (Tahir et al. 2020). Planting pummelo at more than 700 m asl resulted in longer fruit development and produced fruits with thicker rinds and less attractive/paler flesh colour. The main production centre of pummelo in Indonesia is still in Java island. The province of East Java ranks in first with a total harvesting area of 333,339 ha, followed by South Sulawesi at 172,306 ha and Central Java with 111,134 ha in 2022 (Indonesian Agricultural

FIGURE 14.16 Pumello plantation in lowland (50 m asl) at Banjarsari village, Probolinggo Province, East Java (Chaireni Martasari with permission).

Statistics, 2023). The predominant cultivars found are Nambangan, Srinyonya and Magetan (East Java), Nambangan, Srinyonya, dan seedless Bageng Taji (Central of Java) and Pangkep Merah, Pangkep Putih and Maria Sigola-gola (South Sulawesi) (Figure 14.16).

Tangerine has wider environmental adaptations from lowland to highland from 50 to 1200 masl. The plantations of several cultivars like cv. Pontianak & Banjar are found in lowland (<200 masl) in East Java, West Kalimantan, and West Sumatera. Micro-climate modifications through cultural practices are needed on marginal area like swamp or/and peatland to support plant adaptation on the specific location. In some orchards, cultivars like cv. Siam Madu, Gunung Omeh, and Kintamani performed better at medium to highland (400–800 masl). The main production centres of tangerine in Indonesia are the province of East Java, North Sumatera, Bali, West Sumatera, West Kalimantan, and South Kalimantan (Indonesian Agricultural Statistics, 2023) (Figure 14.17).

Unlike tangerine, mandarin citrus was found to be less productive in Iowland. Most of mandarin production centres are located in medium to highlands, like in Batu, Pacitan, Ponorogo (East Java), Solok (West Sumatera), Magelang, Tawangmangu, and Semarang (Central of Java), Nunukan and Bulungan (North Kalimantan), Lebong, Bengkulu Selatan, Bengkulu Tengah, Bengkulu Utara (Bengkulu), Timor Tengah Selatan (East Nusa Tenggara) and other provinces (Figure 14.18).

14.4 SIGNIFICANT TECHNOLOGIES AND PACKAGE OF PRACTICES

Followed in Citrus Farming

Indonesian Government has released a Good Agricultural Practices (GAP) for nursery and orchard management that the farmers could implement. The government, through the Ministry of Agriculture, is focusing on enhancing the quality of planting materials and providing training to farmers. Improved nursery practices and the use of disease-resistant varieties can lead to higher yields and better-quality fruit.

14.4.1 Nursery Management

14.4.1.1 Seed Processing and Seedling Preparation

A qualified seed is required to produce a good rootstock. Most citrus growers used *Japanche citroen* (JC) seeds as a rootstock. The farmers may produce seeds by themselves or buy the seeds from

FIGURE 14.17 Tangerine plantation on swamp area at South Kalimantan (a), and Tangerine plantation in highland (1,200 masl) at Berastagi District, North Sumatera (b) (Chaireni Martasari with permission).

recommended stores. The price for 1 Kg of selected seeds is about Rp. 1.500.000 or equivalent to US$100. In terms of seed preparation, the first step is seed extraction from the fruit by cutting the fruit horizontally and then twisting on halves parts. The extracted seeds are treated with a fungicide solution and then put in sealed plastic bags for storage in a refrigerator at 4 °C. Seeds are usually planted into trays, at a depth of 0.5 to 1.0 cm with a distance of 0.5 x 3.0 cm. The tray medium used is sterilized mixed soil and fine sand with a ratio of 1:2. Seeds start germination 21 days after planting, and within one month after planting almost all seeds germinate. The seedlings are ready to transplant in a poly bag 2.5 3 months after planting. The size of the poly bag is 10 x 30 cm (diameter x height). The medium used in this poly bag is mixed soil, fine sand, and manure in a 1:2:1 ratio. JC seedling for a rootstock is ready to bud 4–6 months after transplanting (Figure 14.19).

14.4.1.2 Scions and Budded Plant Preparation

The success of citrus agribusiness is affected by the quality of budwood used by the farmers. The Indonesian Government set up a Citrus Rehabilitation Program in 1985 to produce virus-free plants. To date, the Ministry of Agriculture via the Indonesian Agency for Agricultural Research and

FIGURE 14.18 Mandarin cv Batu 55 (top) and mandarin cv RGL (bottom) plantation in highland (950 m asl) at Batu City, East Java (Source: Chaireni Martasari and Hardiyanto).

Development (IAARD) has produced and established virus-free budwood production. The standard procedure for producing budwood virus-free materials is mainly via Shoot Tip Grafting (STG) *in vitro* technique. The production of healthy bud wood materials has started from the development of a Nucleic Seed (NS) as a source of shoot materials used for producing Foundation Block (FB). Budwood Multiplication Block (BMB) has been built with materials derived from FB. Meanwhile, materials for producing commercial seedlings have been derived from BMB. Recently, FB and BMB have been built in 29 provinces in Indonesia to provide healthy scion materials for the farmers (Devy et al. 2014; Devy et al. 2019).

The characteristics of qualified twigs used as a source of budwood materials are the optimal age of twigs for scion materials is three months after pruning, the shape of the twigs is round to the triangle is not flat, it is still leafy, and the buds are still prominent but the buds have not yet grown, the twig of the patched eye is still fresh (not wilted), and 2–3 buds at the base of the twigs are discarded, as they tend to be dormant. Scion is budded onto 4–6-month-old JC rootstock through chip budding technique and it was wrapped with plastic ropes. The wrapped plastic rope was opened after 15–21 days (Figure 14.20). The fertilizers used are NPK (15-15-15) and ZA dissolved in a litre of water every 15 days. Pests and disease control are done based on the recommendation. The process of selecting scion materials up to the readiness of the budded citrus plants is monitored by the Seed Testing and Certification Center.

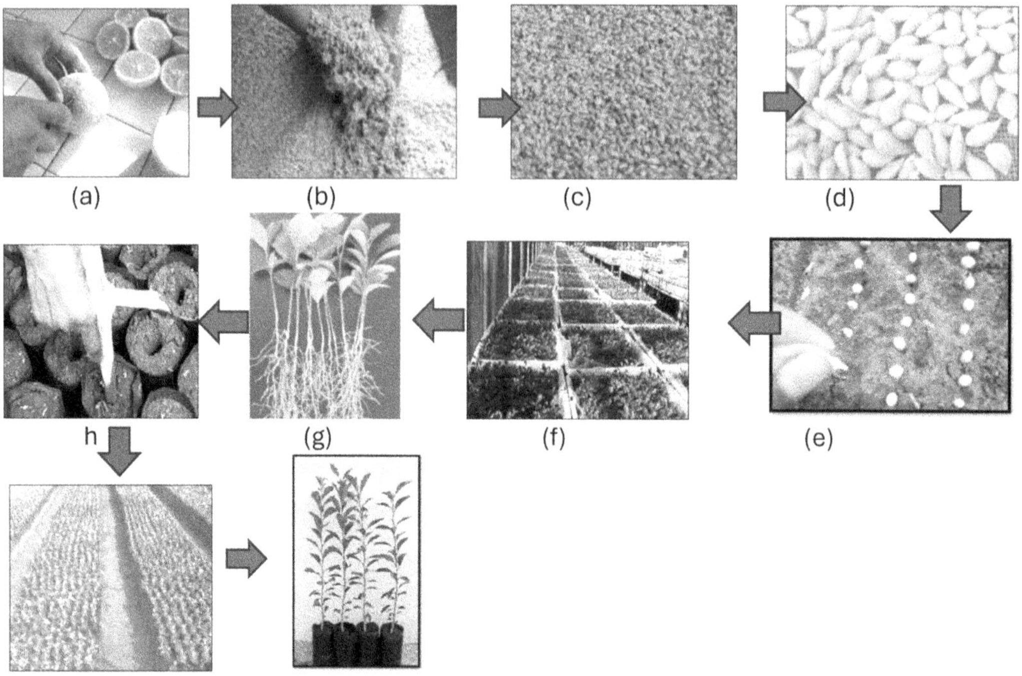

FIGURE 14.19 Flow charts of seed processing and seedling preparation: fruit twisting (a), seed extraction (b), seed washing (c), seed treatment (d), seed planting (e), seedling in trays (f), Seedlings ready to transplant (g), polybag preparation (h), seedling planting in the polybag (i), and JC seedling ready to budding (j) (Pictures obtained from Kusnan and Setiono with permission).

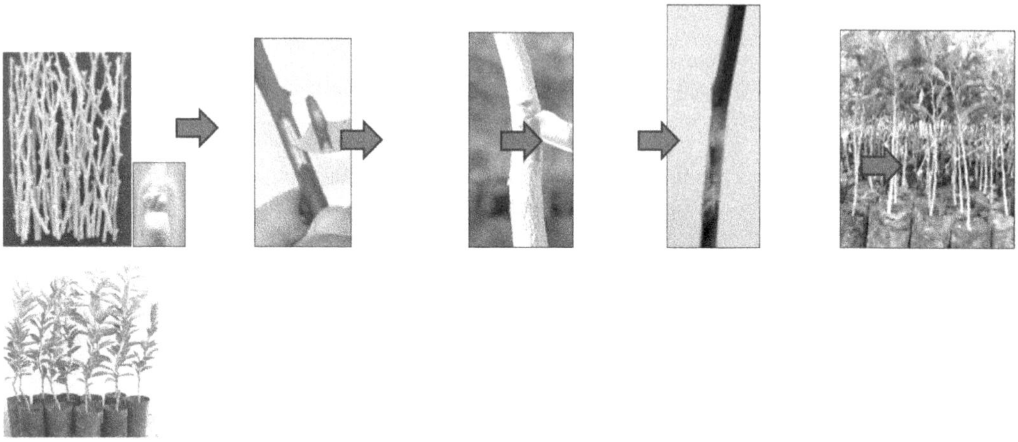

FIGURE 14.20 The process of the certified budded citrus plant production: twig and scion selection, bud wood to be budded, chip budding technique, wrapped with plastic rope, budded seedlings, and 6 to 8-month budded citrus seedlings ready to move to the field. (Pictures obtained from Kusnan and Setiono with permission).

14.4.2 Orchard Management

The criteria for qualified budded citrus seedlings are the use of certified seedlings with a height of 50 to 70 cm and it has experienced two flush periods. The farmers have to implement the healthy orchard management that is developed by IAARD. The "Five" Key Factors in Healthy Orchard Management are using certified (blue labels) nursery stocks, Eradication of infected plants, Vector control of the HLB vector: *Diaphorina citri* Kuw., Optimum maintenance: fertilization, pruning, disease control, others, and Citrus grower group participation/consolidation. The orchard management itself consisted of land preparation, planting, fertilizer application, pests and disease control, irrigation, and harvesting.

Land preparation is usually done one month before planting. The land should be cleaned of weeds or waste trees. The next step is plotting and making planting holes with sizes of 60 x 60 x 60 cm or 1 x 1 x1 m, meanwhile, the distance within a row and among rows is 3 x 3 m, 4 x 4 m, 5 x 5 m, or by using high-density planting system (Figure 14.21). The fertile layer of tiled soil (darker and looser colour) is separated by the layer underneath. The soil material of the cultivated layer is collected near the hole and then mixed with 20–40 kg of manure (compost). If the soil pH is 5.5–6, lime is added to the soil mixture and compost by 0.5 kg or 1 kg if the soil pH < 5.5. A mixture of soil, manure, and lime material is put into the planting hole and left for at least 10 days. The planting hole can be re-covered with a mixture of the soil dug out of the hole + 20 kg of organic matter and if the soil pH < 5.5, the addition of 1 kg of dolomite is highly recommended. After covering the planting hole with media, the existing ajar is installed again in the middle of the planting hole as a marker for the location of the plant. The closure of the planting hole is sought to be higher than the ground level and in the form of a mound of soil. Try to leave the mixed media in the planting hole for 1 week to 1 month.

A planting hole is dug right in the centre of the planting hole, ± 30 cm wide and ±.20 cm deep. As much as 50 g of TSP or 50% of the total dose of P fertilizer a year is included. into the seed hole as a base fertilizer. The planting hole is made into a mountain 20–30 cm high if the soil texture is clay or made flush with the soil surface if the soil texture is sandy or dusty (Andisol). The plastic seed container (polybag) is removed using a knife or scissors. The rooting of the seeds should be carefully examined. Curved or unhealthy root tips should be cut off. Seeds with abnormal root necks, e.g. swollen/purplish, circular, rotten, etc. should not be used (Figure 14.21). The seeds are inserted into the planting holes, then the lateral roots are arranged so that they lead in different directions. Seeds are piled up to 5 cm above the root neck and gently pressed into the soil. The number of leaves can be reduced when it is too dense. The plant support is installed and then the seedlings are watered with water until it wets the planting hole enough.

Good fertilization management is not only oriented to increasing production or maximizing profits alone but also to minimizing environmental damage from fertilizer application activities so that farming is sustainable. In this regard, all knowledge related to plant nutrition and its interaction with soil and plants is an important provision in plant nutrition management. The age/size of the tree is one of the dominant factors that determine the difference in fertilizer needs in citrus plants. In this regard, fertilization recommendations are divided into two, namely fertilization on young plants or plants that have not yet produced fruits and mature plants that have produced fruits. Moreover, the application of fertilization based on plant age applies the principle of as follows:

1. The larger the size/age of the plant, the larger the fertilizer dose
2. The courser the soil texture, the larger the fertilizer dose
3. The higher the rainfall (in wet climates), the greater the fertilizer dose
4. The composition of P and K fertilizers is increased in line with the increase in plant life with N dose remains dominant
5. The older the plant lifespan, the longer the fertilization interval (Sutopo et al. 2021).

FIGURE 14.21 Plotting (a), planting hole making (b), and planting citrus seedlings (c) (Sutopo with permission).

This irrigation is carried out by draining water from the tub and collecting water using pipes or hoses into a soil dish under the canopy. Watering can be done manually by human power or by installing water faucets around trees. To minimize the negative effects of water blows (aggregate crushing soil, compaction, and surface runoff), the height of the water outlet should not exceed 30 cm from the ground level. This method requires less water compared to surface inundation irrigation and line irrigation. In addition, the land will be avoided from erosion caused by waterways. This method is suitable for gardens that are sloped on the plains high, but its efficiency is reduced if the soil is sandy (Sutopo et al. 2021).

Generally, the optimum harvest life for Siamese and tangerine varieties is about 7 to 8 months from the time the flowers bloom. The characteristics of fruit that is ready to be harvested are if the fruit is pressed not too hard, the bottom of the fruit if pressed feels soft and if it is flicked with a finger, it does not sound loud, the colour is attractive (yellow appears for chayote), and the sugar content (PTT) range 8.5 – 13%. Sugar levels can be determined with a hand refractometer in the garden.

14.4.3 Gradual Flowering and Fruiting

Gradual flowering and fruiting technologies, such as the "Bujangseta" method, are being implemented by some farmer growers (Figure 14.22). This approach aims to synchronize flowering and fruiting cycles, enabling farmers to achieve higher yields and better fruit quality. By implementing such technologies, citrus growers can optimize their production schedules and enhance their profitability. Devy et al. (2021) reported that the application of "Bujangseta" changed the flowering pattern and induced flower bud formation, blooming, and fall petals earlier. BS also produced a higher flowering and fruit set percentage, fruit number/10 branches/tree, and total fruit weight.

FIGURE 14.22 "Bujangseta" technology (a), and start of flowering after Bujangseta treatment (b) (Hardiyanto private documentation).

14.5 STATUS OF CITRUS INDUSTRY

14.5.1 Harvesting Time of Major Cultivars of Citrus

The ripeness of citrus fruit is determined by physical changes in the fruit (colour, size, shape, specific gravity) and chemical changes in the fruit (pH, acidity, Total Soluble Solids (TSS)) (Gupta et al. 2021). Citrus fruits variety, including lemon, orange, kinnow, mandarin, pomelo, and others, have distinct growth patterns, flavours, and nutritional profiles but share similar maturation mechanisms and biochemical pathways, which are key to determining their harvest time (Chaudhary et al. 2016; Gupta & Mishra, 2021; Rodrigo et al. 2004). Citrus reticulata var. Mandarin Batu 55, a superior citrus variety from Batu City, East Java, is known for its mildly sweet and tangy flavour, with a sweetness level of 10–12°Brix (Zakiyyah et al. 2022). The Mandarin Batu 55 citrus generally requires about 250–252 days to fully mature after flowering, a timeframe commonly used to estimate its harvest period (Al Riza et al. 2023). The citrus harvesting season in Indonesia typically begins in February and lasts until September, with peak production occurring between May and July (Ratule et al. 2018). The average productivity of tangerine and mandarin citrus at the national level is approximately 20–25 tonnes per hectare (Harwanto et al. 2022).

The harvesting period for major citrus varieties in Indonesia varies depending on the type of citrus, region, and local climate (Table 14.1). Tangerines, mandarins, and sweet oranges typically

TABLE 14.1
Major Harvest Time of Citrus Varieties in Indonesia

Type of citrus	Maturity time
Tangerine, mandarin, orange, etc.*	Juny–September (main season)
	October–January (minor season)
Pamelo/pummelo	February–April
Lemons	All year
Lime	All year
Kaffir lime	All year

Nb: *in some provinces the harvest season can be different.
Source: Hanif with permission.

FIGURE 14.23 The harvest season of citrus in Indonesia, focuses on tangerine and mandarin varieties, with 60% production during the main season (May–August) (Raw data taken from National Statistical Agency was processed by Hanif with permission).

have a peak harvest season (main season) from June to September and a secondary harvest season (minor season) from October to January. Pomelos are generally harvested from February to April and are predominantly cultivated in Java, Sulawesi, and Sumatra. Meanwhile, certain citrus varieties, such as lemons, limes, and kaffir limes, are harvested throughout the year.

Differences in soil and climate types in various citrus production canters in Indonesia mean that citrus fruits can be harvested throughout the year. However, there is a certain pattern in the distribution of the harvest (Figure 14.23). Around 60% of citrus production is harvested in the period from

May to August. During the peak harvest period, the oversupply of citrus in the market often leads to a significant decline in prices. The remaining 40% of the production is harvested between January to April and September to December, resulting in a more evenly distributed harvest throughout the year. However, the citrus industry faces a critical challenge in its distribution system. Inefficiencies in distribution not only hinder the effective marketing of citrus but also negatively impact farmers' incomes and limit the availability of citrus in certain regions.

14.5.2 Marketing Network of Citrus in Indonesia

The marketing network of citrus involves multiple stakeholders, including farmers, farmer associations, wholesalers, and retailers Figure 14.24. The market linkage begins with farmers producing citrus, often organized into cooperatives or farmer associations, which play a vital role in connecting them to larger buyers. Even citrus farmers who are not yet organized will sell directly without going through farmer groups, especially in new development areas. Citrus farmer groups facilitate better access to markets, reduce dependency on intermediaries, and enable collective bargaining to secure fair prices. However, challenges persist, such as price fluctuations due to oversupply during the peak harvest season and inadequate storage and transportation facilities.

Government support has been instrumental in strengthening the citrus marketing network in Indonesia. Policies and programmes, such as subsidies for transportation, grants for cold storage infrastructure, and training for quality improvement, aim to enhance market efficiency. The Ministry of Agriculture collaborates with local governments and farmer organizations to promote export-oriented production and ensure consistent supply chains. Several organizations, including agricultural cooperatives, private sector players, and non-governmental organizations (NGOs), actively participate in the citrus market. These entities help bridge gaps in knowledge, infrastructure, and financing. Farmer associations, such as GAPOKTAN (Farmers' Group Associations), are central to these efforts, enabling farmers to pool resources and achieve economies of scale.

The diagram illustrates the value chain and market linkage pathways for citrus farmers in Indonesia. Farmers can sell their products through various channels, including direct involvement with citrus farmer groups, agro-tourism initiatives (such as fruit-picking tours), or intermediaries like wholesalers (referred to as Bakul). Other selling mechanisms include the Tebasan System, where buyers purchase crops directly from the farm, and the Ijon System, where buyers pay farmers in advance for future harvests (Zainuri & Yamaura, 2021). These channels converge through collector traders, who then distribute the citrus products to retailers, ensuring availability for end consumers. This system highlights the complexity of citrus marketing and the diverse options available for farmers to access markets.

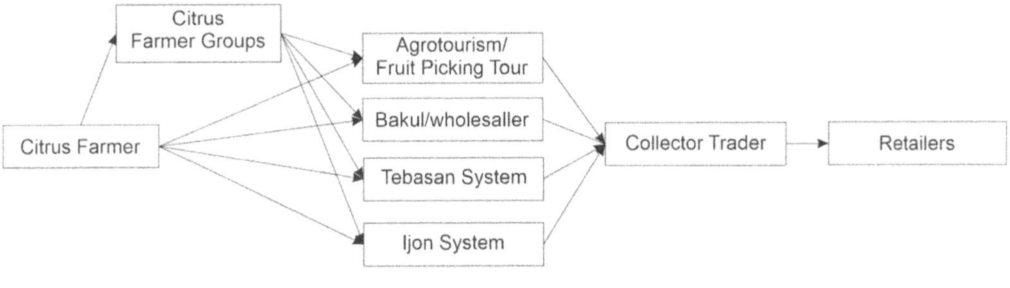

FIGURE 14.24 Citrus value chain in Indonesia and farmers' pathways to deliver products to consumers (Zainuri Hanif with permission).

14.5.3 Status of Export and Import

The import data highlights a consistent reliance on global citrus suppliers, with significant contributions from countries such as China, Pakistan, and Egypt, particularly during peak years like 2019 and 2020. Notably, imports from China dominate the total volume of citrus entering Indonesia (Figure 14.25). In contrast, the export data reflects Indonesia's modest citrus trade, with primary export destinations including neighbouring countries such as Malaysia, Singapore, and Timor-Leste, alongside European markets like France and the Netherlands (Figure 14.26). Despite imports far exceeding exports, the data indicates an uneven yet gradually increasing export trend, suggesting

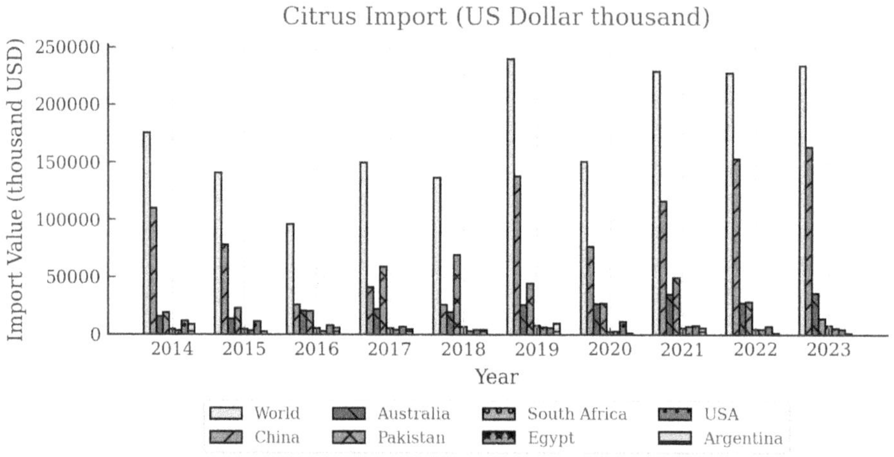

FIGURE 14.25 Trend of imported citrus in Indonesia from 2014 to 2023. (Raw data obtained from International Trade Centre 2024 (https://trademap.org/Index.aspx) was processed by Zainuri Hanif (with permission)).

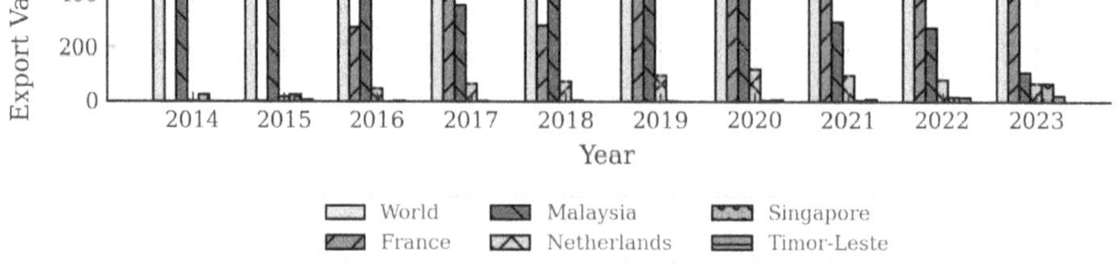

FIGURE 14.26 Trends of exported citrus from Indonesia from 2014 to 2023. (Raw data obtained from International Trade Centre 2024 (https://trademap.org/Index.aspx) was processed by Zainuri Hanif (with permission)).

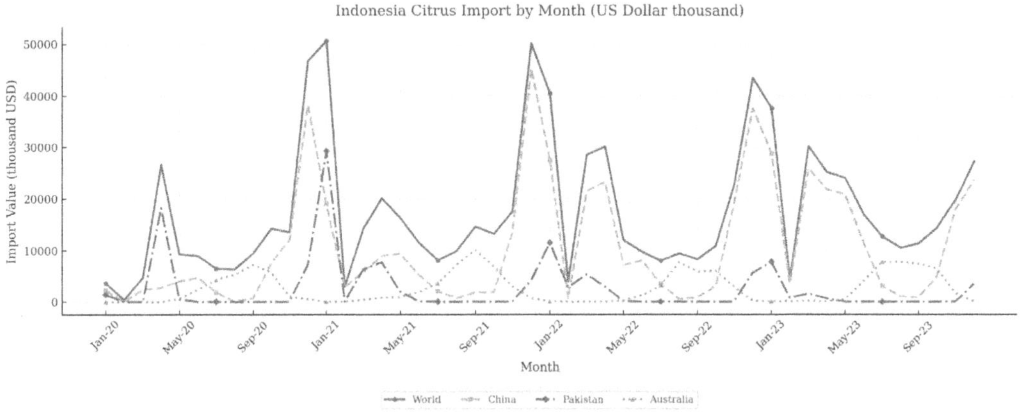

FIGURE 14.27 Monthly trends of Indonesia's citrus imports (in USD Thousand). (Raw data obtained from International Trade Centre 2024 (https://trademap.org/Index.aspx) was processed by Zainuri Hanif (with permission)).

opportunities to enhance Indonesia's citrus export market through advancements in production efficiency and marketing strategies.

Indonesia's citrus imports peak consistently during January and February each year, with additional surges in March, particularly in 2021 and 2022. This trend aligns with the country's minor harvest season, typically from November to February, when domestic citrus production significantly declines, leading to increased reliance on imports. China is the primary supplier during these periods, while contributions from Australia and Pakistan remain relatively smaller. Imports drop sharply between May and Augustus, corresponding to Indonesia's major harvest season, when local production is abundant (Figure 14.27).

These patterns highlight the need for improved domestic production strategies during the off-season, such as adopting greenhouse technologies or cultivating year-round varieties, alongside diversifying import sources to ensure price stability and supply security. In 2023, the growth of kaffir lime exports also occurred to European countries for cooking spices. France accounted for 63.93% of Indonesia's citrus exports for the specified commodity like lime and kaffir lime, making it a key market. France's annual import growth for this commodity from all sources is 3%, while its share in global imports stands at 7.66%, highlighting its significant role in the global market. However, France's import growth from the world outpaces Indonesia's export growth to France, indicating untapped potential for Indonesia to increase its market share. This reliance on France underscores the need for Indonesia to diversify its export destinations while also enhancing its competitiveness and adding value to its products to better capitalize on France's growing demand and strengthen its global presence.

14.5.4 Status of Citrus Processing

The harvesting process plays a crucial role in determining the quality of citrus fruits, both for fresh consumption and processing. In Indonesia, citrus harvesting is predominantly done manually, with workers picking the fruits by hand. Mechanization is considered unnecessary due to the dominant cultivation of tangerine and mandarin varieties, which are primarily consumed fresh as table fruits. This contrasts with orange varieties, commonly processed into fresh beverages or juices. Farmers achieve the highest fruit prices and profit margins when selling citrus fresh, under normal market conditions. However, when market supply is abundant and prices drop, processing becomes a more profitable alternative. Citrus waste can be transformed into a wide range of value-added products, as it contains abundant bioactive compounds such as polyphenols, essential oils, pectin, dietary fibres, and carotenoids (John et al. 2017). These compounds hold significant potential for utilization in the food, cosmetic, and pharmaceutical industries, contributing to the development of innovative products with substantial health benefits (Panwar et al. 2021a; Panwar et al. 2021b).

In Indonesia, citrus research has begun exploring various benefits of tangerine peel, highlighting its potential as a natural preservative to extend fruit shelf life, particularly when processed into essential oil (Ikarini et al. 2023; Utami et al. 2024; Utami et al. 2023). Maintaining fruit quality until it reaches consumers is essential, encompassing both external and internal attributes. External quality refers to visible physical characteristics such as shape, skin colour, size, texture, and smoothness, while internal quality includes sugar content, juiciness, acidity, sugar-acid ratio, juice colour, vitamin C content, and other attributes. These quality parameters are influenced not only by cultivation practices and orchard management but also by proper harvesting and post-harvest handling (Magwaza et al. 2012).

14.6 STATUS OF CITRUS RESEARCH AND EXTENSION

In the last 10 years, government research institutes, state universities, and international private sectors have taken part in citrus research and development. IAARD, through Indonesian Citrus and Subtropical Fruits Institute (ICISFRI), might be recognized as the dominant research executor from 2006 up to 2021 and established the foundation plan for citrus development in Indonesia. The institute has released and been pioneering the implementation of proper technologies for citrus development in Indonesia, starting from adaptive cultivars in the local tropics, qualifiedly certified planting materials, the establishment of an elite planting material production unit, the establishment of an assessment laboratory for citrus diseases, improved cultural methods for pest and disease management, nutrient and water supply, modification of canopy architecture, year-round fruit production and diversification of fruit products. During these periods, ICSFRI had collaborative research activities with state universities and international institutes, like Bogor Agricultural University (IPB Bogor), University of Brawijaya, University of Gadjah Mada, CGN-WUR The Netherlands, SAKATA Ltd from Japan, and others. From 2021 up to the present under BRIN coordination, the citrus research activities have collaborated with Bogor Agricultural University (IPB Bogor), the University of Brawijaya, and several local governments.

Aside from basic and applied research, the framework of national citrus development programmes also employs research extension activities. The research extension provides and facilitates technology service and dissemination from research institutes toward stakeholders, like growers, policymakers, academic institutions, domestic and international private sectors, hobbyists, observers, and other pertinent public sectors. The diverse programmes were designed and customized to the audience and events. The supports of printed and online information sources were constructed in the form of books, flyers, real-time Q&A informative websites, multimedia, and cellular applications. Demonstration plots and supporting laboratories can also be visited for further discussion of problem-solving, short training and field education. The research institute also actively involves and supports wider disseminative activities like training of trainers for extension officers, national and international exhibitions, and symposiums/seminars. At present, a programme under the Research Centre for Horticulture-BRIN that is regularly presented monthly for the horticulture disseminative programme is HortiActive. The several research activities conducted in Indonesia are presented in Table 14.2.

Based on Presidential Regulation No. 78 – 2021, the national research activities were merged and coordinated by a non-departmental agency, called BRIN (National Research and Innovation Agency of Indonesia) which consequently revoked the research and development mandate in all ministries, including Ministry of Agriculture. The research of citrus was then managed by the Research Centre for Horticulture under the coordination of the Research Organization for Agriculture and

TABLE 14.2
Research Activities Conducted in Indonesia

No	Research activities	References
1	Management of citrus germplasms included a collection of edible and Indigenous citrus accessions, characterization of the collected accessions based on morphological and molecular characters, pra evaluation of the accessions to find the potential biochemical containment, features and traits, pre-breeding, and field and *in vitro* conservations and database documentation. Until 2021, ICSFRI had maintained and conserved more than 255 edible citrus accessions collected from many regions in Indonesia	(Anis et al. 2021; Yulianti et al. 2020; Hardiyanto et al. 2023; Devy et al. 2023; Devy and Hardiyanto 2017; Martasari et al. 2023; Indrastuti et al. 2020)
2	Conventional and in conventional breeding programmes through hybridization, mutation, protoplast fusion, and selection. Through these frameworks, more than 10 genotypes have been released as national superior varieties	(Purba et al. 2021; Ylianti et al. 2021)
3	Improved technologies of cultural practices covered mass propagation of citrus for healthy planting materials, rapid detection of systemic disease, integrated pest management, IoT implementation, the use of beneficial microorganisms to trigger plant growth and prevent disease intrusion, management of nutrient, canopy and harvesting periods, and post-harvest technologies	(Devy & Hardiyanto 2015a; Budiarto et al. 2017; Mutia et al. 2021; Devy et al. 2021; Hardiyanto, Devy et al. 2022; Mutia et al. 2023; Shidig et al. 2024; Tuwo et al. 2024; Nugroho et al. 2024; Puspitasari et al. 2019)
4	Socio-economic aspects involved consumer and growers' preferences, feasibility study of pre-implemented technology, supply-chain management, market orientation, and trendsetter perspectives	(Mufidah et al. 2024; Rizka et al. 2024; Hasibuan et al. 2021)

Food-BRIN. BRIN is not the only one conducting research institution citrus and other horticultural crops. BRIN is also designed to facilitate and induce research collaborations among pertinent sectors, like universities, private sectors, international research collaborators, local governments, etc. Up to this moment, research programmes related to citrus are found in four research activities, i.e. morphogenetic and biochemical evaluations of wild Rutaceae, genetic improvement of citrus through genome editing, technology improvement for citrus production in acidic peatland and swamp areas, and registration of local citrus varieties.

14.7 CHALLENGE AND STRATEGIES

Most citrus growers still relied on the traditional culture method, producing/harvesting citrus once a year in huge quantities. With limited capital and knowledge, growers tend to reluctantly remove the infected HLB plants in the field. Technical, knowledge, socio, and economic aspects are the common reasons behind their refusal to eradicate the infected plants. In addition, the availability of qualified and healthy planting materials with the right time, cultivar, and quantity is still a problem for the farmers, particularly at the production centres that are located a little bit far from the certified seed production unit.

The knowledge and skill improvement of growers and extension officers to comprehend HLB handling in the field are the prominent keys to inhibiting the spread of the disease. Moreover, from the domestic and international markets, growers are challenged to produce qualified citrus fruits following consumer preferences. However, several vital aspects are still found and should be paid to attention, i.e. food safety assurance like how to keep fresh fruit fresh when displayed in the market, insufficient packaging and labelling, unacceptable chemical/pesticide containment, costly import duty in some countries, and transportation cost.

These activities need a comprehensive framework involving not only research agencies but dissemination institutes, pertinent logistic-supplying directorates, and local governments as well. The establishment of supervised seed production units near the production centres is needed to guarantee the availability of healthy planting materials of true-to-type varieties at the right time and quantity. This programme consequently needs additional investment in terms of proper production unit facilities including skilled and trained officers. These aspects need support and cooperative work among the pertinent sectors starting from the government regulations and infrastructures like proper fruit storage facilities, packaging technology, assessment laboratories, and quality inspection.

The revitalization of citrus orchards was then expanded to the implementation of GAP/SOP and orchard certification. This programme was based on the concern of the importance of issues of consumer safety to consuming agricultural products in Indonesia. Sufficient food supply, distribution, and ease of access are not the only inevitable cases, product quality also becomes consumer awareness and determines product survival in the market. Therefore, the domestic orchards are pushed to achieve a high certification level which represents the ability to fulfil all the elements in food production requirements, i.e. safe to be consumed, qualified products and the producing processes are environmentally friendly.

Based on the above-mentioned, there are at least four core issues that need to be discussed intensively among the government, research institutes, extension services, stakeholders, and private sectors:

1. Comprehensive technology to boost plant productivity and fruit quality. These include (a) genetic improvement to develop superior cultivars preferred by the market, consumers, and growers, (b) the construction of efficient mass propagation method for true-to-type and healthy planting materials, (c) adaptive IPM implementation in specific area to reduce fruit defections, (d) rapid detection and efficient elimination method for systemic disease, such as HLB, (e) cultural practices related to improvement of plant adaptation toward extreme environments and micro-climate modification to provide conducive environment under suboptimal condition.
2. Organization aspects. The issues are related to improving growers' knowledge and experiences as an entrepreneur in agribusiness. The growers should be induced to realize that their economic activities should be managed for profit-oriented. In many cases, correct decision makings needed to savior, avoid huge economic losses and make the business more competitive. One example of this case was the growers' decision to eliminate the HLB-infected plants when the plants reached the maximum productive ages. The handling of HLB should be carried out collectively in certain covered areas to reduce reinfection incidences. These collective efforts need awareness among the impacted growers to ensure the success of activities.
3. Post harvest and secondary products. The programmes needed in these frameworks are not only related to the classical problem of Cobweb theory for agricultural products but also deal with research activities to increase the additional product values as well. The activities may focus on several aspects, i.e. (a) rapid detection for dangerous chemical containment, (b) efficient methods for preservation to lengthen storage durability, (c) low-cost post-harvest handling, packing, and transportation, (d) exploration of side-products potentials, (e) extraction and mass production of beneficial biochemical containment.
4. Market mastery and intelligence. Market mastery and intelligence are needed to confer growers' insights to penetrate the targeted market segments. Growers must calculate the compensation acceptability of their fruits in terms of quality and quantity toward the market price. On the other hand, growers should also realize that the high quality of fruits is not the only important point. Growers should also be aware of competitors of similar products especially those with cheaper price. The ability to access accurate and dynamic market information

would help the growers maintain their economic business. Market information is also useful for policymakers to take any necessary regulations to open wider market opportunities for local products to be more competitive in terms of quality, supply assurance, and business trust.

14.8 CONCLUSIONS

The development of citrus agribusiness in Indonesia has faced significant challenges, particularly due to the Huang Long Bing disease outbreak. However, the collaborative Citrus Rehabilitation Program initiated in 1987 between FAO and Indonesia has led to the production of disease-free seedlings through the Shoot Tip Grafting technique since 1990. This revival has allowed orange farmers to gradually rebuild their livelihoods. Despite this progress, challenges remain, particularly concerning the consistency of fruit quality and the impacts of climate change, which have sharply decreased both productivity and quality.

To address these issues, collaborative research is essential, alongside training programmes for farmers both domestically and internationally. Furthermore, fostering scientific exchanges between research institutions can facilitate knowledge sharing about citrus cultivation practices globally. Strengthening networks among researchers from various countries is crucial to advancing citrus agribusiness in Indonesia, promoting sustainable practices, and potentially reducing the country's reliance on citrus imports.

The Indonesian Government should create and fund programmes that promote the development and dissemination of climate-resilient crop varieties and farming practices. In addition, subsidies or financial incentives for farmers who adopt sustainable farming practices, such as organic farming and integrated pest management should be taken into consideration.

ACKNOWLEDGEMENT

The authors want to thank Prof. Dr. Kwan Jeong Song and Dr. Dilip Ghosh for giving them access to write a chapter on the Citrus Industry in Indonesia. The authors also thank my colleagues Anis Andrini, Sutopo, Setiono, Kusnan, Sri Andayani, and Agung for providing some pictures for this book.

CONFLICT OF INTEREST

The authors declare that they have no conflict of interest.

REFERENCES

Agisimanto D, Martasari C, Supriyanto A (2007). Perbedaan primer RAPD dan ISSR dalam identifikasi hubungan kekerabatan genetik jeruk siam (Citrus suhuniensis L. Tan) Indonesia. *Journal of Hortikult, 17*, 101–110.

Al Riza DF, Yolanda J, Tulsi A, Ikarini I et al (2023). Mandarin orange (Citrus reticulata Blanco cv. Batu 55) ripeness level prediction using combination reflectance-fluorescence spectroscopy. *Spectrochimica Acta Part A: Molecular and Biomolecular Spectroscopy, 302*, 123061. https://doi.org/10.1016/j.saa.2023.123061.

Anis A, Martasari C, Budiyati E, and Zamzami L (2021). Teknologi Inovatif Jeruk Sehat Nusantara. In Budiarto K, Sugiharto AN (eds). *Klasifikasi Dan Sebaran Jeruk Nusantara*. PT Penerbit IPB Press.

BPS (2023). *Statistik Produksi Buah Indonesia* [Online]. 132–390 pp. Available: www.bps.go.id /subject/8/ekspor-impor.html#subjekViewTab5.

Brillyansyah, D. F., Susanto, S., Fitriana, L., Zaki, M. K., Setyawan, C. & Ngadisih, N (2023). Application of geographic information systems for analysis of rice agricultural land resources potential in Paser Regency as a supporting area for the capital city nusantara. Proceedings of the International Conference on Sustainable Environment, Agriculture and Tourism (ICOSEAT). Advances in Biological Sciences Research, 320–325. https://doi.org/10.2991/978-94-6463-086-2_42

Budiarto, R, Poerwanto R, Santosa E, Efendi D (2017). The potentials of limau (*Citrus amblycarpa* Hassk. Ochse) as a functional food and ornamental mini tree based on metabolomic and morphological approaches. *Journal of Tropical Crop Science*, *4*(2), 49–57.

Cahyaningsih NK, Satriawati DA, Wicaksana IGPAP, Yulita S (2019). Uji aktivitas analgesik ekstrak etanol daun jeruk limau (*Citrus amblycarpa* (Haskk.) Ocshe) pada mencit jantan galur Balb/C dengan metode hot plate. *Jurnal Farmasi Udayana*, *8*(1), 36–43. https://doi.org/10.24843/JFU.2019.v08.i01.p06.

Campos ET, Espinosa MAG, Warburton ML (2005). Characterization of mandarin (Citrus spp.) using morphological and AFLP markers. *Interciencia*, *30*(11), 687–693.

Chaudhary PR, Bang H, Jayaprakasha GK, Patil BS (2016). Variation in key flavonoid biosynthetic enzymes and phytochemicals in 'Rio Red' grapefruit (Citrus paradisi Macf.) during fruit development. *Journal of Agricultural and Food Chemistry*, *64*(47), 9022–9032. https://doi.org/10.1021/acs.jafc.6b02975.

Codoner-Franch P, Valls-Belles V (2010). Citrus as functional foods. *Current Topics in Nutraceutical Research*, *8*(4), 173–184.

Dorji K, Yapwattanaphun C (2011). Morphological identification of mandarin (*Citrus reticulata* Blanco) in Bhutan. *Kasetsart Journal (Nature Science)*, *45*, 793–802.

Devy NF (2014). Aplikasi shoot tip grafting in vitro pada program penyediaan induk jeruk bebas penyakit di Indonesia. *Penelitian Dan Pengembangan Pertanian 33*(3): 105–114.

Devy NF, Hardiyanto, D (2015a). Potential use of in vitro somatic embryogenesis technology for mass propagation of disease-free citrus seeds. *Jurnal Penelitian dan Pengembangan Pertanian*, *34*(4), 169–176.

Devy NF, Hardiyanto, MED (2015b). *Teknologi Shoot-tip Grafting dan Indexing Pada perbenihan Jeruk Bebas Penyakit*. IAARD Press. *80* pp.

Devy NF, Hardiyanto (2017). Citrus diversity of West Sumatra based on morphology, inter-simple sequence repeats and its combined analysis. *Biotika*, *5*(8), 32–39.

Devy NF, Hardiyanto Y (2019). Evaluation and challanges in implementation of citrus Shoot Tip Grafting (STG) technology in Indonesia. *2019 International Symposium on Tropical Fruits*. Liberty Central Saigon Riverside Hotel, Ho Chi Minh City, Vietnam on 24–26 September. International Tropical Fruits Network (TFNet) Box 334, UPM Post Office, 43400 Serdang, Selangor, Malaysia.

Devy NF, Hardiyanto, SA, Harwanto, Mutia DE (2021). Gradual flowering and fruiting technology ("Bujangseta") through fertilizer management in Siam (*Citrus nobilis* l.) Cv. *Pontianak. Emirate Journal of Food and Agriculture*, *33*(9), 764–771.

Devy NF, Yenny, H (2014). The growth performance of citrus derived from somatic embryogenesis plantlet and scion stock. *Indonesian Journal of Agricutural Science*, *15*(2), 71–78.

Dirjen Hortikutura (2024). *Angka tetap Hortikultura 2023*. Kementerian Pertanian, Direktorat jendral Hortikultura.

Gupta AK, Medhi M, Chakraborty S, Yumnam M, Mishra P (2021). Development of rapid and non-destructive technique for the determination of maturity indices of pomelo fruit (Citrus grandis). *Journal of Food Measurement and Characterization*, *15*(2), 1463–1474. https://doi.org/10.1007/s11694-020-00734-4.

Gupta E, Mishra P (2021). Functional food with some health benefits, so called superfood: A review [Article]. *Current Nutrition and Food Science*, *17*(2), 144–166. https://doi.org/10.2174/157340131699920071 7171048

Hardiyanto H, Devy, NF (2011). Status of citrus in Indonesia: Production, research, and sevelopment. *Chronica Horticulturae*, *51(4)*, 14–18.

Hardiyanto H, Devy, NF (2022). Effect of inorganic fertilizer on morphological and physiological responses in 'siam' (tangerine) and 'keprok' (mandarin) seedlings under nursery house. *IOP Conference Series: Earth and Environmental Science*, *985*(1), 012020. https://doi.org/10.1088/1755-1315/985/1/012020.

Hardiyanto, Devy, NF, Mutia DE, Sugiyatno A, Ashari H et al (2023). The fruit peel anatomy, leaf nutrient content, and fruit quality of 'Terigas' mandarin in relation to fruit cracking. *Emirates Journal of Food and Agriculture*, *35*(x), 1–7.

Harwanto, Siregar AS, Mufidah L, Hanif, Zamzami L et al (2022). The role of healthy citrus planting materials production as a major component in the development of the national citrus industry. *IOP Conference: Earth and Environmental Science*, *1160*(1), 012057. https://doi.org/10.1088/1755-1315/1160/1/012057

Hasibuan AM, Gregg D, Stringer R (2021). The role of certification, risk and time preferences in promoting adoption of climate-resilient citrus varieties in Indonesia. *Climatic Change*, *164*, 37. https://doi.org/10.1007/s10584-021-03015-1

Humaira SF, Dewi YSK, Hartanti L., Handojo MAP (2022). Penggunaan Jeruk Sambal (Citrus Amblycarpa) sebagai Bahan Pengasam Alami Terhadap Sifat Fisikokimia dan Sensori Mayones. *Journal Teknologi Pangan, 4*(1), 24–31.

Ikarini I, Suwarda R, Hanif Z, Triasih U, Ashari H (2023). Chemical composition and physical characteristics of orange peel essential oil. *E3S Web Conference*, *425*, 01004. https://doi.org/10.1051/e3sconf/202342501004

Indonesian Agriculture Statistic (2023). Ministry of Agriculture, Republic of Indonesia, 132–203.

Indrastuti NA, Aminah S, Handayani D, Fauziah S (2020). Antioxidant activity of different dried indonesian citrus peel varieties. *Indonesian Journal of Applied Research*, *1*(3), 128–135.

Izza, M. Y. N., Astuty, S., Ramadhan, R. A, Perdani, N. N., Firdaus, M. D., Anwar, M. K., Rosdeawati, N. & Yulihastin, E (2022). Identification of temperature change related to frost phenomenon over Jayapura, Indonesia. ICOPOD-IOP Conference Series: Earth and Environmental Science. *1192*, 012036. https://doi.org/10.1088/1755-1315/1192/1/012036

John I, Muthukumar K, Arunagiri A (2017). A review on the potential of citrus waste for D-Limonene, pectin, and bioethanol production. *International Journal of Green Energy*, *14*(7), 599–612. https://doi.org/10.1080/15435075.2017.1307753

Kandowangko NY, Febriyanti F (2023). Genetic diversity of the genus Citrus in Tomini Bay coastal areas, Indonesia based on morphological characters. *Biodiversitas, 24*(5), 2938–2952. https://doi.org/10.13057/biodiv/d240549.

Magwaza LS, Opara UL, Nieuwoudt H, Cronje PJR et al (2012). NIR spectroscopy applications for internal and external quality analysis of citrus fruit—A review. *Food and Bioprocess Technology*, *5*(2), 425–444. https://doi.org/10.1007/s11947-011-0697-1

Maimunah S, Raihana, Silalahi YE (2020). Aktivitas antibakteri ekstrak daun jeruk purut (Citrus hystrix DC) terhadap bakteri *Staphylococcus aureus*. *Jurnal Pembelajaran dan Biologi Nukleus*, *6*(2), 129–138.

Martasari, C (2012). Characterization of Indonesian "Siam" cultivar (Citrus nobilis Lour.) by morphological and ISSR markers. *ARPN Journal Agricultural and Biological Science*, *7*(10), 830–835.

Martasari C, Karsina H, Reflinur R (2012). Characterization of Indonesian siam cultivars (*Citrus nobilis* Lour.) by morphological and ISSR markers. *ARPN Journal Agricultural and Biological Science*, *7*(10), 830–835.

Martasari C, Yulianti F, Widyaningsih S, Budiyati E, Hardiyanto et al (2023). Genetic diversity assessment of citrus accessions grown in Indonesia using molecular markers. *AGRIVITA Journal of Agricultural Science*, *45*(3), 419–429.http://doi.org/10.17503/agrivita.v41i0.4165

Mufidah L, Devy NF, Wuryantini S, Hanif Z, Sugiyatno A, Endarto O, Hardiyanto et al (2024). Understanding small farmers' technology adoption and its relation to fruit fly attack: A case study from Gunuang Omeh, Lima Puluh Kota, Indonesia. *Ciência Rural, Santa Maria*, *54*(12), e20230393.

Mutia ED, Honestin T, Hardiyanto (2021). Teknologi Inovatif Jeruk Sehat Nusantara. In Bidiarto K, Sugiharto AN (eds). *Pendahuluan*. PT Penerbit IPB Press. 1–7.

Mutia ED, Sugiyatno A, Devy NF (2023). The effect of interstock on the development of Huanglong Bing Disease and vegetative growth of three comercial citrus varieties in Indonesia. *IOP Conference Series: Earth and Environmental Science*, *1172*(2023), 012034. https://doi.org/10.1088/1755-1315/1172/1/012034

Nareswari N, Kuncoro A (2016). Pembuatan salep minyak atsiri daun jeruk limau (*Citrus amblycarpa*) dan uji stabilitas terhadap tipe basis yang digunakan. *Biofarmasi*, *14*(2), 63–68. https://doi.org/10.13057/biofar/f140204

Nugroho K, Santoso TJ, Kosmiatin M (2024). Genetic diversity and population structure of Indonesia's mandarin citrus genotypes using simple sequence repeat and start codon targeted markers. *Genetics Resourses Crop Evolution, 72*, 2939–2957. https://doi.org/10.1007/s10722-024-02130-z

Purba DP, Husni A, Akhidaya A, Kosmiatin M, Purwito A (2021). Effect of gamma rays irradiation and in vitro selection on Citrus nobilis (L.) 'Siam Banyuwangi' to Huanglongbing (HLB) disease. *AGrivita*, *43*(2), 358–366.

Panwar D, Panesar PS, Chopra HK (2021a). Recent trends on the valorization strategies for the management of Citrus by-products. *Food Reviews International, 37*(1), 91–120. https://doi.org/10.1080/87559129.2019.1695834

Panwar D, Saini A, Panesar PS, Chopra HK (2021b). Unraveling the scientific perspectives of citrus by-products utilization: Progress towards circular economy. *Trends in Food Science & Technology, 111*, 549–562. https://doi.org/10.1016/j.tifs.2021.03.018

Puspitasari Y, Sastro Y (2019). Utilization of rapid detection kit in measuring Huanglongbing outbreak level: A case study in Koto Tinggi, West Sumatera, Indonesia. *IOP Conference Series: Earth and Environmental Science, 468*(2020), 012046. https://doi.org/10.1088/1755-1315/468/1/012046

Ratule MT, Supriyanto A, Hanif Z, Hardiyanto (2018). Citrus in Indonesian; production perspective for market development. *International Symposia on Horticulture (ISH) 2018: Emerging Challenges and Opportunities in Horticulture Supporting Development Goals*, 27–30 November 2018, Bali, Indonesia, 132–138 pp. https://doi.org/10.26352/CY27_ISH-2018.

Rizka AN, Hasibuan AM, Novianti T (2024). Impacts of trade policy on the welfare of citrus producers and consumers in Indonesia. *International Journal of Social Economics. 51*(10), 1278–1297.

Rodrigo MJ, Marcos JF, Zacarías L (2004). Biochemical and molecular analysis of carotenoid biosynthesis in flavedo of orange (Citrus sinensis L.) during fruit development and maturation. *Journal of Agricultural and Food Chemistry, 52*(22), 6724–6731. https://doi.org/10.1021/jf049607f.

Rokaya, PR., Baral, DR, Gautam, DM, Shrestha, AK & Paudyal, KP (2016). Effect of altitude and maturity stages on quality attributes of mandarin (Citrus reticulata Blanco). *American Journal of Plant Sciences, 7*, 958–966. http://dx.doi.org/10.4236/ajps.2016.76091

Rusmayadi G, Salawati U, Adriani DE (2022). Adaptasi terhadap dampak iklim ekstrem pada pola tanam jeruk siam Banjar (*Citrus suhuensis*) dengan sistem surjan. *Jurnal Ilmu Pertanian Indonesia, 27*(2), 237–247. https://doi.org/10.18343/jipi.27.2.237.

Shidiq MAA, Widyaningsih S, Wibowo A, Widiastuti A (2024). First report of Colletotrichum queenslandicum and Colletotrichum endophyticum causing citrus anthracnose in Indonesia. *Journal of the Saudi Society of Agricultural Sciences, 23*(2024), 393–403.

Sophia A, Suraini S, Pangestu MW (2021). Ekstrak daun jeruk purut (*Citrus hystrix* D.C) mampu menghambat pertumbuhan *Candida albicans*. *Jurnal Kesehatan Perintis, 8*(2), 159–165.

Sugiyatno A, Setyobudi L, Maghfoer MD, Supriyanto A (2013). Respons pertumbuhan tanaman jeruk keprok Batu 55 pada beberapa interstock melalui metode top working. *Jurnal Hortikultura, 23*(4), 329–338.

Supriyanto A, Ratule, MT, Syakir M (2017). Citrus variety improvement program in Indonesia: Varieties, production and distribution viruses free of citrus nursery. *IOSR Journal of Pharmacy, 1*, 73–78.

Supriyanto A, Setiono (2023). *Teknologi ProduksiBibit Jeruk Premium*. Investasi Masa Depan. UB Press. *128* pp. ISBN. 978-623-296-914-8.

Supriyanto A, Whittle AM (1991). Citrus rehabilitation in Indonesia. *International Organization of Citrus Virologists Conference Proceedings (1957–2010), 11*(11), 409–413. https://doi.org/10.5070/c56hq9h246

Sutopo, Sugoyatno A, Hardiyanto, Buyung AF, Tististyas GA (2021). Teknologi Inovatif Jeruk Sehat Nusantara. In Bidiarto K, Sugiharto AN (eds). *Teknologi Produksi Jeruk*. PT Penerbit IPB Press.

Tahir, M., Suhaenah, A. & Rahim, Y (2020). Potensi aktivitas antioksidan ekstrak etanol dan fraksi n-Heksan buah jeruk pamelo (Citrus maxima (Burm) Merr) asal Kabupaten Pangkep. *Journal Fitofarmaka Indonesia, 7*(2): 18–22.

Tuwo M, Kuswinanti T, Nasruddin A, Tambaru E (2024). Uncovering the presence of CVPD disease in citrus varieties of South Sulawesi, Indonesia: A molecular approach. *Journal of Genetic Engineering and Biotechnology, 22*(2024), 100332.

Utami R, Annisa RR, Praseptiangga D, Nursiwi A, Sari AM et al (2023). Effect of edible coating sodium alginate with addition of siam pontianak tangerine peel essential oil (Citrus suhuinensis cv Pontianak) on the physical quality of strawberries (Fragaria ananassa) during refrigeration temperature storage. *IOP Conference Series: Earth and Environmental Science, 1200*(1), 012058. https://doi.org/10.1088/1755-1315/1200/1/012058

Utami R, Annisa RR, Praseptiangga D, Nursiwi A, Sari AM et al (2024). Preservation effect of sodium alginate-based edible coating enriched with Siam Pontianak tangerine peel oil on chemical and microbiological properties of strawberry during refrigeration storage. *IOP Conference Series: Earth and Environmental Science, 1377*(1), 012035. https://doi.org/10.1088/1755-1315/1377/1/012035

Yassi, A, Demmallino, EB. Sultani, HR (2023). Tropical climate change and its impact on horticultural plants in Enrekang District, South Sulawesi, Indonesia. *Biodiversitas, 24* (6): 3073–3079. https://doi.org/10.13057/biodiv/d24060

Yulianti F, Adiredjo AL, Soetopo L, Ashari S (2020). Short communication: Morphology and genetic characteristics of potential citrus rootstock in Indonesia. *Biodiversitas Journal of Biological Diversity, 21*(11), 5514–5520.

Yulianti F, Adiredjo AL, Soetopo L, Ashari S (2021). Changes in physiological and relative genes expression response of mandarin citrus (Citrus reticulata Blanco) cv Rimau Gerga Lebong (RGL) grafted onto different citrus rootstocks. *Indian Journal of Agricultural Research, I,* 1–7. https://doi.org/10.18805/IJARe.A-615

Yulianti F, Palupi NE, Agisimanto A (2016). Keragaman jeruk fungsional Indonesia berdasarkan karakter morfologis dan marka RAPD. *Jurnal AgroBiogen, 12*(2), 91–100.

Zainuri H, Yamaura K (2021). New farmers' incentives under the new tangerine farming support: The case of Tuban, Indonesia. *International Journal of Fruit Science, 21*(1), 392–399. https://doi.org/10.1080/15538362.2021.1895032

Zakiyyah A., Hanif Z, Indriani DW, Iqbal Z, Damayanti R, Al Riza DF (2022). Characterization and classification of citrus reticulata var. Keprok Batu 55 Using Image Processing and Artificial Intelligence [Article]. *Universal Journal of Agricultural Research, 10*(4), 397–404. https://doi.org/10.13189/ujar.2022.100409

Zamzami L, Widyaningsih S (2020). Peningkatan Kapasitas Penangkar Benih Jeruk Terhadap Teknologi Produksi Benih Jeruk Bebas Penyakit Melalui Metode Pelatihan Di Kabupaten Purworejo, Jawa Tengah. *Jurnal Ekonomi Pertanian Dan Agribisnis, 4*(1), 191–99.

Index

D

I

J

K

N

O

P

Q

R

S

T

U

V

W

X

Y

Z

For Product Safety Concerns and Information please contact our EU
representative GPSR@taylorandfrancis.com
Taylor & Francis Verlag GmbH, Kaufingerstraße 24, 80331 München, Germany

www.ingramcontent.com/pod-product-compliance
Ingram Content Group UK Ltd.
Pitfield, Milton Keynes, MK11 3LW, UK
UKHW061913111225
465993UK00010B/1276
9781032894669